Adhesives Technology for Electronic Applications: Materials, Processing, Reliability

Dale Swanson dedicates this book to his son, Keni

Adhesives Technology for Electronic Applications: Materials, Processing, Reliability

Second edition

James J. Licari and Dale W. Swanson

ELSEVIER

Amsterdam • Boston • Heidelberg • London
New York • Oxford • Paris • San Diego
San Francisco • Singapore • Sydney • Tokyo
William Andrew is an imprint of Elsevier

William Andrew is an imprint of Elsevier
225 Wyman St, Waltham, MA 02451, USA
The Boulevard, Langford Lane, Kidlington, Oxford, OX516B

First edition 2005
Second edition 2011

Library of Congress Cataloging in Publication Data
A catalog record for this book is available from the Library of Congress

British Library Cataloguing in Publication Data
A catalogue record for this book is available from the British Library

ISBN: 978-0-12-810370-8

Typeset by TNQ Books and Journals

Transferred to Digital Printing in 2012

Contents

Preface

Since the publication of the first edition of this book in 2005, significant changes have occurred in technologies, economy, and government regulations all of which have impacted adhesives generally and specifically those used for electronic applications. The biggest impact has been the consolidation of the industry. Larger firms have acquired many of the smaller specialty adhesives' companies and many large firms have merged. During this process, adhesives that were considered as duplications or where sales were considered too small were dropped, forcing users to requalify the alternate available adhesives for their specific applications. Further, the emphasis on developing new adhesives and adhesive formulations has been placed on new or emerging high-volume low-cost consumer and commercial electronics. Associated with the wide use of adhesives in commercial/consumer applications has been the emergence of stricter government regulations regarding flammability, toxicity, outgassing, and environmental hazards.

From a technology standpoint, a reassessment of adhesive requirements for new applications such as LED, OLED, security devices, and nanotechnology is in progress to improve the properties such as thermal conductivity, optical, tamperproofing, and electrical conductivity.

This second edition has been updated to reflect the business, technology, and environmental changes that have occurred; for example, all the tables have been reviewed and updated, new sections have been added to cover formulation additives, LED and OLED applications, market trends, physics of failure, and reliability prediction. However, most of the basic chemistry and processes remain current.

Although other books have been written on adhesives, we believe this continues to be the only one that focuses entirely and comprehensively on the use of adhesives for the multi-billion-dollar electronics field. The rapid advancements that have been made and continue to be made in microelectronic devices and packaging have been the drivers for the changes that have occurred in adhesives materials and processes. Polymer adhesives such as epoxies, silicones, cyanate esters, and polyimides have, to a large extent, replaced metallurgical and vitreous attachment methods for the very fine pitch, high-density flip-chip, chip-scale, chip-stack, and ball-grid array packages. Polymer adhesives have advantages over solder because of their lower cost, precision placement of devices, low-temperature processing, and ability to be reworked.

An understanding of the electrical, thermal, and environmental requirements of the electronic devices and circuits is important for materials and process engineers and for chemists in formulating and developing new adhesive compositions. At the same time, electrical design engineers require a better understanding of the basic chemistry and properties of adhesives in order to select and specify reliable adhesives for specific applications. Manufacturing engineers require knowledge of the various cleaning, deposition, curing, and rework processes and equipment. Finally, quality assurance personnel must be familiar with potential failure modes and mechanisms and the reliability aspects of adhesives.

This book brings together the entire technology of adhesives for electronics, starting with an Introduction that defines the various devices and packaging approaches; then progresses to Functions and Theory, Chemistry and Properties, Bonding and Manufacturing Processes, Applications, Reliability, Test Procedures, and Specifications. In addition to discussions on general applications such as die, substrate, and lid attachments, specific applications such as those for automotive, consumer, medical, space, and military are covered. Special treatment has been given to relatively new developments including anisotropic adhesives, electrically conductive adhesives as replacements for lead-bearing solders, and underfill adhesives for flip-chip devices and ball-grid array packages. Mindful of current environment and energy requirements, we also discuss and compare formulations and curing methods that will meet these requirements such as UV curing, microwave curing, moisture curing, and 100% solids' adhesives.

The reader will find an abundance of tables listing commercially available adhesives, adhesive properties, suppliers, equipment, and test methods that should save considerable time in selecting the most reliable material and process for a specific application. The tables provide examples, but are not intended to be comprehensive. The reader should consult each supplier for the latest product list and for the most recent data sheets.

It is our hope that this book will be of value to a broad group of personnel working in the field of electronics including design engineers, materials and process engineers, chemists, quality assurance and reliability personnel, manufacturing engineers, and marketers.

Note: Throughout the book we have vacillated between the metric and English systems, especially for dimensions. At first, we wished to be consistent, but this was not practical because of the wide variation in dimensions from small-pitch devices to large circuit boards. Thus, mils, micrometers, and nanometers are used for the smallest dimensions while inches or millimeters are used for the large dimensions. For convenience, conversion tables are provided in the Appendix.

James J. Licari PhD

AvanTeco, Whittier, California

Dale W. Swanson MS

Yorba Linda, California

Acknowledgments

Besides those already acknowledged in the first edition, the authors thank the following for their contributions to this edition: Christopher M. Carpenter (ReliaSoft), Mark Costigliola (Advanced Lumonics, EarthLED Div.), Roberta J. Foster-Smith (Nordson Asymtek), James Frick (Technic Inc.), Lisa Geiffer (Philips Lumileds Lighting Co.), Jackki-Morris Joyner (International Microelectronics and Packaging Society), Dr. Ewen Kellar (TWI, UK), Phyllis Swanson (Boeing Co.), and Dr. Larry Tannas (Tannas Electronic Displays).

In addition, the authors are grateful to Matthew Deans (Senior Publisher Science and Technology Books, Elsevier), Frank Hellwig (Editorial Project Manager, Elsevier), Renata Corbani (Project Manager), Andrea Sellers, Naomi Robertson and the rest of the Elsevier staff for their guidance and support in bringing this book to publication.

1 Introduction

Chapter Outline

Adhesives Technology for Electronic Applications. DOI: 10.1016/B978-1-4377-7889-2.10001-4

Adhesives are used extensively and are vital in the assembly and packaging of electronic devices, especially in the current proliferation and mass production of electronic hardware. Adhesives are used in assembling semiconductor die, both in single-chip packages and in multichip assemblies. Both bare-chip devices and prepackaged components are attached and electrically connected with adhesives to produce electronic circuits such as printed-wiring assemblies (PWAs), thin- and thick-film hybrid microcircuits, and multichip modules. Adhesives, as pastes or as solid films, are also used in fabricating high-density multilayer interconnect substrates, flexible circuits, flat-panel displays, and a host of other emerging applications including optoelectronics; high-speed, high-frequency circuits; sensors; and smart cards. Due to their low cost, ease of rework, and low-processing temperatures, polymer adhesives have replaced many traditional interconnect materials such as solder, eutectic alloys, and wire, especially for most commercial and consumer electronics. However, because of the large number of adhesives available and the variety of polymer types, forms, and formulations, a basic understanding of adhesives and their properties is necessary in their selection and application to assure performance and to avoid subsequent reliability problems. The prime objective of this book is to provide this basic understanding and define guidelines for selecting and qualifying adhesives based on the product and the conditions the adhesive is expected to meet.

1.1 Adhesive types and classifications

Adhesives used in assembling electronic circuits may be classified based on their physical form, polymer type, molecular structure, formulation, curing method, function, or intended application.

1.1.1 Classification by form

In terms of their physical forms, adhesives may be liquids, solids, or pastes. Liquids may be used where excessive flow or spreading is not a problem. In most electronics applications, however, adhesive flow needs to be contained and controlled during cure especially when used in the assembly of microcircuits. Pastes are generally highly filled with mineral fillers and thixotropic compounds such as fine silica powder to produce semi-solid properties having non-Newtonian flow behavior. They are easily dispensed by forcing them through a needle or by pressing them (squeegeeing) through a screen or stencil. Paste or liquid adhesives, in turn, may be *one-part* or *two-part adhesives*.

A one-part adhesive may contain a polymer resin system that cures directly on exposure to moisture, oxygen, UV light, or elevated temperature. Two-part adhesives consist of a separate resin portion, referred to as Part A and a hardener or catalyst portion, Part B, either of which may contain fillers and other formulation ingredients. The two parts are packaged and stored in separate containers. When ready for use, the parts are weighed, mixed, deaerated, dispensed, and cured. Some two-part systems can be formulated as one part by admixing the resin portion with a hardener or catalyst that is unreactive under normal conditions (latent) but decomposes at some elevated temperature to release a reactive curing agent as, for example, the dicy (dicyandiamide) curing of epoxy resins.

Two-part adhesives may also be converted to one-part systems by quickly mixing weighed amounts of Parts A and B, deaerating, freezing, packaging and then storing at temperatures of −40 °C or lower. These one-part systems, known in the trade as *frozen adhesives*, may be purchased from several suppliers in tubes or syringes of various sizes and amounts. Frozen adhesives, generally epoxies, are highly desirable from a manufacturing standpoint since they obviate the risk of human errors by the electronics assemblers of weighing, mixing, and processing small batches of adhesives.

Solid one-part adhesives may also consist of *preforms* (B-staged or partially cured polymers) that, with additional temperature and pressure, will bond two surfaces by flowing, wetting, and fully curing as with copper-epoxy prepreg laminates. Other solid adhesive forms include *film* and *tape* adhesives. Film and preform adhesives often contain a carrier or "scrim" consisting of glass or plastic mesh or cloth for easier handling and bondline thickness control during cure. Film and preforms may be precut to the exact size and configuration needed to attach components to printed-circuit substrates, lids to packages, and insulation to circuit substrates for 3-D module assemblies.

Solid adhesives may also be "hot melt" *thermoplastic films* or pellets that liquify on heating and resolidify on cooling and are useful in the fabrication of high-density multilayer circuits.

1.1.2 Classification by polymer or chemical type

Adhesives are often referred to by their polymer type. The major polymer types used in electronics packaging are epoxies, silicones, acrylics, polyurethanes, polyimides, and cyanate esters. Some generic properties apply to each type but, in generalizing, one must be careful since there are hundreds of formulations of each type on the market, each with minor or major differences in properties. Even for a specific formulation, an adhesive's processing conditions, such as its cure schedule, can affect the final properties. Chapter 3 provides a discussion of the major polymer types and their properties.

Polymer adhesives may be further classified as either *thermoplastic* or *thermosetting*, depending on whether their molecular structures, after curing, are linear or crosslinked. The linear polymers may be straight chains or branched chains (Fig. 1.1). Thermoplastic adhesives melt and flow at a specific temperature or within a narrow temperature range and then quickly resolidify on cooling. Thermoplastic adhesives are sometimes called *hot-melt adhesives* and are particularly advantageous in their

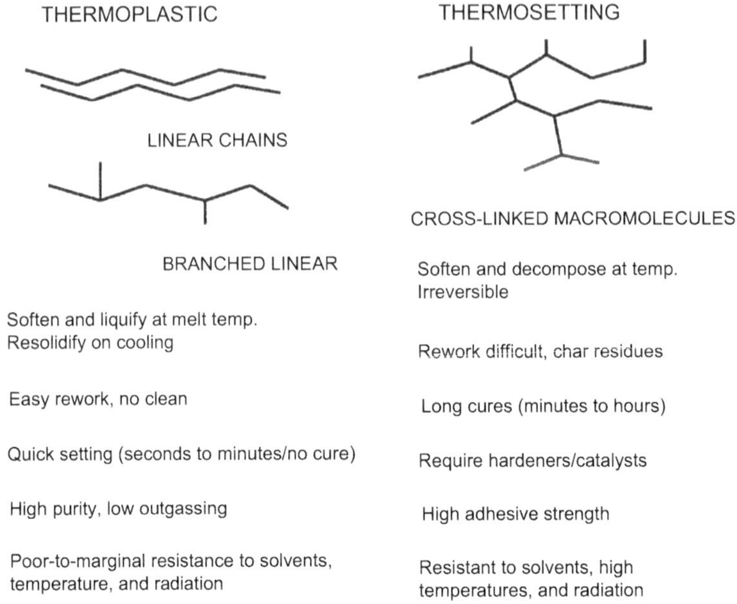

Figure 1.1 Representations of thermoplastic and thermosetting polymer structures.

ease of processing and reworking. Most polyurethanes and polyamides fall in this category. Although thermoplastic polymers are linear, not all linear polymers behave as thermoplastics. Many linear polymers have a high aromatic or heterocyclic content in their structures, thus increasing their thermal stabilities to a temperature that, instead of melting, char and decompose like thermosetting polymers. Among these are polyimide pastes. Other thermoplastic film adhesives that are fully polymerized, high-molecular-weight resins behave in the same way; examples are polyetheretherketone (PEEK®, a trademark of Victrex), polyetherimide (ULTEM®, a trademark of General Electric), fluorinated ethylene propylene (FEP), polysiloxane imide, and some epoxies.

Thermosetting adhesives, on the other hand, soften as the temperature is increased to the glass-transition temperature, but do not melt and resolidify. Instead, because of their highly cross-linked macrostructures, they decompose and char on reaching their decomposition temperatures. Epoxies, cyanate esters, and phenolics are generally thermosetting types.

A unique adhesive type is *silver-glass*; a paste consisting of a glass matrix blended with silver particles, a polymer binder, and a solvent. Silver-glass adhesives are not considered polymer types since the polymer binder is burned off during final processing. The polymer resin is added to the formulation to give the paste, the thixotropic properties needed during dispensing or screen-printing. Silver-glass adhesives are processed at much higher temperatures (400–450 °C) than either polymer adhesives (75–175 °C) or solders (180–300 °C). These high temperatures are required to burn-off the polymer binder, melt the glass, and fuse the glass and oxides.[1]

1.1.3 Classification by formulation

In many adhesives formulations, the resin portion is the same or similar and what determines its properties is the hardener or catalyst that is used to cure the resin. Thus, depending on the hardener, epoxy adhesives may be referred to as amine-cured, anhydride-cured, polyamide-cured, or novolac-cured. Polyurethanes may be polyol-cured or hydroxypolyester-cured.

Formulations may also be either *solvent-based* or *100% solids*. One-hundred percent solids adhesives contain no solvents and are rapidly replacing the solvent-based types because of environmental concerns over volatile organic compounds (VOC), government regulations restricting their use, and reliability concerns.

1.1.4 Classification by curing method

The curing process consists of transforming low-to-moderate molecular-weight resins (monomers or oligomers), generally liquids, into high molecular-weight solid polymers. Most curing mechanisms involve polymerization requiring a catalyst or hardener and are initiated by some form of energy.

1.1.4.1 Heat-cured adhesives

Heat curing by convection in box ovens is the most widely used and simplest method for curing adhesives. Heat is applied until a temperature is reached at which polymerization occurs and continues until the polymerization is complete. Time–temperature curing schedules are developed for each adhesive for which optimum properties are achieved. For each of their adhesives, suppliers provide several cure schedules that are considered equivalent relative to final properties. Users may also develop curing schedules to achieve properties specific for their application and maintain production flow. This can be done experimentally by measuring a desired parameter such as electrical resistivity, dielectric constant, hardness, or bond strength as a function of cure temperature and time. Optimum cures may also be determined from the degree of polymerization as measured by infrared spectroscopy, thermomechanical analysis (TMA), or differential-scanning calorimetry (DSC).[2,3] The cure times and temperatures required are largely dictated by the hardener or catalyst used. With the rise in mass production of consumer electronics and the need to reduce energy costs, epoxy adhesives known as *snap-cured adhesives* have been developed that cure in less than 1 minute at temperatures of 160–170 °C. Snap-cured adhesives are generally cyanate ester or modified cyanate esters that form high-temperature-stable triazine structures on curing.[4,5]

1.1.4.2 Ultraviolet (UV)/visible-light-cured adhesives

Adhesives that cure on exposure to UV radiation are also finding wide use because of their rapid cures (several seconds) and because they generally do not require heat energy. Their chemistry is such that exposure to UV light in the range of 2,000–4,000 angstroms initiates polymerization and effects cures in less than 1 minute. However,

UV curing must often be followed by a short heat cure to fully polymerize portions of the adhesive that are shadowed from UV light by other components or portions of the circuit. Ultraviolet-light-curing adhesives are generally acrylics or include the acrylic, vinyl, or ethylene groups as part of the polymer structure. UV-cured adhesives are of particular interest in positioning and staking precision parts such as glass fibers in optoelectronic circuits or staking components on one side of a double-sided printed-wiring board (PWB) while the other side is being assembled, for example, by solder dispensing, attaching components, and reflowing solder.[6]

1.1.4.3 Microwave curing

Variable-frequency microwave (VFM) curing is yet another rapid-curing method for adhesives. It differs from conventional microwave in using swept frequencies rather than a fixed frequency. The main advantages of VFM curing are cure times that are 2–10 times faster than conventional heat curing and uniform heating of large sections because of an even distribution of energy.[7,8]

1.1.4.4 Moisture curing

Moisture-cured adhesives require only ambient moisture to cure. These adhesives contain functional groups that have been purposely blocked and rendered unreactive until exposed to humidity. Moisture reacts with the blocked groups releasing the functional (active) group, allowing it to polymerize. Moisture-curing adhesives may take days to attain complete curing and may have to be supplemented with a short heat cure. Both silicones and polyurethanes are commercially available as one-part moisture-curing adhesives.

1.1.5 Classification by function

Adhesives are often classified according to the function they are to perform. There are four main functions: mechanical attachment, electrical connections, thermal dissipation, and stress dissipation.

1.1.5.1 Mechanical attachment

Adhesives that are used to bond dissimilar materials must produce sufficient shear strength and tensile pull strength and maintain this strength for the life expectancy of the part and under the environmental conditions that the part will encounter. A reduction in bond strength generally occurs on aging at elevated temperatures and under moist conditions, but the reduced strength may still be sufficient for the intended application.

1.1.5.2 Electrical connection

Adhesives may substitute for wire or solder in forming electrical connections; for example, in connecting capacitors, resistors, or semiconductor devices to a printed-wiring board or to a thin-film or thick-film hybrid microcircuit. The best and most widely used electrically conductive adhesives are silver-filled epoxies. Because of the large amount of silver required in electrically conductive adhesives (60–80% by weight), they also provide a high level of thermal conductivity, allowing components to dissipate heat and operate in their normal temperature range. Electrically conductive adhesives may be further classified as either *isotropic* or *anisotropic*.

Isotropic adhesives conduct current equally in all directions and are the most common and widely used in industry. The anisotropic types, also referred to as *z-directional adhesives* or *anisotropic-conductive adhesives* (ACAs), although filled with metal particles, are filled at much lower levels (0.5–5% by volume) than isotropic types (filled 25–30% by volume). The volume fraction of filler is well below the percolation threshold at which the adhesive becomes highly conductive in all directions. Because of the low volume fraction of metal particles, there are no continuous electrically conductive paths in the x–y plane. During the connection process, the anisotropic adhesive, either as a film or paste, is positioned between a flip-chip bumped die or a tape-automated bonded (TAB) die and the corresponding pads of an interconnect substrate. Pressure and heat are applied so that the conductive particles contact and form z-direction connections. The connections are held in place by the surrounding nonconductive cured adhesive. The lack of continuous particle connections in the lateral direction provides electrical insulation around the connections (Fig. 1.2). Conductive particles, ranging in size from 5 to 20 μm, may be pure metals such as silver, gold, or nickel; solder; plated-metal cores; or plated cores of plastic or glass.

Although dating as far back as the 1960s, anisotropic adhesives have only recently been rediscovered and advanced to a state where they are being used in place of solder to interconnect fine-pitch, high I/O-count flip-chip devices, TAB devices, and high-density interconnect substrates. Anisotropic adhesives are also used to interconnect edge connectors and flex cable to PWBs and flat-panel displays (FPDs). ACAs are

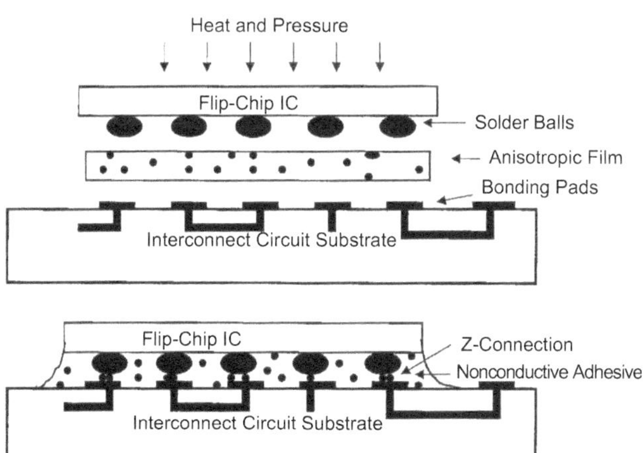

Figure 1.2 Representation of anisotropic conduction.

commercially available in either paste or film form and as either thermoplastic or thermoset types.

Nonconductive or *electrically insulative* adhesives are used where electrical shorting may be a problem or where electrical isolation is required. Nonconductive adhesives are available in either paste or film form and are generally filled with silica or alumina to provide optimum flow properties for dispensing and to improve their thermal conductivities.

1.1.5.3 Thermal dissipation

Thermally conductive adhesives may be filled with metal, ceramic, or inorganic particles. Silver-filled epoxies have high thermal conductivities, but may not be used where there is a risk of electrical shorting. In such cases, epoxies or other polymers filled with electrically resistive, but thermally conductive materials such as aluminum nitride, boron nitride, alumina, or beryllia must be used. Some applications for thermally conductive adhesives include attachment of power devices, heat sinks, large components such as capacitors and transformers, large ceramic substrates, and edge connectors.

1.1.5.4 Stress dissipation

Adhesives that reduce and dissipate stresses consist of low-modulus elastomeric types. Low-stress adhesives are required in attaching large die, large substrates, and flex circuitry. Low-modulus elastomeric adhesives are also used to attach and fillet large components to dampen vibration and shock.

1.1.6 Classification by intended use

Adhesives may also be designated according to the electronic application for which they are intended and formulated. Some of these applications are discussed below.

1.1.6.1 Die attachment

Adhesives that are used to attach bare die* (unpackaged semiconductors such as transistors, diodes, and integrated circuits) or chip* devices (unpackaged capacitors, resistors) into a single package or onto an interconnect substrate are referred to as die-attach or chip-attach adhesives. They may be either electrically conductive, used where ohmic contacts are required as in connecting capacitors, resistors, and transistors, or electrically insulative, used in attaching ICs, stacked chips, connectors, heat sinks, and substrates. Die-attach adhesives are widely used to attach semiconductor die (chips) onto leadframes (chip-on-lead [COL] or leadframes onto die (lead-on-chip [LOC]) prior to plastic molding or hermetic sealing in ceramic or metal packages (Fig. 1.3). Die-attach adhesives are also used in the assembly of hybrid microcircuits and multichip modules.

* The words *die* and *chip* are often used interchangeably, although chip is a broader term that includes bare semiconductor die, but also chip capacitors and resistors.

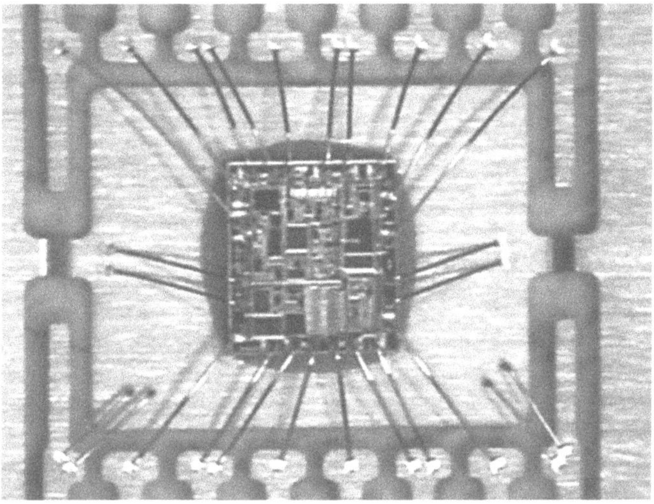

Figure 1.3 Single chip, adhesive-attached and wire bonded to leadframe.
Courtesy: Analog Devices.

Bare die and other chip devices are attached with electrically conductive or nonconductive adhesives to ceramic substrates having defined circuit patterns produced by thin-film vapor deposition and photoetching of metals or by screen-printing and firing of thick-film pastes.[9] With recent advancements in fine-line printed-circuit boards, adhesives are also finding use in attaching bare die to PWBs, a technology known as chip-on-board (COB).

1.1.6.2 Substrate attachment

In assembling hybrid microcircuits or multichip modules, ceramic interconnect substrates fabricated using thin-film or thick-film processes are attached to the inside base of a ceramic or metal package. Generally, film adhesives that have been cut to size are used to attach large substrates (greater than 1-inch square) while either paste or film adhesives may be used for smaller substrates. Substrates may be alumina, beryllia, aluminum nitride, or silicon.

1.1.6.3 Lid attachment

Adhesives in both paste and film form have been used to attach and seal ceramic or metal lids to cavity packages that may also be ceramic or metal.[10] These packages cannot be considered hermetic in the sense that seam-sealed, welded, or soldered lids are, but they are adequate for most commercial and consumer electronics where extreme moisture-temperature environments are not encountered. Packages sealed with Epo-Tek H77 epoxy adhesive are reported to provide near-hermetic sealing of hybrid microcircuits, integrated-circuit devices, and MEM packages withstand 100 hours of 95% relative humidity.[11]

1.1.6.4 *Surface mounting*

Surface mounting of components onto PWBs is commonly referred to in the industry as *SMT* or *surface-mount technology*. It consists of the attachment of electronic components by automated dispensing or screen-printing of solder paste and adhesive onto a PWB, placement of the components, then curing the adhesive and reflowing the solder to form the electrical connections (Fig. 1.4). For SMT, the components are generally prepackaged as opposed to bare die and have pins, leads, or terminals that are solderable or can be connected using electrically conductive adhesives. Adhesives have been developed that will cure during the same operation in which the solder is reflowed, thus avoiding a separate cure step and accelerating production. Solder may also be replaced with an electrically conductive epoxy. Insulative adhesives may be used to reinforce large capacitors, inductors, or transformers.

Surface-mount technology replaces previous methods of inserting component leads into plated through-holes of PWBs, wave soldering from the back side to flow solder into the holes, and simultaneously forming both mechanical and electrical connections.[12] SMT is highly automated and currently the most widely used production process for the assembly of single-layer, double-sided, and multilayer circuit boards.

1.1.6.5 *Underfill for flip-chip-attached devices*

With the availability and increased use of flip-chip devices, a need has arisen to reinforce the connections and dissipate stresses resulting from mismatches in expansion coefficients among the solder, the substrate, and the device. This need has

Figure 1.4 Printed-wiring assembly (4 × 8 in.) with surface-mounted components. *Courtesy:* The Boeing Co.

been met by underfilling the space between the solder connections with specially formulated underfill adhesives (Fig. 1.5). Underfill adhesives are generally electrically insulative epoxies or modified epoxies formulated to have good wetting properties and rapid flow and, once cured, to have low moisture absorption and stress-dissipating properties.

1.1.6.6 *Underfill for ball-grid-array (BGA) packages*

After attaching the BGA packages to a PWB and reflowing the solder, an underfill adhesive similar to that used for flip chips may be dispensed and allowed to fill the spaces between the solder balls, thus reinforcing the connections and relieving stress in the entire structure.

1.1.6.7 *Solder replacements*

Because of the toxic nature of lead in lead-alloy solders and health concerns from being exposed to hot solder fumes during circuit board assembly, interest has developed in replacing lead-containing solders with no-lead solders. Since most of the no-lead solders evaluated to date require higher reflow temperatures than the currently used tin–lead solder, a concurrent effort has been the investigation of electrically conductive adhesives as solder replacements. The feasibility of forming silver-filled epoxy bumps on the die at the wafer stage has been demonstrated for flip-chip devices, and adhesive formulations have been made commercially available.[13]

1.1.6.8 *Particle getters*

A minor, but in some cases, essential use of adhesives is to getter particles entrapped in a sealed microcircuit package. Some microcircuits have been found difficult to clean and to remove all metallic and nonmetallic particles from their assembly prior to hermetic sealing. Others, even after cleaning and sealing, have been found to slough off particles during vibration. In these situations, a soft, tacky adhesive may be attached to the inside of the lid prior to sealing. Particle getters are primarily silicones.

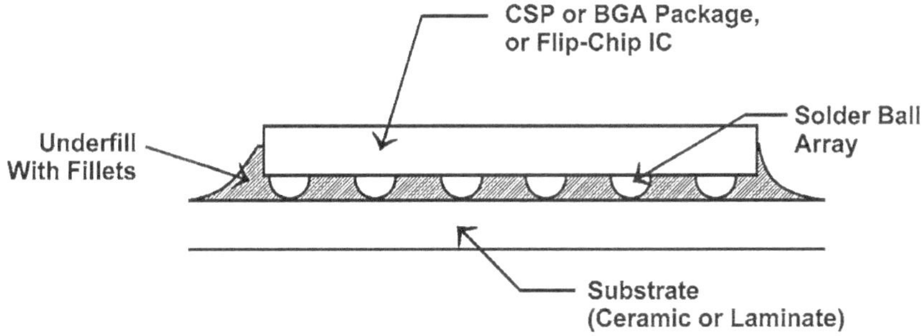

Figure 1.5 Cross-section of flip-chip-attached die showing underfill.

1.2 Summary of packaging technologies

A brief description of the main electronic packaging technologies is provided to assist the reader in understanding how and where adhesives are being used. The large number of packaging approaches and variations for each approach precludes a thorough discussion here. The literature, however, is replete with hundreds of technical articles and several books for each approach.[9,14–21]

1.2.1 Single-chip packaging

After mechanical attachment to a substrate, a leadframe, or to the inside of a package, bare die or chip devices are electrically connected by one of five methods: wire bonding, flip-chip bonding, TAB, solder attachment, or attachment with electrically conductive adhesives. Fig. 1.6 shows some of these methods.

Wire bonding is still the foremost method used. Aluminum or gold wire, generally 1-mil diameter, is used for most applications. High current power circuits require thicker wire up to 10-mil diameter. Die can be wire bonded at high speed with automatic thermocompression, thermosonic, or ultrasonic equipment.

In *flip-chip bonding*, the normal wire-bond pads at the perimeter of the die are replaced by solder or metal bumps. Solder bumps may also be formed throughout the die area (area arrays) to achieve up to 1,000 I/Os per die. The bumped die are flipped face down to mate with corresponding solder pads formed on the interconnect substrate that may be a PWB, a thin- or thick-film ceramic substrate, a chip carrier, or a ball-grid-array package. The solder is then reflowed to form the *z*-axis electrical connections. Flip-chip bonding has major advantages over wire bonding in the large number of connections that can be made per unit area and their short electrical paths, important for high-speed, high-frequency circuits. Variations of flip-chip designs include flip-chip-on-board (FCOB), flip-chip-on-flex (FCOF), and flip-chip-in-package (FCIP).

In *tape-automated bonding* (TAB), the semiconductor die, fabricated with bumped interconnect pads, are positioned in apertures of a polymer-film tape, generally polyimide (Kapton®, a registered trademark of DuPont), then gang-bonded to cantilevered beams that are formed by photodelineated metal pads on the tape. The tape (film) has sprocket holes along its sides similar to movie film and, like movie film, is handled on reels in automatic equipment for high-speed assembly.[17]

Once the die are attached and electrically connected, a large variety of packages and packaging approaches is available ranging from low-cost transfer-molded plastic

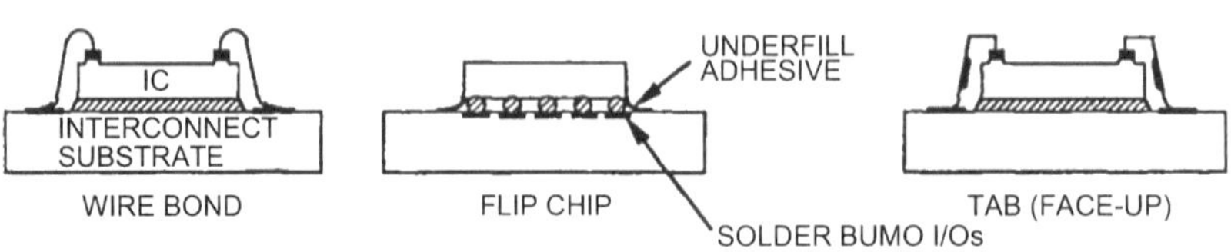

Figure 1.6 Cross-sections of basic die bonding and chip-connection methods.

or plastic-encapsulated devices to devices that are hermetically sealed in metal or ceramic packages. Initially, transistors were welded shut in small metal cans of various sizes referred to as TO (transistor outline) cans. With the introduction of ICs, followed by the progressive increase in functionality to large-scale integration (LSI) and very-large-scale integration (VLSI), increasing numbers of I/O connections (up to 1,000 per device) were required. To meet these needs, new cavity and non-cavity (plastic and ceramic) packages have been designed and built that have a higher number of leads of smaller pitch. These packages can be used for both insertion mounting and surface mounting. Among packages introduced were the dual-in-line packages (DIPs), quad-flat packs (QFPs), pin-grid arrays (PGAs), and chip-scale packages (CSPs) and, along with them, numerous variations and a host of new abbreviations and acronyms.

1.2.1.1 Dual-in-line package (DIP)

The dual-in-line package is a very popular rectangular package in which the leads emanate from two parallel sides of the package (Fig. 1.7). DIPs may be constructed of molded plastic (plastic dual-in-line package [PDIP]), primarily epoxy, or of ceramic (ceramic dual-in-line package [CERDIP]) and may have 8–64 leads with a pitch of 2.54 mm (100 mils). In one process, the die is attached with adhesive or eutectic solder to a leadframe, wire bonded and then transfer molded with plastic. The leads are formed (bent vertically), solder coated and then are ready for insertion in plated through-holes of a PCB. CERDIPs consist of a preformed ceramic body with one or two cavities containing the bonding pads to which the IC chips are bonded. The leads are brazed to connecting pads at the side edges.

1.2.1.2 Quad-flat packs (QFPs)

Quad-flat packs may be metal or ceramic cavity types or plastic-molded packages where the leads emanate from all four sides (Fig. 1.8). Packages may be of various sizes with lead pitches as low as 12 mils and up to 370 leads. QFPs may be used to package either single chips or multiple chips in one package. The leads of QFPs may be formed in the shape of "gull wings" so that the packages can be surface mounted onto an interconnect substrate.

Figure 1.7 Examples of dual-in-line packages (DIPs).

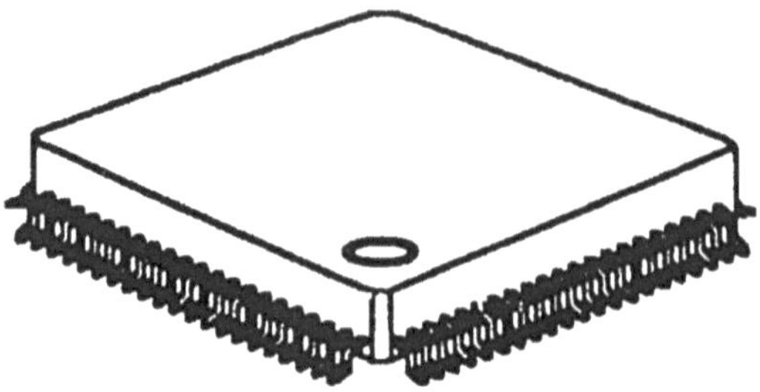

Figure 1.8 Diagram of a quad-flat pack (QFP) showing "gull-wing" leads for surface mounting.

1.2.1.3 Area-array devices and packages

ICs, chip carriers, and packages may have their I/O connections formed throughout the surface area (area arrays) as opposed to peripheral I/Os as with wire-bond pads on die or axial leads on packages. Area arrays include flip-chip, BGA, and pin-grid arrays. Area-array devices provide very high I/O counts per surface area, which are necessary for increased functionality.

1.2.1.3.1 Flip-chip devices

Flip-chip devices have solder bumps, other metal bumps, or even conductive adhesive bumps on the face of the device for I/O connections. During assembly, the devices are flipped face down, then mated and bonded to corresponding solder or metal pads on the package or interconnect substrate. In the quest to eliminate tin–lead solders, electrically conductive epoxy adhesives are beginning to be used for the bumps.

1.2.1.3.2 Ball-grid array (BGA)

Ball-grid arrays are analogous to flip-chip devices except that the solder balls are formed or attached at the next level: the package or the chip carrier (Fig. 1.9).[15] Thus, one can have flip-chip or wire-bonded devices or combinations in a BGA package. BGAs were developed because other packaging approaches such as the QFP had reached their limit in the number of I/Os that were practical and producible. Reducing the lead pitch for QFPs had increased the number of leads but, since the leads were at the periphery, such packages became quite large and risked damage to the closely spaced leads. With bumps spread throughout the area of a BGA package, even though the I/O pitch is greater, the overall size is smaller and up to 1,000 I/O bumps are possible. Fig. 1.10 shows a flip-flip configuration where flip-chip die are attached and connected to a BGA interconnect substrate.

1.2.1.3.3 Pin-grid array (PGA)

PGA packages contain vertical solder-plated or gold-plated Kovar® (a registered trade name of Carpenter Technology Corp.) leads on the bottom of a ceramic or plastic package (Fig. 1.11). Several rows of pins may emanate from the bottom perimeter of the package or there may be a full array of pins throughout the area.

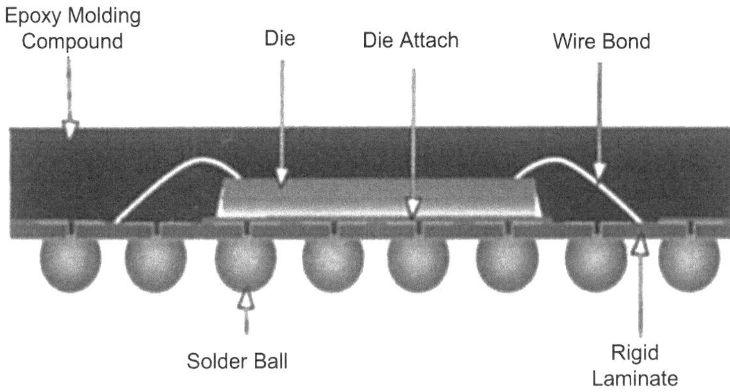

Figure 1.9 Ball-grid-array package.
Source: Amkor.

PGA packages have a lead pitch of 100 mils and are designed for insertion mounting in PWBs. Fig. 1.12 shows the construction of a ceramic PGA (CPGA) package whose base is a Cu–W slug that provides high heat conduction and heat dissipation from the device.[22]

1.2.1.4 Chip-scale packages (CSPs)

Bare die such as integrated circuits may be assembled in plastic or ceramic carriers, called chip-scale packages, whose dimensions are slightly larger than the chip. CSPs are often defined as packages that are no larger than 1.5 times the area of the die or no more than 1.2 times the width or length of the die.[23,24] If the carrier is a BGA type, an alternate definition is that the solder-ball pitch be less than 1 mm.[25] The pitch of

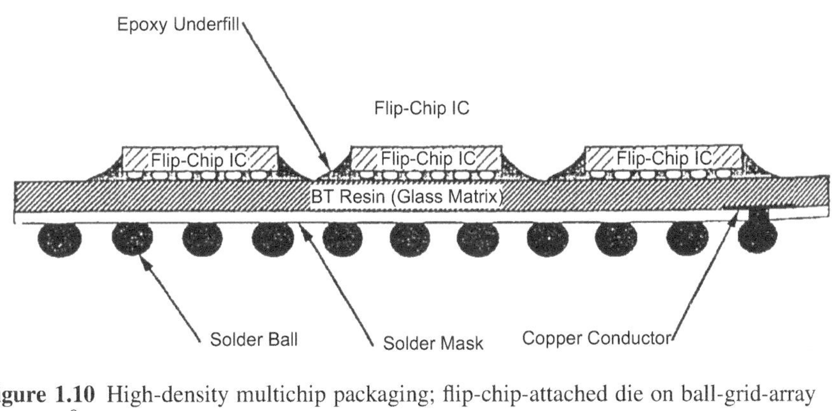

Figure 1.10 High-density multichip packaging; flip-chip-attached die on ball-grid-array substrate.[9]

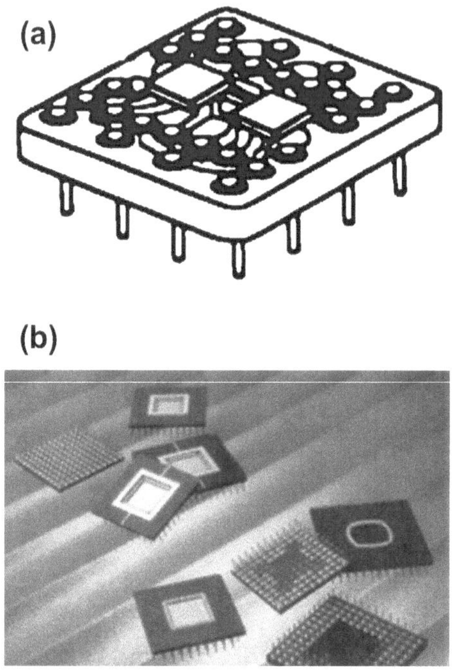

Figure 1.11 Pin-grid array: (a) construction and (b) examples.

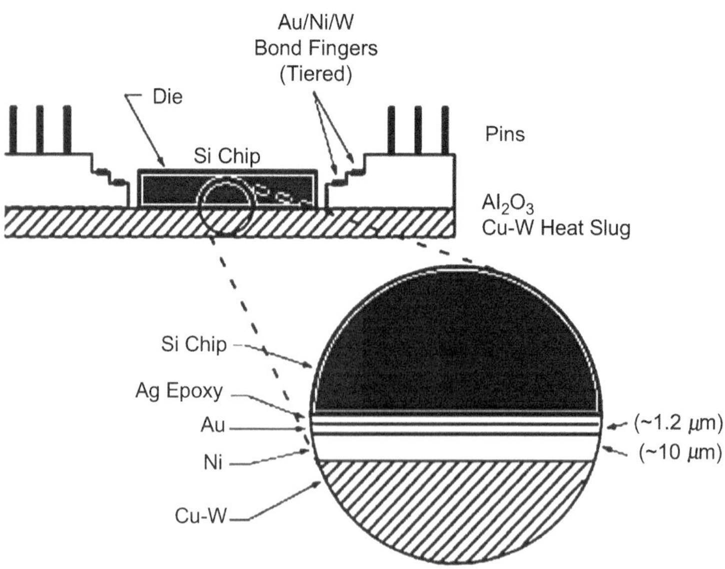

Figure 1.12 Silicon die, epoxy attached and wire bonded in a CPGA package.[22]

interconnects varies by I/O count. Parts having fewer than 100 I/Os have pitches of 1 mm, decreasing to 0.5 mm as the number of I/Os increases to 500.[26]

Chip-scale packages have many advantages over bare-chip attachments especially in their ease of handling, ability to be burned-in and electrically tested, and compatibility with automated soldering and epoxy attachment equipment. Chip-scale packages also have advantages over QFP and PGA packages in the higher levels of silicon density that are obtained. While the silicon density (ratio of silicon-to-package area) of a QFP is 10–60%,[27] that for a single chip in a CSP may be as high as 60–100%.[28] CSPs may be bumped for flip-chip connections or wire bonded. Adhesives are used to attach the chip to the carrier and the carriers to an interconnect substrate. In the past decade, there has been a proliferation of designs and configurations for CSPs, and they have become the preferred method of packaging ICs having up to 500 I/Os. One design is shown in Fig. 1.13. Chip-scale packages are classified as four general types.[24]

1.2.1.4.1 Leadframe-based CSPs

Leadframe CSPs are most closely related to conventional, single-chip packages. According to this approach, the die is attached and wire bonded to a copper leadframe using multiple rows of gold wire bonds. The completed assembly is overmolded, usually with an epoxy molding compound. Small size and weight as well as improved thermal and electrical performance are the key benefits of this technology.

1.2.1.4.2 Rigid substrate-based CSPs

The die may be attached and wire bonded or flip chip bonded to a small rigid interconnect substrate such as a ceramic or a plastic laminate, for example, bismaleimide triazine (BT). Conventional wire bonding and overmolding processes, as in leadframe CSPs, are also used.

1.2.1.4.3 Flex interposer-based CSPs

This CSP type uses a flexible circuit having solder balls or metal bumps as an interconnect interposer between the chip and the next circuit board level. The bare

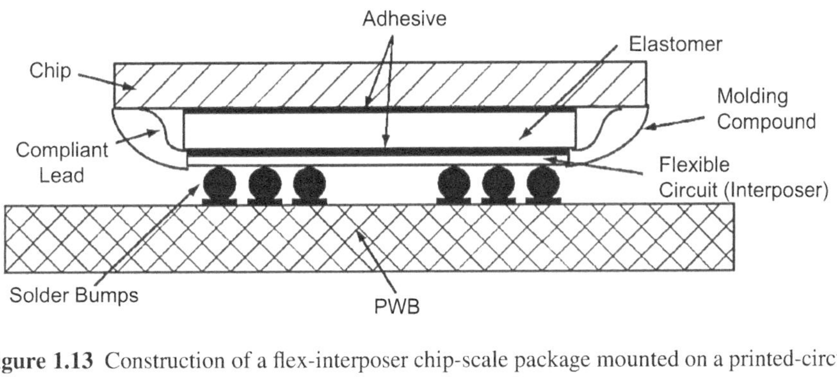

Figure 1.13 Construction of a flex-interposer chip-scale package mounted on a printed-circuit board.

chip is attached face down and wire bonded to the interposer. A thin elastomer, sandwiched between the chip and interposer, cushions the chip and the solder-ball interconnects, relieving stresses (see Fig. 1.13). The interposer generally consists of a metallized, flexible polyimide tape on which are formed electrical connections by photolithographic processes. As a final step, the exposed wire bonds and edges of the chip are molded with epoxy.

1.2.1.4.4 Wafer-level CSPs

CSPs can be batch fabricated at the wafer level by the semiconductor manufacturer resulting in lower costs and higher yields. In their production, through several more steps, the I/O bonding pads are redistributed to facilitate the formation of bumps. The bumped die are then encapsulated and singulated. In redistributing the pads and forming the bumps, photolithographic processes similar to those used in the original fabrication of the ICs are used. Cyclotene® (a benzocyclobutene dielectric coating developed by Dow Chemical Co.), polyimide, or other temperature-stable coating is used as an interlayer dielectric. A beam-leaded variation is also available.

1.2.1.5 Lead-on-chip (LOC)

Lead-on-chip is a variation of CSP in which the leadframe is attached with adhesive to the face of the chip so that wire can be bonded within the chip area. The leadframes are attached to the chip with thermoplastic polyimide film or with a paste adhesive, and the metallized bonding pads on the chip are wire bonded to the leadframe (Fig. 1.14). Leadfames may consist of rigid or flex cable. Parts are subsequently epoxy molded on both sides of the leadframe except that the bottom surfaces of the bottom leads are left exposed so they may be solder plated. The leads are trimmed very close to the edge of the package. Advantages of LOC include high packaging density, ease of pretesting, improved power distribution, low parasitics, and high clock rates. Several variations of LOC have been developed that improve on the lead configuration for better standoff height, better heat transfer, increased compliance, and better inspectability of solder joints.[29] LOC chip-scale packages are used extensively for dynamic random-access memory (DRAM) and other memory chips, providing packages with small form factors, high speed, and low profiles.

1.2.1.6 Chip stacks

A recent approach to increasing component densities is to connect the die or chip devices vertically as a chip stack instead of assembling them horizontally on

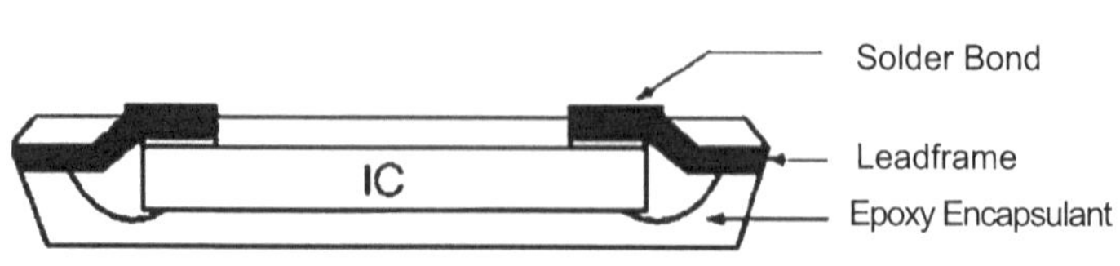

Figure 1.14 Example of a lead-on-chip (LOC) construction.

a substrate. The silicon chips are first thinned to several mils (2–4 mils), then bonded together. Where the chips are of different sizes, such as an application-specific integrated-circuit (ASIC) chip stacked on a static random-access memory (SRAM), the chips may be connected by wire bonding or flip-chip bonding, first to each other, then to a base substrate. Where the chips are of the same size, they must be staggered so that all the wire-bond connections can be made from one side of each device to the base substrate. Cross-sections of wire-bonded, two-chip and three-chip stacks are shown in Fig. 1.15. Chip stacks incorporating up to four chips having a height of only 1.3 mm have also been produced (Fig. 1.16). Chip stacks are widely used in cell phones.

1.2.2 Surface-mount technology

Surface-mount technology (SMT) is basically a component assembly technology related to printed-circuit boards wherein the components are attached and connected on the surface of the board using batch solder-reflow processes. SMT differs from other PWB methods where the component leads are inserted into plated through-holes and wave-soldered from the bottom to fill in the holes and interconnect the components. SMT has the advantages of achieving higher packaging densities, higher reliability, and reduced cost than the plated through-hole insertion process. SMT is currently the process most widely used for low-cost, high-production consumer electronic assemblies.

1.2.3 Multichip packaging

Multichip packaging involves the attachment and interconnection of a variety of chip devices on an interconnect substrate that may be single layer or multilayer. There are two basic types of multichip packaging: hybrid microcircuits and multichip modules.

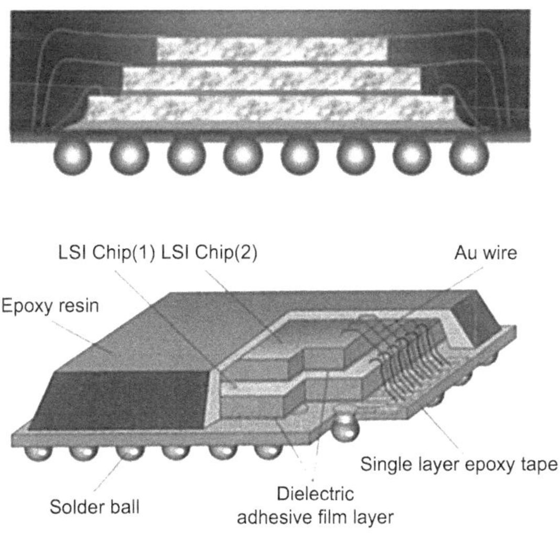

Figure 1.15 Cross-sections of chip stacks.[30,31]

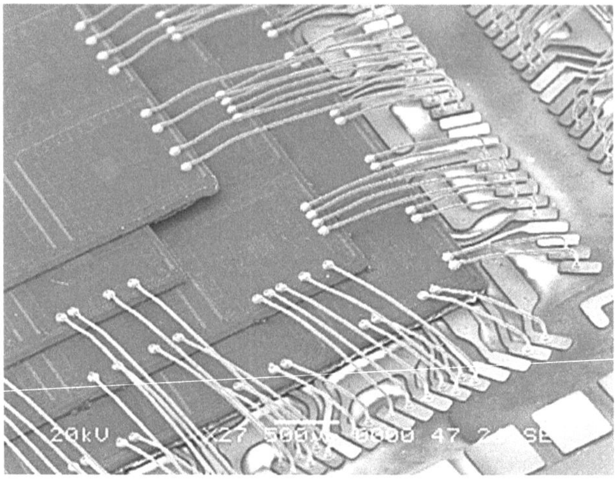

Figure 1.16 Stacked chip-scale package configuration.
Source: Amkor.

1.2.3.1 Hybrid microcircuits

Hybrid microcircuits are high-density circuits produced by directly attaching and connecting bare chips to a fine-line interconnect substrate. The word *hybrid* derives from the large variety of devices, connection techniques, and interconnect substrates that can be used in their assembly. Bare semiconductor die (ICs, transistors, and diodes) and passive devices (chip capacitors, resistors, inductors, and transformers) may be mechanically attached with insulative adhesives and electrically connected using conductive adhesives, gold or aluminum wire, flip-chip soldering, or tape-automated bonding (TAB). The interconnect substrates may be constructed of single or multilayer thin- or thick-film circuits (Fig. 1.17). Additionally, prepackaged devices such as plastic-molded ICs may be assembled alongside the unpackaged chips. Because bare die, fragile wire bonds, and potentially corrosive thin-film metallizations are used, hybrid microcircuits must be well cleaned, vacuum baked,

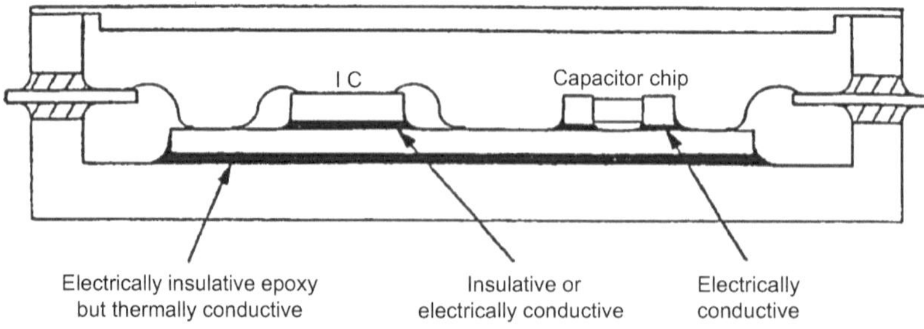

Figure 1.17 Cross-section of a hybrid circuit showing IC and capacitor attachments.

and hermetically sealed in an inert ambient. Packages may be gold-plated or nickel-plated Kovar® with glass-sealed leads emanating either laterally from the sidewalls or vertically as pins from the base. Packages may also be constructed from cofired ceramic in which the leads are formed as an integral part of the package.[9]

Thin-film hybrid microcircuits may be assembled on ceramic substrates such as alumina or beryllia where interconnect conductor lines and bonding pads have been produced by vapor depositing or sputtering thin films of metals, then etching the patterns using standard photolithography processes. Conductor metallizations are generally gold or aluminum. Conductor line widths and spacings are typically 3–5 mils. Thin-film resistor patterns may also be integrated with the conductors by depositing and photoetching nichrome or tantalum nitride. A key attribute of thin-film hybrids is the ability to produce precision resistors that, after laser trimming and annealing, have temperature coefficients of resistance (TCRs) as low as 0 ± 3 ppm/ °C and resistor drifts of less than 100 ppm after aging 1,000 hours at 150 °C.

Thick-film hybrid microcircuits are also assembled on ceramic substrates such as alumina, beryllia, or aluminum nitride, but the conductors and resistors are applied by screen-printing and firing conductor and resistor pastes (Fig. 1.18). Conductor line widths and spacings may be as narrow as 3–5 mils but, for high yields, are generally 10–20 mils. The ability to form multilayer circuits by screen-printing dielectric layers between the conductor/resistor patterns and simultaneously forming vias in the dielectrics to interconnect the layers is a distinct advantage of thick films over single-layer thin films. A further advantage of thick-film over thin-film circuits is the lower cost of materials, processing, and equipment.

1.2.3.2 Multichip modules

Multichip modules (MCM) are basically extensions of hybrid microcircuits, the differences being in their higher degree of density and improved electrical

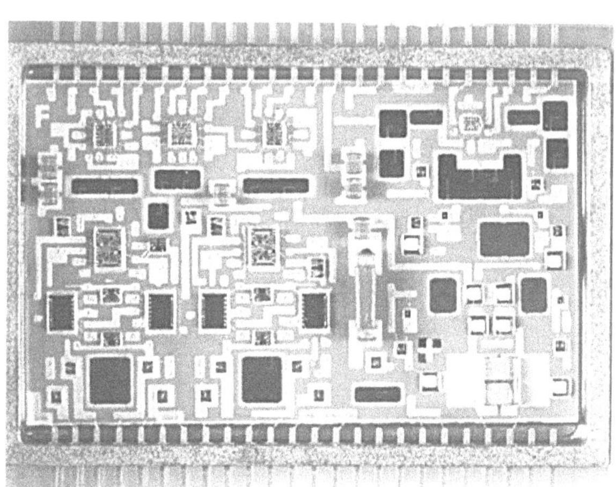

Figure 1.18 Thick-film hybrid microcircuit.

performance. In general, an MCM is defined as a microcircuit that has a silicon-to-substrate density greater than 30%. Three basic MCM types, defined according to the process used in fabricating the interconnect substrate, have been developed. They have been classified by the Institute of Interconnecting and Packaging Electronic Circuits (IPC) as MCM-D (deposited thin film), MCM-C (ceramic thick film), and MCM-L (plastic laminate).[32] MCM-D, the first MCM type to be developed, is produced by a vapor deposited or sputtered thin-film photoetch process similar to that used for hybrid thin-film substrates, except that multilayering has now been introduced. High-performance dielectric layers such as polyimide or benzocyclobutene (BCB) coatings are deposited over the etched conductors, the coatings are cured, and vias are formed by plasma or laser etching. On metallizing the dielectric and photoetching the next conductor layer, *z*-direction vias are also metallized and form electrical connections between the layers. Of the three MCM types, MCM-D provides the finest line widths and spacings (25 μm or less), and smallest vias (25 μm or less).[16] MCM-Ds may also be multilayered to form up to eight conductor layers. Fig. 1.19 shows the construction of an MCM-D interconnect substrate consisting of five conductor layers.

MCM-C interconnect substrates are produced from either low-temperature cofired ceramic (LTCC) or high-temperature cofired ceramic (HTCC). Either process can produce multilayer substrates having high numbers of conductor layers (up to 100), although for most applications 2–20 layers are sufficient. The fabrication of MCM-C involves thick-film processes that have wider lines and spacings (5–20 mils) than MCM-D, but are lower in cost.

MCM-Ls are multichip modules whose interconnect substrates are produced from plastic laminates similar to PWBs, but having much finer dimensions.

1.2.3.3 High-density interconnect (HDI) overlay process

The HDI overlay packaging process, developed by General Electric, is a unique variation of MCM-D. In it, the bare chips are attached first, face up, in a closely packed manner, to a bare substrate and the interconnect layers are then formed over the chips. A polyimide film is next bonded to the array of chips using a thermoplastic adhesive and vias are formed through the polyimide directly over the bonding pads on the devices using a programmed laser beam. The film is next metallized and photo-delineated to generate the fine-line circuitry and *z*-direction interconnections.

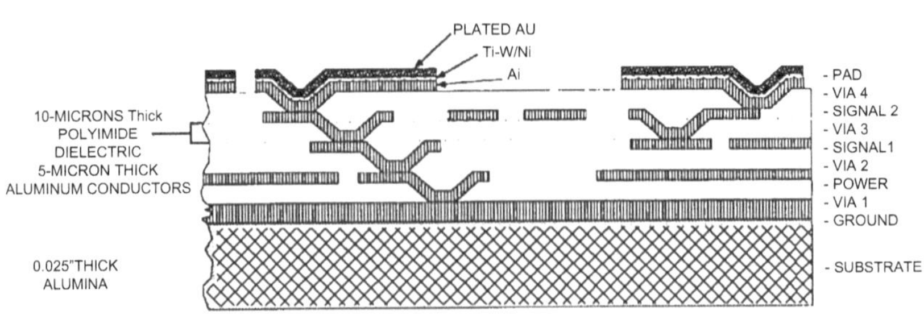

Figure 1.19 Cross-section of an MCM-D multilayer construction.

Subsequent interconnect layers may be formed by repeating the steps[33,34] (Fig. 1.20). A thermoplastic polyetherimide adhesive (GE's ULTEM®) was used to bond the devices to the substrate and also to bond the Kapton® film to the tops of the devices. The process has many advantages over the normal "chips last" approach used for hybrid microcircuits and most multichip modules. Among these advantages are the extremely short electrical paths (no wire, TAB, or flip-chip bonding required), high packing density of devices, and improved thermal conduction from the bottom of the die directly to the substrate which may be alumina, aluminum nitride, beryllia, or other material having high thermal conductivity. Plastic substrates are often used for low-cost consumer applications.

1.2.3.4 Chip-on-board

Chip-on-board (COB), also known as chip-on-substrate (COS), is a relatively new technology that has evolved as a cross between high-density hybrid microcircuits and low-cost PWAs. Recent advances in producing fine lines and spacings and closely spaced bonding pads on PWBs have allowed the assembly of active and passive chip devices as in hybrid and multichip modules. Chip-on-board has resulted in much higher densities than PWBs that were assembled with prepackaged components. As with hybrid circuits, the chips are attached with electrically conductive or insulative die-attach adhesives, or they may be wire bonded or flip chip attached with solder. Also, as with hybrid circuits, the bare-chip devices must be protected from ambient moisture and other contaminants. For most commercial and consumer electronics, it is sufficient to selectively encapsulate ICs and other semiconductor die with a plastic encapsulant specially formulated for low moisture absorption and low stress. These encapsulants are known as *glob tops* and are mainly thixotropic epoxy formulations (Fig. 1.21).

1.2.3.5 Flexible circuits

Flexible circuits (flex circuits) are analogous to rigid printed-circuit boards except they are fabricated from a thin flexible dielectric film to which a thin copper foil is

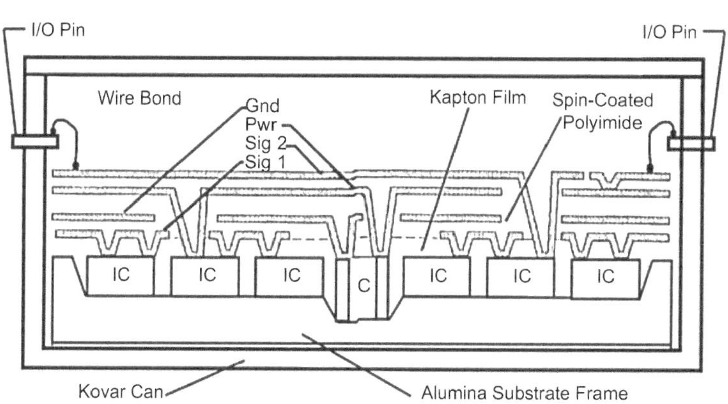

Figure 1.20 HDI overlay, "chips first" packaging.

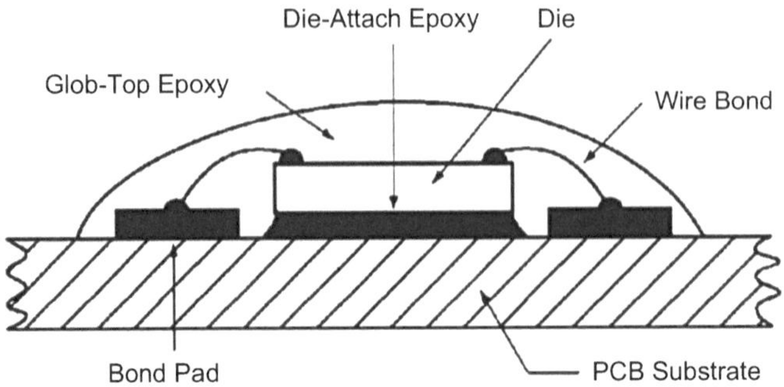

Figure 1.21 Chip-on-board cross-section showing glob-top protection for IC chip.

adhesive bonded. The copper is then photoetched to form a circuit pattern using normal photolithography processes. A plastic film (coverlay) is then adhesive bonded to the etched copper circuitry except that open areas are left for subsequently attaching components. The coverlay provides moisture, contaminant, and handling protection much as a conformal coating does for a PWA. Dielectric films most widely used are polyimide, for example, DuPont's Kapton® H, polyester terephthalate (DuPont's Mylar®), epoxy polyester, and various fluoropolymers such as Teflon® FEP or Teflon® PTFE. Polyimide, because of its very high-temperature stability, is best suited where subsequent assembly involves solder attachments. A variety of adhesives may also be used including acrylics, epoxies, and epoxies modified with flexible resins such as phenolic butyral. Adhesives for flex circuits must meet several requirements, some of which are as follows:

- Retain adhesion while the circuit is flexed
- Wet and bond well to both the polymer film and the copper foil
- Have low moisture absorption
- Have good dielectric properties
- Have low or no-flow characteristics
- Be thermally resistant to solder-reflow temperatures

As with PWBs, flex circuits may be single layer, double sided, or multilayer.

1.2.3.6 Chip-on-flex (COF)

Chip-on-flex is similar to chip-on-board except that the die, chips, or CSPs are wire bonded, flip chip attached, or epoxy connected to a flexible circuit instead of to a rigid interconnect substrate such as a PWB.

1.3 History of adhesives in electronic applications

The basic polymer resins used in adhesives formulations are not new. Epoxy resins, for example, now the most widely used in adhesive formulations, were first introduced in the late 1930s by Ciba.[2] Other early producers of epoxy resins included Shell Chemical Co., Dow Chemical Co., Union Carbide, Celanese, Reichold, and DuPont.[35] Polyurethane resins were introduced in the 1940s and produced by several large chemical companies including DuPont, Mobay Chemical Co., and Monsanto. Polyimides were developed by DuPont in the 1950s, and silicones, the newest of the second generation of electronic materials, were developed almost simultaneously at Dow Corning and General Electric in the 1960s.[36] The assembly of devices and components started shortly after the invention of the transistor by using solder (primarily tin–lead), eutectic alloys (gold-silicon, gold-tin), and wire (copper, aluminum, or gold). In the early 1960s, polymeric adhesives and, in fact, any organic materials in hermetically sealed electronic packages were not allowed in military programs, and rightfully so, since adhesives first used were off-the-shelf commercial products not intended to meet the reliability standards of aerospace programs. Failures due to corrosion, metal migration, and outgassing initially gave organic materials a bad name. At the same time, the government and industry saw the need to use polymer adhesives because of their low cost, low-processing temperatures, and ease of rework. During the next decade, with improvements in purifying resins, reducing the amounts of ionic contaminants, and defining the requirements for qualification, polymer adhesives began to be accepted.

Although metallurgical methods are still widely used today, in the drive toward lower costs, higher densities, and further miniaturization, polymer adhesives have displaced them in many applications. Polymer adhesives such as electrically conductive epoxies have many advantages over solder, eutectic, or glass attachments[37] as listed in Table 1.1. However, the introduction of polymer adhesives in the 1960s was not without problems. Among problems encountered were: high levels of ions and other impurities that were present in the commercially available adhesives of the period; high outgassing in sealed packages; voids due to retention of solvents, absorption of water, or entrapment of air; and stresses due to mismatches in coefficients of thermal expansion (CTE). An awareness of the requirements of the emerging electronics industry led some large chemical firms to conduct research and development to solve these problems and to address the specific needs of the electronics industry. Among these companies were DuPont, General Electric, Dow Corning, Dow Chemical Co., Shell Chemical Co., and 3M. Following suit, during the 1960s and 1970s, many new companies emerged that were dedicated to developing electronic-grade adhesives and other polymer materials for the rapidly growing electronics industry. Among these new-start companies were Ablestik Laboratories, Amicon, Epoxylite, Furane Plastics, Epoxy Technology, AI Technology, Hysol, Loctite, and Emerson & Cuming.

Epoxies became popular for the first generation of electrically conductive adhesives introduced in the mid-to-late 1960s.[38,39] The early adhesive formulations contained organic solvents, which often became trapped in the bondline during cure and resulted in voids, especially under large chips and substrates. These early adhesives also contained large amounts of ionic impurities, notably chloride ions, which, in the case of nonconductive adhesives, resulted in reduced electrical resistivity, ion migration, and corrosion, especially in humid environments. Early epoxies were also

Table 1.1 Comparison of metallurgical, polymeric, and vitreous attachment methods

Attachment/ connection method	Advantages	Limitations
Tin–lead soldering	• Batch process • Automated dispensing or screen-printing • Long history of use/mature • Easy rework • Good electrical connections • Good thermal conduction	• Requires flux • Risk of flux residues and corrosion • Concern over toxicity of lead • Corrosion of solder in humidity, if not overcoated • Risk of voids under large components • Stress due to large CTE mismatches • High-temperature exposure during solder reflow (200–240 °C) • Limited wetting and adhesion to some surfaces (e.g., glass)
Polymer adhesive bonding	• Easy rework • Low cost • Low-processing temperatures (<165 °C) • Automated dispensing or screen-printing • Wide variety of commercially available products to choose from • Ability to relieve stresses • Excellent adhesion to most surfaces • Snap-cure types cure in seconds (160–200 °C) • Directional conductivity with anisotropic forms	• Risk of outgassing in enclosed packages and for space hardware • Most require moderate to long cures (1–2 hours) • Limited thermal stabilities • Finite absorption of water presents "popcorn" problem during reflow soldering
Silver-glass bonding	• High thermal conductivity (50–90 W/m-K) • Good electrical conductivity • Suitable for single-chip packaging	• Long cure schedule to remove binder and fuse glass/oxide • Very high processing temperatures (up to 400 °C) • Not practical to rework • Not suitable for multichip modules with temperature-sensitive materials and devices

cured with aliphatic amines, which outgassed ammonia. Yet, in spite of all this, the first generation of epoxy adhesives performed well for the small chips and devices, conventional PWAs, and low-density hybrid circuits of the time. Among the first electrically conductive die-attach adhesives was DuPont's 5504, a silver-filled, solvent-based epoxy-anhydride system.[40]

Film adhesives (preforms) were introduced in 1968.[41] Film adhesives, originally developed for substrate attachment, are now also used for conductive die attach, lead-on-chip (LOC) bonding, and in fabricating multilayer flex and rigid interconnect substrates. During the 1960s, both one-part and two-part room-temperature vulcanizing (RTV) silicone adhesives were introduced by General Electric and Dow Corning which met a need for higher temperature resistance and higher purity than epoxies. Some of these silicones were convenient to use because they were one-part systems that cured at room temperature on exposure to air moisture.

During this early period, polyurethanes were used where flexibility and vibration damping of large components were required. Cyanoacrylates and anaerobic acrylic ester adhesives that cured in the absence of air were formulated in the 1950s and marketed by Loctite.[42] They were studied for bonding electronic components and heat sinks to printed-wiring boards in the early 1960s. Both cyanoacrylates and anaerobic adhesives provided very rapid tacking and curing in the absence of air, but were soon abandoned for electronic applications because of their poor resistance to elevated temperature-humidity environments, especially where they had to meet military specifications. However, anaerobic adhesives found wide use in locking threaded fasteners, thread sealing, and in impregnating porous surfaces.

The next generation of adhesives (1970s) saw further improvements in purification of epoxy adhesives in which the chloride, potassium, sodium, and other ions were reduced to levels below 20 ppm. Outgassing was also reduced by changes in the hardeners used to cure the epoxies and by optimizing the cure schedules. The first specification for the qualification of adhesives was issued during this period by NASA's Marshall Space Flight Center in 1978.

In the 1980s, a third generation of high-density, high-performance IC devices and interconnect circuits having fine lines and spacings was developed and, concurrently, advances were made in high-purity (low ionic content) and non-solvent (100% solids) epoxies, polyimides, cyanate esters, and other adhesives. Interest was renewed in flip-chip devices for high-speed circuits and for high I/O devices, and underfill adhesives were introduced to fill the spaces between solder-ball connections and relieve stresses. Also, in the 1980s, high-density multichip modules were developed based on multilayer thin-film materials and processes. Thermoplastic film adhesives were used in some designs such as the overlay process developed by General Electric. Other thermoplastic films were used in constructing flex circuits and cable. In 1987, DuPont introduced Ditac®, a thermoplastic preform for die attachment.

A fourth generation of electrically conductive adhesives based on cyanate esters and modified cyclo-olefin thermoset (MCOT) resins was introduced in the 1990s. These resins were specially formulated for low moisture absorption, high thermal stability, and low stress to be compatible with solder-reflow conditions. They were developed to prevent or reduce the so-called *popcorn* effect, a failure mode attributed to the rapid release of absorbed water in adhesives and other plastic materials on exposure to the high solder-reflow temperatures. Researchers were prolific in the 1990s in the development of new approaches to meet the high demand for miniaturized consumer electronics. The requirement for high-throughput automated assembly resulted in the wider use of snap-cure adhesives, UV and microwave rapid-curing adhesives, and silver-glass adhesives. With the greater use of flip-chip and area-array devices and new package designs, faster capillary flow and faster cure underfill adhesives were formulated. At the same time, all these developments had to be moderated by strict government regulations to reduce emissions of volatile organic compounds (VOC). A progression of improvements and advances in adhesives is summarized in Table 1.2.

Table 1.2 Historical developments for adhesives used in electronics

Decade Time-frame	Development
1960s	• First use of electrically conductive epoxies • Premixed, frozen adhesives introduced by Ablestik Laboratories, Epoxy Technology, and Amicon • Reduced ion content epoxies and polyimides • Introduction of epoxy preforms • First high-purity RTV silicone adhesives from G.E. and Dow Corning
1970s	• First specification to qualify die-attach adhesives (NASA MSFC-SPEC-592) • Low bleed-out adhesives • Further improvements in high-purity, low outgassing adhesives
1980s	• MIL-STD-883 Method 5011 coordinated and released • Thermoplastic film adhesives • Silver-glass die-attach adhesives introduced • First underfills for flip-chip devices • Snap-cure adhesives for high production, rapid throughput • Thermoplastic film adhesives for multichip modules (G.E.) • Introduction of anisotropic adhesives for fine-pitch assemblies
1990s	• UV/visible curing adhesives • Film adhesives for lead-on-chip packaging • High thermal conductivity die-attach adhesives • First generation of low-stress adhesives (paste and film) • Low-firing, silver-glass adhesives • Fast-cure, high-flow underfills • No-flow, fluxing underfills • Cyanate ester and MCOT adhesives introduced for low moisture absorption, compatible with solder-reflow temperatures (popcorn resistant) • "Skip cure" or "on-bonder" cured adhesives • Variable-frequency microwave curing (Lambda Technologies)

Table 1.2 Historical developments for adhesives used in electronics—*cont'd*

Decade Time-frame	Development
2000 to present	• Application-specific adhesives • Chip carriers and 3-D stacking • Wafer-level packaging; silicon through vias • LED and OLED lighting • MEMS • Liquid crystal and LED displays

1.4 Comparison of polymer adhesives with metallurgical and vitreous attachment materials

Metallurgical methods for die attachment or for die and component electrical connections consist primarily of soldering using tin–lead solder. The application of adhesive and solder paste by automated dispensing or by screen-printing, placement of the components, and reflowing the solder are the main process steps for surface mounting and connecting components onto PWBs. The solder-reflow temperatures are high (220–240 °C) and may not be compatible with other materials or components in the assembly. In addition, with the trend toward eliminating lead in solder pastes, new no-lead solders are being introduced that require even higher reflow temperatures. These high temperatures have also created a new problem for plastic-encapsulated microcircuits (PEMs) where absorbed moisture in the plastic rapidly vaporizes producing stresses and delaminations within the module. This phenomenon is called *popcorning* because of the crackling sound heard during the reflow step. Since most solder pastes either contain flux or must be used together with flux, the effective cleaning and removal of flux residues has always presented a problem.

Polymer adhesives avoid most of these problems, but do not possess the high thermal or high electrical conductivities of metals or metal alloys. Adhesives also are limited in their temperature stabilities, decomposing, and outgassing above a certain temperature.

Vitreous materials comprise a third type of attachment material. Glasses may be used to seal lids to packages for hermetic enclosures. Silver-glass pastes have been developed for die attachment. They have high thermal and electrical conductivities and are used extensively for single die attachment in single packages.

A comparison of the three main attachment methods was given in Table 1.1.

1.5 Specifications

As the adhesives technology matured, specifications were needed to assure the quality and properties of the incoming material and of the cured material. To this end, users and suppliers of adhesives, together with government agencies, collaborated in defining the critical requirements, the tests to be performed, and the parameters that the material had to meet to qualify and assure a reliable part. This assignment, undertaken by NASA MSFC in 1974, led to the first document for the procurement of adhesives to be used in hybrid microcircuits. After further coordination and revisions, this initial document was formally released as a specification in 1978 and revised in 1982 as NASA MSFC-SPEC-592, "Specification For The Selection and Use of Organic Adhesives in Hybrid Microcircuits." For the first time, the critical requirements for ionic contaminants, electrical resistivity, die-shear strength, corrosivity, and outgassing were specified. Work on the development and evaluation of adhesives continued into the mid-1980s and, as a result, several commercially available adhesives were qualified.[43]

Shortly after the initial release of the NASA specification, other government agencies followed suit with the release of their own specifications modeled after the NASA specification. Ultimately, all government agencies agreed on one specification to be administered by the Air Force at Rome Air Development Center. Previous specification requirements became incorporated in MIL-STD-883 as Method 5011[44,45] which is still in use today for the qualification of adhesives, as well as for other polymer materials, to be used in high-reliability electronics. MIL-STD-883 is now administered by the Defense Supply Center, Columbus (DSCC), Ohio.

The government-sponsored programs and the requirements and standards that were established had a pronounced effect on the development of new and improved adhesives that met the needs of the entire electronics industry and resulted in the start-up of many adhesives' companies.

1.6 Market and market trends

The adhesives market for electronics naturally follows the growth in the electronics market as a whole where increases are projected in consumer electronics, communication systems, industrial controls, autos, aerospace, and medical devices. Two factors are expected to drive the adhesives market higher in the next 5–10 years: further miniaturization of existing products and continued development of new solid-state semiconductor technologies and applications. With the general trend toward further integration and miniaturization, for example, going from printed-circuit board systems to multichip modules and on to wafer-level packaging, the amount of adhesive used per system will decrease. On the other hand, miniaturization generally leads to lower cost and expansion of the global market resulting in a net increase in the amounts of adhesives used. Second, major new applications for solid-state devices continue to emerge, two of which are expected to be high-volume consumer/industrial uses. These are: information processing devices such as smart cards and security electronic detection systems, and lighting and displays based on LEDs and possibly OLEDs. In fact, LED lighting is forecast to be one of the fastest growing fields at 10.9% annually through 2013.[46]

For automobiles, the electronics content has increased dramatically over the past 10 years and is expected to continue with greater use of sensors for collision avoidance and engine status, power electronic modules, fuel cells, power-train controls, LED packaging for lights and displays, and anti-theft systems.

During the next 3–5 years, electronics based on innovative technologies and developments addressing issues such as health, energy conservation, global warming, safety, and security are expected to dominate. In general, specialty adhesives such as formulations based on modified epoxies, silicones, cyanoacrylates, and UV- or radiation-cured acrylics for medical and automotive markets are expected to show strong growth at approximately 4.4% per year through 2013.[47] Electrically conductive epoxies are expected to replace much of the lead-containing solder for assembling and interconnecting micro- and macro-electronic modules while continued developments in nanofillers to improve electrical conductivity and adhesive strength are anticipated.

As the world population and urbanization increase at rapid rates, heightened fears of crime and terrorism will drive the development of microelectronic devices and systems to detect, report, prevent, and analyze crime. These devices may consist of miniature flexible sensors, imaging devices, chemical detectors, and explosives detectors. Growth of approximately 7.5% per year through 2012 is expected for adhesives used in these fields. There is also a trend to develop small radio frequency wireless devices such as cryptographic radio frequency identification (RFID) modules.[48] Smart cards are expected to continue their tremendous growth and become "smarter" by incorporating more microprocessing and data analysis functions especially benefiting the healthcare industry.

Wafer-level packaging (WLP), where specialty adhesives are used at the wafer level before devices are singulated, and 3-D packaging showed growths in 2009 in spite of the economic downturn.[49] Both WLP and 3-D packaging are expected to continue to increase by eliminating or combining steps in the packaging process, thereby reducing costs.

The broad field of "green" technology has already had a large impact on the materials used in formulating adhesives, on the processes used to prepare and clean surfaces and on curing adhesives. Major thrusts have already started and are expected to continue, for example: replacing toxic materials (brominated flame retardants and lead solders), reducing or eliminating organic solvents, and reducing energy consumption to alleviate global warming. Advancements are expected in low cost, low-energy, light-emitting diodes (LEDs) through developments in high thermal conductivity adhesives to replace solder. Even higher efficiency is projected by replacing LED chips with organic light-emitting devices (OLEDs).[50]

Table 1.3 presents a summary of the expected market trends and associated adhesives requirements.

Table 1.3 Market trends and challenges

Trend	Requirements for adhesives
High production, low cost to meet continued demands of existing commercial electronics (mobile phones, texting, video and personal communication, and entertainment devices)	Rapid dispensing, rapid curing without altering physical and electrical properties, high moisture, temperature, and mechanical shock resistance. Meet increasing safety regulations

(*Continued*)

Table 1.3 Market trends and challenges—*cont'd*

Trend	Requirements for adhesives
"Green" materials and processes	Solvent-less, 100% solids adhesives UV-light or radiation curing, low outgassing during/after curing, low-temperature curing
Increased safety regulations	Non-toxic, non-flammable, elimination of organic solvents, elimination of halide (bromo, chloro) flame retardants
New security electronic devices and systems; new sensors, ultra-thin flexible detectors and imaging systems, biosensors, explosives detectors, smart cards	Flexible adhesives for wrap-around ultra-thin electronics, stress free, tamper-proof, non-removable without destroying encrypted devices and circuits
New developments in nanotechnology	Improved thermally and electrically conductive nanofillers, higher adhesion strength
Increased use of LEDs; rapid developments in OLED for low energy, low-cost solid-state lighting, and TV displays	High thermal conductivity LED attach adhesives; improved moisture resistance, hermetic sealing, extremely low moisture and air permeation/penetration for OLEDs
Increased developments and use of 3-D and STV	Stress-free electrically conductive adhesives for filling small vias
Increased use of electronics in auto and aerospace, LED lighting and displays	High- and low-temperature stability, high moisture resistance, outdoor global weather extremes, high thermal conductivity
Advanced medical diagnostic modules, on/in body electronic sensing and recording devices	Biocompatibility with human tissue, anticoagulant properties, high-purity adhesives
Replacement of lead-containing solders	High electrically conductive thermoplastic adhesives as replacements for solder, investigate nano-electrically conductive fillers
Expanded applications for flat-panel displays	Improved sealing from moisture and oxygen especially for OLEDs; low-stress adhesives, low internal outgassing

References

1. Dietz RL, Winder L. New die attach material for hermetic packaging, *Proc. Mfg. Eng. EMTAS Conf.* Phoenix, AZ: EE83-145; 1983.
2. Lee H, Neville K. *Handbook of Epoxy Resins*. McGraw-Hill; 1967 [chapter 6].
3. Chen Y, Chew F, Goodwin C. Determination of curing schedule for advanced packaging materials. *Int J Microcircuits Electron Packaging* 1997;20(4):540-544 [4th Quarter].
4. Galloway DP, et al. Reliability of novel die attach adhesive for snap curing, *Proc. IEMT Symp*. IEEE; 1995.
5. Burkhart A, Grosse M, Nguyen M. Novel snap cure die attach for in-line processing. *Solid State Technol,* Jun. 1995.
6. Fisher E, Hinerman JB, Dixon D. SMT adhesive deposition: the line to success. *Surface Mount Technol,* Dec. 2000.

7. Fathi Z, Tucker D, Ahmad I, et al. Innovative curing of high reliability advanced polymeric encapsulants. *Proc. NEPCON West,* Anaheim, CA; 1998.
8. Zou Y, Harris J, Ahmad I, Tucker D, Fathi Z. Comparison of die level stresses in chip-on-board packages processed with conventional and VFM encapsulant curing, IMAPS. In: *Proc. High-Density Packaging MCM Conf.*, Denver, CO; Apr. 1999.
9. Licari JJ, Enlow LR. *Hybrid Microcircuit Technology Handbook.* 2nd ed. Noyes Publications; 1998.
10. Epo-Tek H77 Technical Data Sheet, Epoxy Technology, Rev. IV; Apr 2010.
11. http://www.lewrac.co.uk/index_files/a-h77.pdf.
12. Coombs CF. *Printed Circuits Handbook.* 4th ed. McGraw-Hill; 1996.
13. Kuleza FW, Estes RH. Solderless flip-chip technology. *Hybrid Circuit Technol,* Feb. 1992.
14. Lau JH. *Flip Chip Technologies.* McGraw-Hill; 1995.
15. Lau JH, ed. *Ball Grid Array Technology.* McGraw-Hill; 1995.
16. Messner G, Turlik I, Balde JW, Garrou PE. Thin Film Multichip Modules. *ISHM Monograph, Int Soc Hybrid Microelecton*; 1992.
17. Licari JJ. *Multichip Module Design, Fabrication, and Testing.* McGraw-Hill; 1995.
18. Brown R. Materials and Processes for Microwave Hybrids. *ISHM Monograph, Int Soc Hybrid Microelecton*; 1991.
19. Harper CA, Sampson RM, eds. *Electronic Materials and Processes Handbook.* 2nd ed. McGraw-Hill; 1993.
20. Lau JH. *Chip on Board.* Dordrecht: Kluwer Academic Publishers; 1994.
21. Wong CP. *Polymers for Electronic and Photonic Applications.* Academic Press; 1993.
22. Tan N, Lim K, Chin B, Bourdillon A. Engineering surfaces in ceramic pin grid array packaging to inhibit epoxy bleeding. *Hewlett-Packard J.* Aug. 1998.
23. DiStefano T, Fjelstad J. Chip-scale packaging meets future design needs. *Solid State Technol.* Apr. 1996.
24. Bauer, CE, Let's talk chip scale, Circuits Assembly, IPC Supplement (Apr. 1997).
25. Ngulty TA, Ekere NN. Chip scale vs flip chip: issues to consider. *Circuits Assembly.* Nov. 1999.
26. Schueller, R.D, Portable CSV, *Adv Packaging* (May 1998).
27. Baliga J. High-density packaging: the next interconnect challenge. *Semicond Int.* Feb. 2000:91-94.
28. Karnezas M. Advantages of system on a package and system on a chip. *Electronics and Production Packaging.* Jul. 2001.
29. Francis D, Jardine L. Lead-on-chip package extends the use of wire bonding and lead-frame technology. *Chip Scale Rev.* Jul./Aug. 1999.
30. Chien D, Hellmold S, Yee M, Kilbuck K. Stacked multichip CSP standards: a push forward. *Adv Packaging.* Jun/Jul. 2000.
31. Gilleo K, Solberg V. No place to go but up. *Electronics and Production Packaging.* Oct. 2001.
32. *Guidelines for Multichip Module Technology Utilization, IPC-MC-790.* Lincolnwood, IL: Institute of Interconnecting and Packaging Electronic Circuits; Aug. 1992.
33. Fillion RA, Neugebauer CA. High density multichip packaging. *Solid State Technol.* Sep. 1989. Hybrid Supplement.
34. Levinson LM, et al. High density interconnects using laser lithography. *Int. Soc. Hybrid Microelectronics.* 1998 *West.* 1989.
35. Licari JJ. *Coating Materials for Electronic Applications.* Noyes Publications; 2003.

36. Lau J, Wong CP, Prince JL, Nakayama W. *Electronic Packaging Design, Materials, Processes, and Reliability*. McGraw-Hill; 1998.
37. Bolger JC, Morano SL. Conductive adhesives: how and where they work. *Adhesives Age*. Jun. 1984.
38. Seymore RB, Carraher CE. *Structure Property Relationships in Polymers*. Plenum Press; 1984.
39. Burkhart A. New epoxies for hybrid circuit evaluation. *Hybrid Circuit Technol.* Oct. 1988.
40. Pietrucha BM, Reiss EM. The reliability of epoxy as a die attach in digital and linear integrated circuits. In: *Proc. 12th Annual Reliability Phys. Symp.* Apr. 1974:234-238.
41. Shenfield DM. Microcircuit adhesives tutorial. *Hybrid Circuit Technol.* Oct. 1987.
42. Rich RD. Anaerobic adhesives. In: Pizzi A, Mittal KL, eds. *Handbook of Adhesive Technology*. Marcel Dekker; 1994.
43. Kraus HS. Chip adhesives update. *Solid State Technol.* Mar. 1987.
44. Ebel GH, Englekey HA. Polymeric materials and the new mil specs: a hybrid users view. In: Proc. ISHM; 1984:189–193.
45. Microcircuits, Test Method Standard, Method 5011, MIL-STD-883F, DSCC Columbus, OH.
46. Advanced Lighting, Freedonia Group, FR 2486; Apr. 2009.
47. Specialty Adhesives to 2013, Freedonia Group, FR 2561; Oct. 2009.
48. Bono SC, Green M, Stubblefield A, Juels A, Ruben AD, Szydlo M. Security analysis of a cryptographically-enabled RFID device. In: *14th USENIX Security Symp.* Baltimore, MD; 2005.
49. Vardaman J. Sixth Annual International Wafer-Level Packaging Conf., Santa Clara CA; Oct. 27–30, 2009.
50. Chanchani R. "An overview – advanced packaging for solid state light emitting diode (SSLED) lighting,". *Advancing Microelectronics*. Jan/Feb. 2010.

2 Functions and theory of adhesives

Chapter Outline

Adhesives used in the assembly of electronic circuits serve four main functions:

- Mechanically attach parts such as semiconductor die, components, substrates, packages, and heat sinks.
- Electrically interconnect components on a substrate or in a package.
- Dissipate heat from components and circuits.
- Dissipate stresses.

Adhesives Technology for Electronic Applications. DOI: 10.1016/B978-1-4377-7889-2.10002-6

The prime function of adhesives is to mechanically attach or bond devices, components, heat sinks, wire, connectors, and other parts onto a circuit board or an interconnect substrate. Adhesives are also used as pastes or films to attach lids in sealing cavity packages and as dielectric films in fabricating multilayer interconnect substrates. The most important consideration in obtaining a reliable adhesive bond is the ability of the adhesive to flow and wet the surfaces. For a reliable bond, strong adhesion to both surfaces and strong cohesion within the adhesive are necessary.

Besides this prime role of attachment, electrically conductive adhesives are widely used to form electrical contacts between components and the printed-wiring board or other interconnect substrate, such as thin-film or thick-film ceramic substrates or flexible cable. For this function, adhesives serve as low-cost alternates to wire, solder, and other metallurgical connections.

The third function, one that is becoming increasingly important, is to draw heat away from components. With increasing circuit densities and power generation, it is essential to avoid excessive operating temperatures to prolong device lives and avoid premature failures. High heat-generating devices are generally cooled by conduction of heat through metal heat sinks, metal tabs, metal or ceramic substrates or packages, and through liquid inert coolants. The conduction of heat-to-heat sinks demands that adhesives have high thermal conductivities and are applied as thinly as possible without generating voids. In their normal unfilled state, polymer adhesives are poor thermal conductors. They must be highly filled with metal or ceramic particles to increase their thermal conductivities. Heat dissipation by convection is also used to cool electronic devices and assemblies; an example of which is blowing cool air or other gas across the parts while operating.

Lastly, adhesives are used to dissipate stresses that may be generated from thermal excursions, mechanical shock, vibration, or moisture. Specially formulated adhesives are effectively used as underfills for flip-chip devices and ball-grid-array packages to compensate for mismatches of expansion coefficients among the solder, the silicon chip, and the ceramic or plastic-laminate substrate. Low-stress adhesives are also used to attach fragile devices such glass diodes and to dampen stresses due to vibration.

2.1 Mechanical attachment

Mechanical attachment of components, devices, and other parts of an electronic assembly is the prime function of adhesives. Although adhesives are expected to bond a wide variety of materials for electronic applications, they do not need to be structural. They should, however, meet minimum tensile and shear strengths in order to withstand mechanical shock, thermal shock, thermal cycling, and vibration as specified for the intended application. For consumer and commercial products, these stresses may be minimal. For high reliability aerospace and medical systems, more severe tests as defined in MIL-STD-883 and other documents must be used.

Cohesive and adhesive strengths of adhesives develop during the curing or polymerization process and reach maximum values at some optimum time and temperature conditions, usually determined empirically by measuring shear strength or

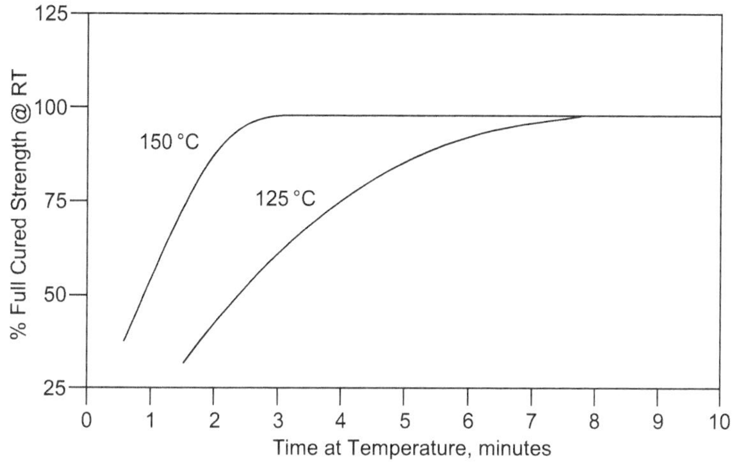

Figure 2.1 Torque strength versus cure schedule for Loctite® Chipbonder® 3607 (strength measured on 1206 capacitors @ 22 °C per IPC SM-817).[1]

another desired property as a function of increasing cure schedules. Figure 2.1 shows how torque strength was used to establish two optimum cure schedules for a surface-mount adhesive.[1] The degree of polymerization or completeness of cure may also be measured by infrared spectroscopy by the amount of reduction of absorption of specific groups as they are converted by crosslinking through addition or condensation polymerization. For example, the oxirane (epoxy) group of epoxy resins absorbs strongly in the 912 cm^{-1} frequency, but this absorption band decreases as the epoxy group is consumed during polymerization with a hardener or catalyst.[2]

Besides the degree of cure, the ability of an adhesive to form strong bonds between different surfaces depends on its ability to wet the surfaces, its flow or rheology properties, and the nature of the bonding mechanism.

2.1.1 Wettability theory

Adhesives must effectively wet and completely contact the surfaces to assure a strong bond. The ability to wet a surface, *wettability*, is related to the ease with which a liquid spreads on a solid surface and is essential in maximizing coverage and minimizing voids in the bondlines.[3] Wettability is measured by the equilibrium contact angle, θ, which is defined by balancing surface-tension forces in Young's equation:

$$\gamma_{sv} - \gamma_{sl} = \gamma_{lv} \cos \theta \tag{2.1}$$

where γ_{sv} is the surface energy at the solid–vapor interface, γ_{sl} is the surface energy at the solid–liquid interface, and γ_{lv} is the surface energy at the liquid–vapor interface. Adhesive wetting is based on the concept of a work function for adhesion as derived from the first law of thermodynamics:

$$W_a = \gamma_{sv} + \gamma_{lv} - \gamma_{sl} \tag{2.2}$$

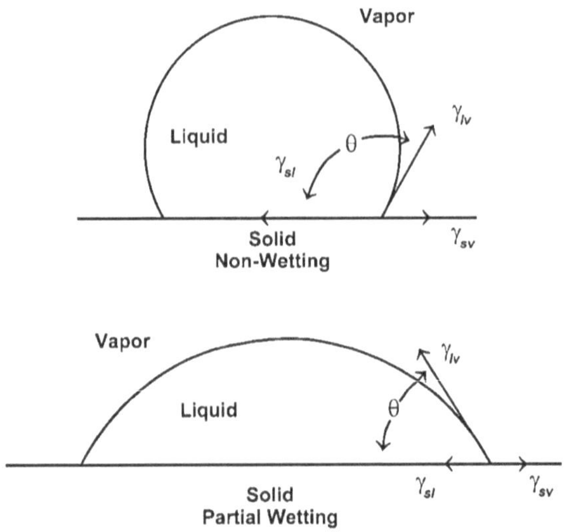

Figure 2.2 Wettability of adhesives and contact angle.

Combining Eq. (2.2) with Young's equation yields the Young–Dupre equation:

$$W_a = \gamma_{lv}(1 + \cos\theta) \tag{2.3}$$

The spreading coefficient or work function (W_a) in Eq. (2.3) shows that good wetting ($\theta < 90°$) is realized when W_a is greater than zero. Good wetting also requires using a liquid having a high surface tension. As W_a decreases, θ increases from 90° to 180° and conditions go from partial wetting to non-wetting.[4] Based on this equation, conditions of wetting versus non-wetting are as follows (Fig. 2.2):[3]

$\theta = 0°$: Complete wetting; liquid spreads spontaneously; high surface attraction
$0° < \theta < 90°$: Partial wetting
$90° < \theta < 180°$: Non-wetting; liquid beads up

Some surfaces that wet well ($\theta < 90°$) include:[5]

- Metal oxides
- Metal plus native oxide films
- Metal plus foreign oxides (alumina, silicates)
- Metal plus adsorbed chemical films (wetting agents, surfactants)
- Hydrated silica (Si–OH surface bonds)
- Ceramics such as alumina
- Epoxy-glass laminates

Among surfaces that do not wet ($\theta > 90°$) are the following:[5]

- Gold
- Metals free of oxides
- Organic nonpolar polymers such as Teflon®
- Surfaces with hydrophobic contaminants such as silicones, oils, and greases
- Metals having adsorbed ions such as fluoride

- Silicon etched with hydrofluoric acid, forming Si–F surface bonds
- Strongly heated silicon, forming Si–O–Si surface bonds

These observations are consistent with the high surface energies displayed by certain metals, metal oxides, ceramics, and plastics.[6] Equations (2.2) and (2.3) also show that high surface tensions of adhesives as liquids require high surface energies of the surfaces to be bonded to produce low contact angles and good wettability. The surface tension of most epoxies is approximately 80 dynes/cm. As a general rule, if water, with a surface tension of 72 dynes/cm, wets a surface, epoxy adhesives will also wet that surface.[7,8] Wetting also occurs if the surface tension of the liquid adhesive is close to or slightly less than that of the substrate.[9]

Contact angles are modified for rough surfaces. In general, high surface roughness degrades wettability if θ is greater than 90° and improves wettability if less than 90°. The contact angle is greater on rough surfaces compared to planar surfaces where θ is greater than 90°. For example, highly roughened surfaces such as sandblasted silicon display poor wetting while mechanically polished silicon has improved uniformity and superior wetting to solder.[3] On the other hand, slightly abraded surfaces, as for thick-film gold conductors, are necessary to improve adhesion. Thus, Eq. (2.3) may be modified by introducing a "contact coefficient" that is a function of surface roughness. Another modification introduces a strain energy γ' function that results during the cooling down of the adhesive after heat curing.[3]

$$W_a = k[\gamma_{sv}(1 + \cos\theta - \gamma')] \tag{2.4}$$

2.1.1.1 Factors affecting wettability

In addition to the contact angle, surface energy, and surface-tension considerations just discussed, the degree of wetting depends on the pot life and gel time of the adhesive. *Pot life* is the useful life of an adhesive after a two-part system has been mixed or after a one-part system has been removed from a container or thawed, in the case of frozen adhesives. Beyond its pot life, the adhesive increases in viscosity to an extent that it no longer flows readily and is difficult to dispense. Pot life is most often measured in terms of a viscosity increase that varies according to the product and the supplier. One supplier, for example, uses a 25% increase in viscosity as the limit.[10] Hannafin et al. considered a 10% increase suitable for automated dispensing but a 100% viscosity increase is acceptable for adhesives that are manually dispensed.[11] Material that has exceeded its pot life may be partially gelled or partially polymerized, a condition also called *B-staged* or *B-staging*. At the gelation point, the adhesive will no longer wet the surface. As an example, the viscosity of Epoxy Technology's Epo-Tek® E3001 is stable over the first 48 hours of use, but increases significantly from less than 6,000 cP to 16,000 cP during the next 24 hours, at which point gelation has occurred.[12] Table 2.1 shows the effect of temperature on gel time for several adhesives. Adhesives are sometimes dispensed at slightly elevated temperatures to reduce their viscosity and permit better wetting, but at the expense of a reduction in pot life since elevated temperatures also accelerate polymerization and gelation.

Table 2.1 Effect of temperature on gel time[13]

Material	Pot life (10 g sample)	Temperature (°C)	Gel time
Epibond® 7002	1–2 days	525	>2 days
		150	5 minutes
Epibond® 7200	1–2 days	25	>2 days
		150	5 minutes
Uralane® 7760[a]	3 hours	25	>3 hours
		100	5 minutes

[a] Available from Specialty Polymers and Services, Inc.
Data courtesy of Vantico Corp.

2.1.1.2 Excessive wetting and bleedout

Excessive wetting may result in the separation and migration of adhesive constituents, either the resin or hardener portion, a phenomenon known as *bleedout*. Bleedout is sometimes visible as a slight discoloration of the surface around the adhesive joint or on a wire-bonding pad and may occur prior to or during cure. Bleedout is especially prevalent on surfaces that are super clean and have high surface energies. Bleedout may be considered analogous to chromatography where the substances in a multi-component mixture absorb, wet, and migrate along the chromatography medium at different rates, resulting in separation into the individual components. Bleedout has been a problem especially with two-part epoxy adhesives where one ingredient migrates and contaminates an adjacent wire-bond pad. This contaminant interferes with the ability to form a wire bond, and if formed, results in a bond of marginal strength. The problem has been largely solved by cleaning the circuits with an oxidizing plasma prior to wire bonding. An oxygen–argon plasma effectively removes the traces of organic residues.

As a better understanding of the bleedout phenomenon was gained, other techniques were used to reduce bleedout. Vacuum baking the interconnect substrate prior to attaching devices has been reported to reduce bleedout. Several theories have been proposed to explain this. One is that vacuum baking removes adsorbed moisture and other polar surface residues that have a high affinity for the equally polar resin or hardener constituents of the epoxy. This theory is reinforced by experiments that showed higher contact angles for vacuum baked than for non-vacuum baked substrates.[4] A second theory proposes that thin films of oil from the vacuum pump may backstream and deposit onto the surface. The oil film, being hydrophobic, increases the contact angle of the adhesive and reduces its flow. Auger spectroscopy has been used to identify carbon, nitrogen, oxygen, and sulfur on gold-plated surfaces after vacuum baking, confirming the presence of oily residues.[4] A more controlled method to reduce bleedout that does not rely on vacuum baking involves treating the surface with a surfactant that leaves known concentrations of hydrocarbon films. These films behave the same way as the oils backstreaming from the vacuum pump. With controlled amounts of hydrocarbon films, adhesive die-shear strengths still met the minimum requirements of MIL-STD-883.[14] Thus, there is a tradeoff and a fine line between achieving good wettability to insure high adhesive strength and reduced wettability to prevent bleedout.

Lastly, and more importantly, suppliers have reformulated their standard adhesives to reduce or prevent bleedout. For example, the widely used Henkel's Ablebond® 84-1 LMI (Low Mobile Ions), a silver-filled epoxy, was reformulated as a no-bleed version: Ablebond 84-1 LMI NB (No Bleed). This NB version, however, was found to have reduced wettability to capacitor terminations and, as a result, yet another version (NB1) was formulated as a compromise.[15] Adhesives that are 100% solids are more prone to bleedout than solvent-based types since they contain low-molecular-weight reactive diluents that have a greater tendency to migrate prior to complete curing. Some surface-related and material-related observations concerning bleedout mechanisms may be summarized as follows:

- Formulation ingredients that tend to bleed consist of low-molecular-weight resins, reactive diluents, and some curing agents such as imidazoles.[16]
- Extremely clean surfaces increase bleedout.
- Smooth, uniform surfaces bleed less; grainier, porous surfaces promote bleedout due to capillary action.
- Vacuum baking can reduce bleedout by removing adsorbed moisture and other polar contaminants, and it can also reduce bleedout due to pump oil backstreaming onto the surfaces.
- Solvent-based adhesives are less likely to bleed than 100%-solids adhesives.
- Adhesives on etched leadframes bleed more than on stamped leadframes; this is believed to be due to the presence of residual cutting oils on the stamped versions.
- Selenium on leadframes was found to increase bleedout.[17]
- Residues from electroplating baths also increase bleedout.

2.1.2 Rheology

The rheology, or flow properties, of adhesives is important not only to achieve adequate coverage and wetting, but also to develop the process parameters necessary to dispense the adhesive (see Chapter 4). Flow properties are determined by measuring viscosity and thixotropy.

Viscosity, η, is a measure of the internal friction of a material and its resistance to flow. It is represented by the proportionality constant between shear stress and shear rate.[18]

$$\eta = \text{Shear stress} \div \text{Shear rate} \tag{2.5}$$

The viscosity of liquid adhesives decreases with increasing temperature until polymerization begins to occur, then increases. The viscosities of epoxies are reported to decrease 20% to as much as 80% for a 10 °C rise in temperature above room temperature.[19–22] Figure 2.3 shows this viscosity reduction as a function of temperature for two anhydride-cured epoxy adhesives. For low-molecular-weight polymers, the temperature dependence of viscosity follows the Arrhenius equation:

$$\eta = Ae^{-(E/RT)} \tag{2.6}$$

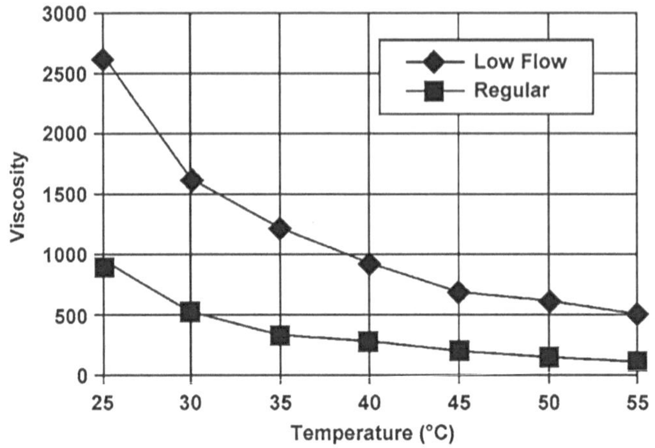

Figure 2.3 Viscosity (in cP) as a function of temperature for two epoxy-anhydride adhesives.

where E is the activation energy for viscous flow, A is a constant, T is temperature, and R is the Boltzmann constant.

2.1.2.1 Newtonian and non-Newtonian flow

When the shear stress of a liquid is directly proportional to the strain rate, as in Fig. 2.4a, the liquid is said to exhibit ideal viscous flow or Newtonian behavior. Most unfilled and capillary underfill adhesives[23] are Newtonian fluids. Materials whose viscosity decreases with increasing shear rate are said to display non-Newtonian behavior or *shear thinning* (Fig. 2.4b). Non-Newtonian fluids are also referred to as pseudoplastic or thixotropic. For these materials, the shear rate increases faster than the shear stress. Most filled adhesives that can be screen printed or automatically dispensed for surface-mounting components are thixotropic and non-Newtonian.[12,18] A second deviation from Newtonian behavior is *shear thickening* in which viscosity increases with increasing shear rate. This type of non-Newtonian behavior, however, rarely occurs with polymers.[24]

Thixotropy occurs when clusters of filler particles break up and fluidity increases. At rest, fillers exist in clusters and their strength, size, and shape determine the static viscosity of the adhesive. The total surface area of the filler particles also contributes to static viscosity. Shear thinning occurs when the clusters, such as silver flake in silver-filled epoxies, break apart and viscosity decreases; then, as stress is removed, the clusters form again and viscosity increases.[21,25]

The thixotropic property of an adhesive may be defined by its *thixotropic index*, the ratio of viscosities measured at two shear rates, generally a decade apart, for example, viscosities measured at 0.5 rpm and at 5 rpm (1.9 s^{-1} and 19.2 s^{-1}).[21] Thixotropic indices are valuable in selecting the best method to dispense an adhesive. Adhesives that have an index of 2–5 are considered suitable for automatic dispensing. Materials with indices close to 1 are generally Newtonian types and suitable for capillary-flow underfills.

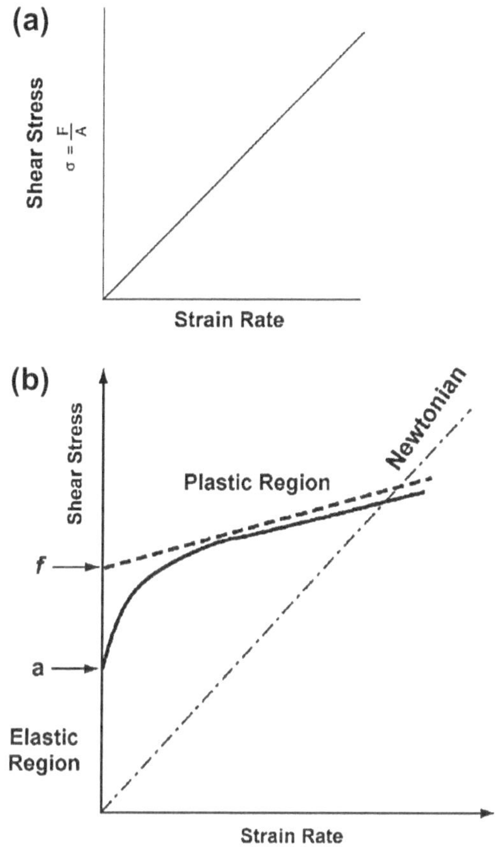

Figure 2.4 Viscosity behavior of adhesives: (a) Newtonian and (b) non-Newtonian.

2.1.2.2 Capillary flow

Adhesives used to fill spaces underneath a flip-chip device or BGA package must flow readily and rapidly when dispensed at the edges of the chip in order to completely fill the gap and free space surrounding the solder bumps. To be effective, the adhesive must be drawn into the narrow space by capillary action. The theory behind underfilling is based on capillary-flow behavior between closely spaced parallel plates (Fig. 2.5) by considering the drop in pressure (p) across a liquid–vapor interface:[26,27]

$$\frac{\partial p}{\partial x} = \frac{-12\mu}{h^2}\frac{\partial L}{\partial t} \tag{2.7}$$

where μ is viscosity of the underfill material, h is the gap height, and L is the length of the die surface traveled by the underfill. The equation is solved by separating the distance (x) and time (t) variables, resulting in:

$$\Gamma = -\frac{2\gamma\cos\theta}{hL} \tag{2.8}$$

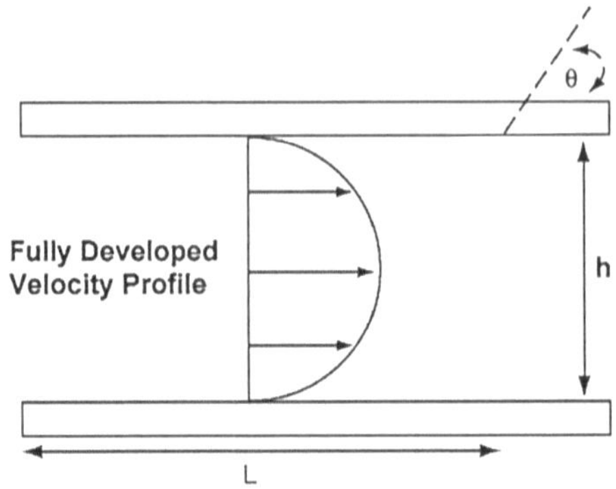

Figure 2.5 Capillary-flow diagram showing the velocity of an advancing front of underfill material as it flows between the two surfaces.
Copyright: PennWell and Advanced Packaging, 1999, reproduced with permission.

where γ is the surface tension of the liquid–vapor interface and Γ is a constant. Application of boundary conditions resulted in the time, t_f, to reach a distance, L_f between the parallel plates, of:

$$t_f = \frac{3\mu L_f^2}{h\gamma \cos\theta} \tag{2.9}$$

where μ is the absolute viscosity, L is the flow distance at time t_f, h is the gap under the part, θ is the wetting contact angle, and γ is the surface tension of the underfill-to-air interface.

2.1.2.2.1 Material factors affecting capillary flow

Material, process, and physical factors all affect capillary flow. Material characteristics that affect the flow rate of underfill adhesives include viscosity, the contact angle that the adhesive makes with the surface, and ability to wet the surfaces. Viscosity and contact angle are related; increasing the viscosity increases the contact angle, thus reducing wettability and flow. The amount and nature of the filler in adhesive formulations also affect viscosity and contact angle, the higher the filler concentration, the greater the viscosity. Of three underfill adhesives, having contact angles of 5–10°, 18–20°, and 10–15° the one with the highest filler content also had the largest contact angle.[28] However, a high filler concentration, although undesirable for capillary flow, is important in reducing the coefficient of thermal expansion (CTE) of the cured material. Thus, a tradeoff is required between a low CTE and high capillary flow by managing the type and amount of filler. The particle sizes of fillers also affect the ability to underfill parts effectively. As a rule, the gap height should be twice the size of the filler particles. The gap should be three times the filler size when die are to be bonded to PWBs because of the greater surface roughness of plastic-laminate boards and the fact that larger particles limit the flow through the gap.[23,29]

In practice, even for highly filled formulations, the viscosity can be lowered during the dispensing operation by heating the material or the substrate. For example, viscosity can be lowered from a typical value of 4,000 cP at 25 °C to about 500 cP by increasing the temperature to 70–90 °C.[19,30] As an example, one underfill, whose viscosity is 4,000 cP at 25 °C, is reduced to 200 cP at 80 °C while another falls from 6,000–9,000 cP at room temperature to 400 cP at 80 °C.[31] It is estimated that over 90% of all underfill materials are dispensed at elevated temperatures at two adjacent edges of a chip.[32] A practical limit, however, is reached when gelation or expiration of the pot life occurs at the dispensing temperature.[23,33]

Besides viscosity, the surface-wetting ability of underfills is critical to capillary flow. For capillary flow to occur, the underfill material must wet the surfaces so that the advancing contact angle is less than 90°.[34] Also, for capillary flow, the intramolecular forces of attraction among adhesive molecules must be weaker than the intermolecular attraction of the adhesive for the die, the substrate, and the solder surfaces.[35] This occurs when the surface tension of the underfill is lower than the surface energy of the surfaces being bonded. As a rule, suitable wetting for capillary flow results when the surface energy of the substrate is 10 dynes/cm greater than the surface tension of the underfill.[9] The surface energy of most PWBs is 35–40 dynes/cm while underfill surface tensions are about 25 dynes/cm.[33]

2.1.2.2.2 Physical factors affecting capillary flow

The gap height, size of the die, and free volume beneath the die all have a bearing on the efficiency of capillary flow, and the ability to completely underfill a device in a few seconds. Small gaps are harder to fill and require low viscosity adhesives to keep the underfill times as short as possible. To fill a 0.5-mil gap, the viscosity of the underfill should be approximately 2,000–4,000 cP at 25 °C. As the gap increases to 1 mil, viscosities can be 6,000–45,000 cP and for a 3-mil gap, they may be 5,000–25,000 cP. These ranges overlap but, in practice, viscosities of underfills that differ by a factor of two exhibit about the same flow rates. This inconsistency is explained by the fact that when measuring viscosities using a Brookfield viscometer, the shear stress applied is too high to simulate capillary forces. Brookfield viscometers are better suited to determine the flow characteristics of screen printable pastes or highly filled, syringe-dispensible adhesives.[36] An accurate test method for capillary-flow viscosity utilizes a controlled-stress rheometer* where the shear stress is controlled to 10 dynes/cm and the temperature held at 90 °C. Viscosities measured using the HaakeRheo equipment correlate well with the time required to underfill devices having different gap heights.[36] Figure 2.6 shows the flow times required to underfill a 3-mil gap with various viscosities of a highly filled material.

The size of the die is a further consideration in the time it takes for the underfill adhesive to traverse the distance and completely fill the gap. A plot of the square of the distance as a function of time produces a straight line as expected from Eq. (2.9)[23] (Fig. 2.7). For a given die size, increased viscosity of the underfill material also increases the time to fill the gap. For high production rates, the fill time must be as short as possible. With increasing sizes of IC chips, this has presented

* For example, the HaakeRheo Stress Rheometer from ThermoFisher Scientific.

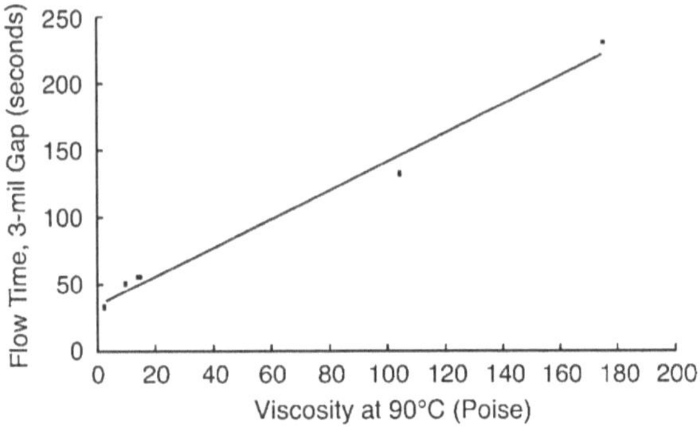

Figure 2.6 Viscosity versus underfill time using controlled-stress rheometer.
Source: Loctite.

a challenge. For example, the die size cannot be greater than 15-mm square for a 15-second fill time using Henkel's E-1216 underfill adhesive (6,000 cP, 20-μm filler size). However, the die can be as large as 30-mm square if a 45 second time is acceptable.[29]

The volume of underfill must be controlled to assure complete filling beneath the die and some filleting around the die. A complex combination of factors (gap height, contact angle, viscosity, wettability) affects the amount of underfill required for complete coverage. The volume of underfill (V) is calculated based on the volume of free space under the die (V_c), the volume of the solder bumps (V_b), and the fillet volume (V_f) or: $V = V_c - V_b + V_f$.[37]

The volume of the fillet, as shown in Eq. (2.10) is a function of the die length and width and the fillet height and width, as well as a shape factor (Fig. 2.8).

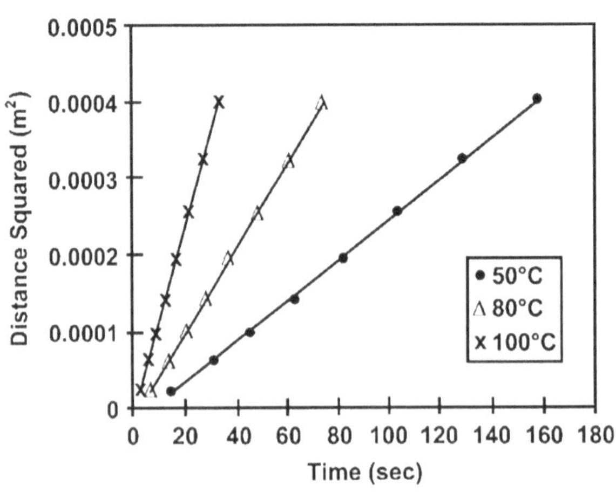

Figure 2.7 Underfill time as a function of distance.
Copyright: IEEE, 1988, reproduced with permission.

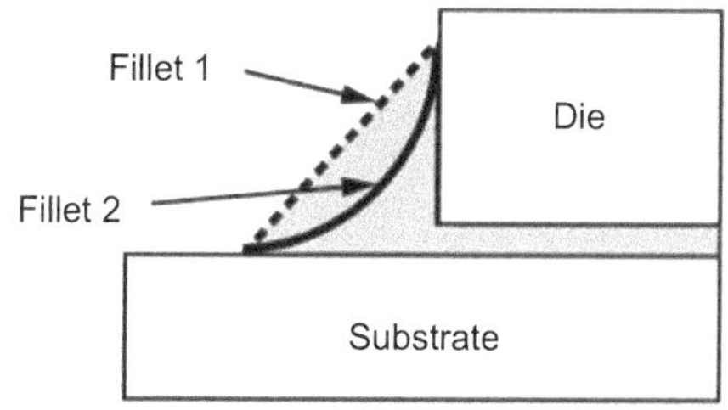

Figure 2.8 Shape factors for underfill fillets.
Source: Electronics Engineer, 1998.

$$V_f = \Big\{\Big[\text{Fillet height} \times \text{Fillet width} \times 2 \times \Big(\text{Die length} + \text{Die width}\Big)\Big] + \pi/3 \times (\text{Fillet width})^2 \times \text{Fillet height}\Big\} \times \text{Shape factor} \quad (2.10)$$

The shape factor is related to the type of underfill and its wetting characteristics. Materials with large contact angles have a high shape factor close to that of Fillet 1 in Fig. 2.8; a larger reservoir remains due to reduced wetting. Allowable filleting for dispensing takes into account the tolerances of chip size and underfill gap. A large gap allows more material to flow under the device and less material remains for the fillet (shape factor Fillet 2). A small gap leaves a larger reservoir and produces a higher fillet (shape factor Fillet 1). Examples of the relationship between shape factors and contact angles are given in Table 2.2.[37]

2.1.3 Adhesion mechanisms

Bonding of electronic components and parts may be metallurgical using solder, other metal alloys; or eutectics, organic using polymer adhesives; or vitreous using

Table 2.2 Contact angle and fillet shape factors for underfills[37]

Contact angle (°)	Shape factor
1	0.1109
5	0.3202
10	0.4739
15	0.5849
20	0.6742
25	0.7505
30	0.8187
35	0.8817
40	0.9416

Source: Electronics Engineer (1998).

glass-phase adhesives. Polymer adhesive bonding may result from direct-chemical reactions, hydrogen bonding, or mechanical interlocking. In reality, bonding is quite complex and involves more than one mechanism.

2.1.3.1 Metallurgical bonding

Metallurgical bonding and interconnections dominated the early electronics field. Tin–lead solder has been used for decades and is still in wide use today both for mechanical attachment and electrical connections. Examples include surface mounting of components on PWBs, soldering the leads of cable assemblies, edge connectors and transformers to PWBs, and attaching leadframes to substrates and packages. Eutectic alloying of silicon with gold was especially useful initially for bonding small transistor die and electrically connecting them. Although solder attachments have been supplanted to a large extent by polymer adhesives, they are still widely used in the large-scale production PWAs, flip-chip devices, and BGA packages. Major limitations of solder attachment that have plagued the industry from the beginning include the difficulty in thoroughly removing flux residues, most of which are corrosive; the high temperatures (220–240 °C) required to apply and reflow the solder; and, in some cases, solder fatigue and cracking. The limitations in using eutectics for die attachment have been the high temperatures required to interdiffuse the gold with the silicon and the stresses generated when attaching large die.

2.1.3.2 Direct bonding

Direct-chemical bonding occurs when functional groups of a polymer react directly with functional groups of a substrate.[38,39] Strong covalent bonds can thus be formed between the functional groups of a polymer adhesive and those of a substrate such as a plastic-laminate circuit board during the curing of the adhesive. The functional groups have to be highly reactive and high temperatures may be required for this type of mechanism to occur. For this reason, and because most other surfaces to be bonded are ceramic or metal, direct-chemical bonding is not a major type. One exception is the use of organosilane adhesion promoters, also known as silane coupling agents. Organosilanes such as aminomethyl-triethoxysilane form chemical bridges between a polymer and a surface containing hydroxyl functional groups.

2.1.3.2.1 Adhesion promoters

For some surfaces, adhesion must be augmented using an adhesion promoter either as an ingredient in the adhesive formulation or separately as a primer applied to the substrate prior to dispensing the adhesive. *Adhesion promoters* are difunctional or multifunctional monomers or oligomers where one or more reactive groups chemically bond to the substrate and the other reactive groups form bonds with the adhesive, thus creating a molecular bridge. Because of this bridging mechanism, adhesion promoters are also known as coupling agents. The most widely used adhesion promoters are the aminoalkoxysilanes (such as 3-aminopropyltrimethoxysilane) and mercaptoalkoxy silanes. The amino or mercapto groups form chemical bonds with the polymer molecule while the alkoxy groups, after conversion to hydroxyl groups when

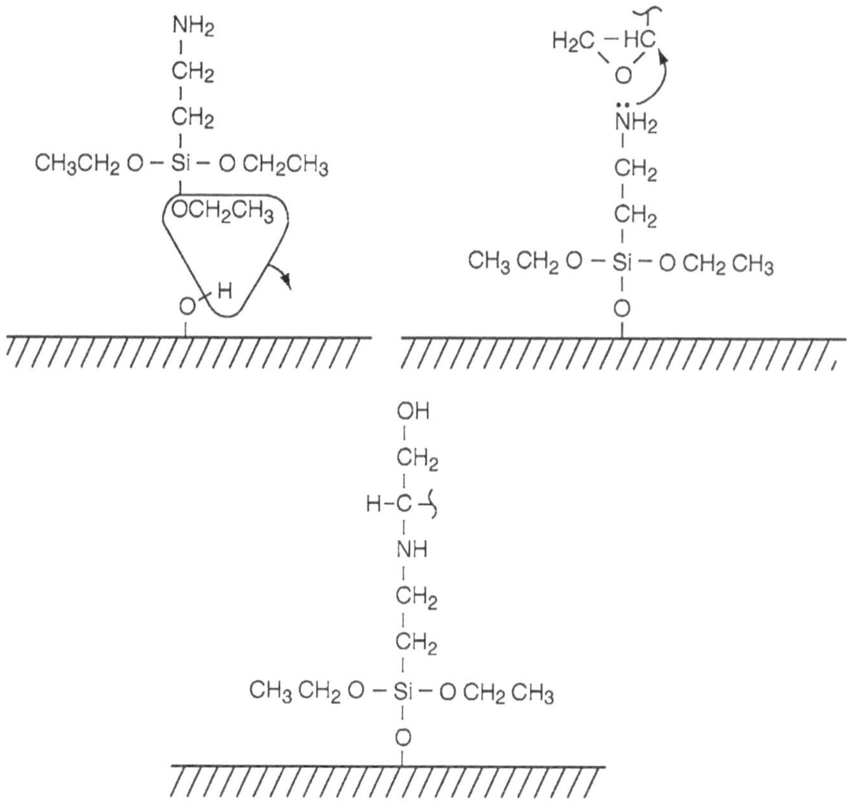

Figure 2.9 Proposed mechanism for adhesion promotion using a silane.

exposed to small amounts of moisture, bond chemically to the oxide or hydroxyl groups of ceramic or metal substrates. The more chemically active sites the substrate has, preferably hydroxyl groups, the better the adhesion will be. A proposed mechanism is shown in Fig. 2.9. Adhesion promoters consist of very dilute solutions, for example, 0.1–1% by weight of the organosilane in a mixture of 95% methanol and 5% water. Organosilane primers are especially effective in enhancing the adhesion of polyimides to silicon, silicon oxide, and silicon nitride surfaces.[40] Organosilanes are also used to coat filler particles to enhance their adhesion to the resin matrix in formulating metal-filled epoxy or polyimide adhesives.

2.1.3.3 Hydrogen bonding

One mechanism for bonding highly polar polymers to equally polar sites on substrate surfaces involves hydrogen bonding (H-bonding). *Hydrogen bonding* occurs between the hydrogen atoms of one molecule and the highly electronegative atoms of other molecules or of a substrate.[41] H-bonding is the attractive force between atoms or groups of opposite polarities without actually forming a chemical bond. Hydrogen

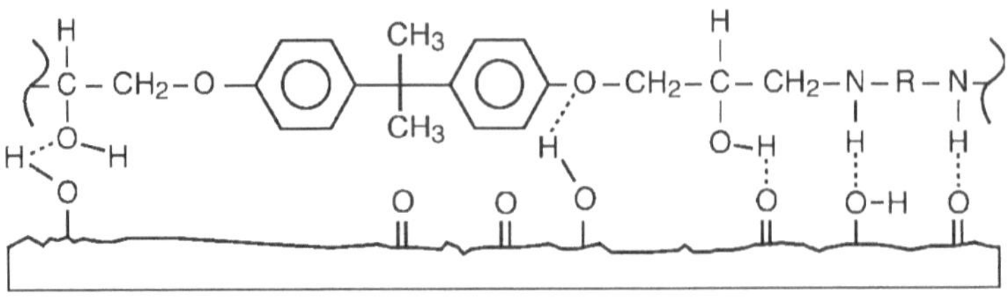

Figure 2.10 Representation of H-bonding of epoxy to an oxide surface.

atoms bond strongly to highly electronegative atoms such as fluorine, oxygen, nitrogen, and sulfur. Thus, the hydrogen of hydroxyl groups, present in most epoxy, polyurethane, and polyimide resins and polymers, readily bonds with the oxide and hydroxyl groups of metal oxides present on metal and ceramic surfaces. Hydrogen bond interactions between epoxy adhesives and epoxy-laminate or other plastic-laminate circuit boards also result in strong adhesion. Figure 2.10 shows the authors' concept of the role that H-bonding plays in the adhesion of a partially cured epoxy to an oxide surface such as a silicon oxide layer on a silicon substrate or an aluminum oxide of an alumina ceramic. The figure shows several ways in which hydrogen, oxygen, and nitrogen can participate in forming H-bonds.

2.1.3.4 Mechanical interlocking

Together with hydrogen bonding and, in some cases direct-chemical bonding, mechanical interlocking plays a key role in forming strong bonds. Mechanical interlocking is a mechanism whereby a low viscosity liquid or a thixotropic paste fills the porous surfaces of some materials or encapsulates the peaks and valleys of a rough surface. In both cases, the greater exposed surface areas compared with smooth non-porous surfaces provide greater sites for attachment. Examples include the nascent oxides that are present on the surfaces of metals such as aluminum or purposely formed oxides such as the anodizing of aluminum or the chemical oxidation of copper. Rough surfaces also enhance mechanical interlocking. Adhesives form stronger bonds to a 96% alumina substrate with a surface roughness of 15–25 μin than to a smooth, high-purity $99^{+}\%$ alumina with a roughness of only 1–4 μin.

2.1.3.5 Vitreous bonding

Silver-glass paste adhesives are reported to bond by a chemical reaction between the glass oxides and the oxides contained in a ceramic substrate such as alumina. A simultaneous mechanism may involve melting of the lead borosilicate glass phase and fusion with the glass phase and oxides of a ceramic substrate or with oxides of a silicon substrate. Adhesion between the silver-glass and the backside of a die is reported to occur by out-diffusion of silicon through the backside chromium–gold metallization.[42] Bonding occurs during the final temperature step in the processing of the silver-glass. At the 325–340 °C step, the glassy phase softens followed by liquifying, flowing, and sintering at temperatures of 410–430 °C.[42–44] Glass die bonding is performed in an air or oxidizing ambient to avoid reducing the lead oxide of the glass to metallic lead, which then hinders the oxide bonding mechanism.[45]

2.2 Electrical connections

A second key function for adhesives is to form ohmic contacts or electrical connections between active and passive devices, connectors, leadframes, I/O leads, or other electronic parts to form a circuit. Other applications for electrically conductive adhesives include electrostatic dissipation and electrical grounding connections.

To function as electrical conductors, epoxies and other polymer resins, because they are inherently insulators, must be filled with electrically conductive particles such as metals. The selection of electrically conductive or insulative adhesives is based largely on their conductivities or, conversely, on their volume resistivities. Electrically conductive adhesives should have low resistivities initially and retain those values on aging, moisture exposure, thermal cycling, and other operating and test conditions. The resistivities of metal-filled epoxy adhesives can range from 10^{-3} to 10^{-5} ohm-cm. Electrically insulative adhesives should have high volume resistivities ranging from 10^{8} to 10^{15} ohm-cm.

2.2.1 Conductivity and resistivity

According to Ohm's Law, *electrical resistance* is the ratio of applied voltage to the current between two electrodes in contact with a material. Resistance is directly proportional to the length and inversely proportional to the cross-sectional area of the sample as follows:

$$R = \rho l/A \quad \text{and} \quad \rho = Rwt/l \qquad (2.11)$$

where R is the insulation resistance in ohms, l is the length of the sample in cm, A is a cross-sectional area in cm^2, w is the width of the sample in cm, t is the thickness of the sample in cm, and ρ is a proportionality constant called the resistivity, in ohm-cm.

It is convenient to compare and select adhesives in terms of their resistivities because these values reduce resistance measurements to a common denominator. *Volume resistivity*, which is the ohmic resistance of a cube of bulk material 1 cm per side, is frequently used to compare the electrical properties of both electrically conductive and electrically insulative adhesives.

The reciprocal of resistance is *conductance*, expressed as ohm^{-1} or mhos, and the reciprocal of resistivity is *specific conductance* or *conductivity* expressed as ohm^{-1}-cm^{-1}.

2.2.2 Electrical conductivity and fillers

Electrically conductive adhesives may be isotropic (conduction equally in all directions) or anisotropic (conduction in the *z*-direction only). Both types are widely used in the assembly and packaging of electronics.

2.2.2.1 Isotropic conduction

To obtain the lowest electrical volume resistivities (in the 10^{-3} to 10^{-5} ohm-cm range) and the lowest contact resistances, polymer resins must be highly filled with 70–80% by weight (25–30% by volume) of metal particles and must be thoroughly cured. Silver and silver alloys are the most widely used fillers but, in specific applications, other metal fillers including gold, copper, nickel, and even carbon are used. Carbon-filled polymers are often used to dissipate static electricity or to protect from electromagnetic interference (EMI).

The resistivities of the best silver-filled epoxies used for die and chip attachment are in the 10^{-4} to 10^{-5} ohm-cm range. Although these values are still several orders of magnitude worse than pure silver metal, they are adequate for most circuit applications.

In adding conductive filler to an insulating resin, the volume resistivity changes slowly until a critical level of filler is reached, called the percolation point. The *percolation point* occurs when the resistivity drops abruptly, then continues to drop slowly (Fig. 2.11).[25,46,47] Almost continuous linkage of metal particles occurs at the percolation point where typical filler volumes for silver flakes are 25–30%. According to the percolation theory, there is a minimum critical volume of filler required for electrical conductivity in a polymer at which each filler particle must contact two other particles. A misconception in the use of silver flakes is that increasing the number of contacts lowers volume resistivity.[25] Actually, the converse is true because, once the percolation point has been reached, each additional contact adds resistance. Thus, increasing the particle size can increase conductivity since the total number of contacts for a fixed volume decreases.[46]

Electrical conduction in filled epoxies occurs by current flowing through and between the metal particles. Shrinkage of epoxies during cure (approximately 3%) results in a compressive stress on the silver flake particles promoting better

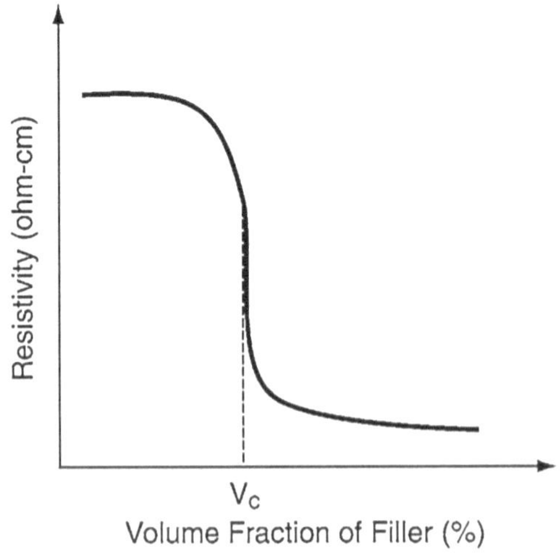

Figure 2.11 Percolation point (V_c) for loading conductive fillers in epoxy resin.

Table 2.3 Volume resistivity versus cure conditions for a silver-filled epoxy adhesive[51]

	Volume resistivity (ohm-cm)			
	Time (hours)			
Temperature (°C)	**8**	**16**	**24**	**48**
25	–	66	0.349	0.071
	Time (hours)			
Temperature (°C)	**1**	**2**	**3**	**4**
50	–	111	0.148	0.041
65	0.086	0.010	0.016	0.007
100	0.005	0.004	0.004	0.004
120	0.003	0.003	0.003	0.003

Data Provided by Thermoset, Lord Chemical Products.

particle-to-particle contact and improved conductivity.[48] Filler particles come in various shapes and sizes, for example, spheres, flakes, fibers, or granules. Silver is normally used as flakes along with anisometric particles added to provide a higher number of contact points.[49] The optimum particle sizes and geometries are those that require the minimum filler concentration to provide the highest conductivity.[46] In an evaluation by Jost et al.,[50] the effect of filler size on volume resistivity was investigated. Adhesive formulations in which the fillers consisted of large flakes (6–11 μm) and low surface area (0.3–0.5 m^2/g) had volume resistivities two orders of magnitude lower than formulations with flakes nearly half the size (3–7 μm) and surface area 0.6–0.9 m^2/g.

Optimum curing conditions are also important in attaining the highest conductivity. The improvements in electrical conductivity resulting from increased time and temperature cure conditions for a commercial silver-filled epoxy adhesive are found in Table 2.3.[51] This low temperature-curing adhesive is used in assemblies having temperature-sensitive components.

Electrical conductivities of fully cured, silver-filled epoxies are quite stable, decreasing only slightly when measured at elevated temperatures (Fig. 2.12, Table 2.4). Extended cures and aging at 150 °C for 1,000 hours have even improved their conductivities.

2.2.2.2 *Anisotropic conduction*

Electrical conductivity in anisotropic adhesives occurs by a mechanism different from that of isotropic adhesives. Although metal fillers are also used, they are used in much lower amounts (0.5–5% by volume) so that the adhesive is essentially an insulator in the x–y directions. On inserting the adhesive between the electrodes (for example, the metal bumps of a flip-chip device with metal pads on a flex circuit) of two parts and applying pressure and heat, the metal particles form a z-direction electrical connection between the electrodes while the surrounding material remains insulating. The

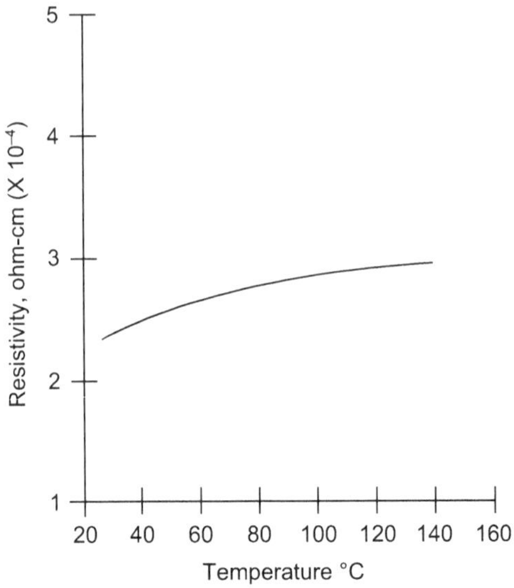

Figure 2.12 Volume resistivity versus temperature for silver-filled epoxy paste adhesive.

Table 2.4 Volume resistivities for silver-filled epoxy adhesives (ohm-cm)

Adhesive	At 25 °C	At 60 °C	At 150 °C	At 25 °C after 1,000 hours at 150 °C
Ablebond® 36-2[a]	2.4×10^{-4}	2.6×10^{-4}	2.9×10^{-4}	1.4×10^{-4}
Ablebond® 84-1- LMI[a]	7.0×10^{-5}	8.0×10^{-5}	9.0×10^{-5}	5.0×10^{-5}
Amicon C868-1	3.4×10^{-4}	3.8×10^{-4}	4.8×10^{-4}	2.1×10^{-4}
Epibond® 7002	8.0×10^{-5}	4.8×10^{-4}	5.0×10^{-4}	5.5×10^{-4}
Epo-Tek® H35-175M[b]	2.0×10^{-4}	2.5×10^{-4}	2.5×10^{-4} (at 125 °C)	2.7×10^{-4}

[a] Henkel.
[b] Epoxy Technology, Inc.

shrinkage of the polymer that occurs during curing generates a compressive force on the joints that sets them in place. The dependence of the volume fractions of a nickel filler on the *x–y* and *z*-directions of an anisotropic film adhesive is shown in Fig. 2.13.

2.3 Thermal dissipation

Thermal dissipation is becoming more important as component densities increase and devices operate at higher speeds and consume more power. Semiconductor junction temperatures must be kept low to assure long-term reliability. This is especially critical for gallium-arsenide ICs because the thermal conductivity of gallium arsenide (GaAs) is about half that of silicon and because GaAs devices operate at higher

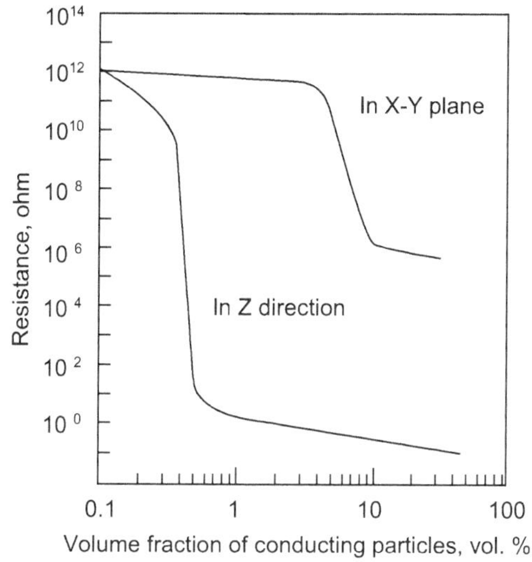

Figure 2.13 Dependence of volume fractions of conductive filler on the electrical conductivity of an anisotropic adhesive.[52]

frequencies than silicon devices. The main methods for removing heat are conduction through solids or liquids, convection by flowing cooled air or other gas over the devices, and radiation.[53] Of the three, conduction and convection are the methods most widely used for cooling electronic devices and assemblies. Adhesives play a key role in conducting heat from the devices through several interconnect substrate materials and interfaces and ultimately to a heat sink, a heat spreader, or a heat exchange unit. In high-density multichip modules, there are at least three adhesive interfaces: from die to substrate, substrate to the inside of a package base, and package case to a PWB. Because the thermal conductivities of unfilled polymer adhesives, and even of filled adhesives are low compared to metals and ceramics, they are often the limiting factor in the transfer of heat.

2.3.1 Thermal conductance and thermal conductivity

Heat transfer by conduction is governed by Fourier's Law, which states that the flow of heat through a material is directly proportional to the cross-sectional area, the temperature difference across the material, and the time, but inversely proportional to the thickness of the material as follows:

$$Q = \mathrm{k}A(T_2 - T_1)t/x \tag{2.12}$$

where Q is the rate of heat flow or thermal conductance, A is the cross-sectional area, $T_2 - T_1$ is the temperature drop, t is the time, and x is the thickness.

The proportionality constant, k, is the thermal conductivity or coefficient of thermal conductivity, a value specific for each material and generally reported as W/m K by electrical engineers or as cal/second cm °C by chemists and materials engineers (cf. Appendix Table A-1 for conversion factors).

Materials having the highest thermal conductivities are the metals, metal alloys, and some ceramics while plastics, polymeric materials, and glasses have the lowest thermal conductivities.

The flow of heat from the junction of a semiconductor device through the device material and on through several layers of an interconnect substrate may be treated as a series of temperature drops or thermal resistances at each interface. The junction-to-case thermal resistance θ_{JC} expressed in °C/W is widely used in thermal analysis. The junction-to-case thermal resistance is proportional to the drop in temperature from the semiconductor junction to the case and inversely proportional to the power dissipation, thus:

$$\theta_{JC} = \left(T_j - T_{case}\right)/Q_{chip} \tag{2.13}$$

where T_j is the junction temperature, in °C, T_{case} is the case temperature, in °C, and Q_{chip} is the power dissipated from the chip, in W. For efficient heat flow and removal, both the temperature drop and the θ_{JC} should be low. To achieve this, the best approaches are:

- Selecting adhesives that have high thermal conductivities.
- Using as thin a bondline as possible, since thermal conduction is inversely proportional to the bondline thickness.
- Avoiding the generation of voids in the bondline due to air or solvent entrapment.

The thermal conductivities of various metals, alloys, and ceramic materials used in electronic packaging are given in Tables 2.5 and 2.6.

Table 2.5 Thermal conductivities of metals and alloys commonly used in electronic devices and assemblies

Material	Symbol or composition	Thermal conductivity (W/m K)
Alloy 42	58% Fe, 42% Ni	15
Aluminum 6061	Al	237
Chromium	Cr	66
Copper	Cu	401
Copper–molybdenum	Cu (20%)–Mo	155
Copper–tungsten	Cu (20%)–W	248
Gold	Au	318
Gold–tin (20%)	80% Au, 20% Sn	57
Gold–platinum	Au–Pt	130
Gold–silicon eutectic	96.85% Au, 3.15% Si	27
Indium	In	78
Invar	64% Fe, 36% Ni	11
Kovar® (iron–nickel–cobalt)	53% Fe, 29% Ni, 18% Co	17
Lanxide	Al-SiC	200
Lead	Pb	34
Molybdenum	Mo	130–146
Nickel	Ni	90–92

Table 2.5 Thermal conductivities of metals and alloys commonly used in electronic devices and assemblies—*cont'd*

Material	Symbol or composition	Thermal conductivity (W/m K)
Palladium	Pd	70
Platinum	Pt	72
Silicon	Si	114–125
Silicon–germanium	Si–Ge	114
Silver	Ag	429
Silver–palladium	Ag–Pd	150
Tantalum	Ta	54
Tin	Sn	63
Tin–lead solder	Sn–Pb (63/37)	53
Tin–lead solder	Sn–Pb (37/63)	42
Tin–lead solder	Sn–Pb (5/95)	36
Tungsten	W	160

Note: values vary depending on the test conditions and the purity of the sample. Ranges are based on values from several sources.

Table 2.6 Thermal conductivities of ceramic and inorganic materials commonly used in electronic devices and assemblies

Material	Symbol	Thermal conductivity (W/m K)
Alumina (92%)	Al_2O_3 (92%)	18
Alumina (96%)	Al_2O_3 (96%)	20–35
Alumina (99%)	Al_2O_3 (99%)	40
Aluminum nitride	AlN	170–260
Beryllia	BeO	250
Boron nitride single crystal powder, hexagonal platelets	BN	300+
Boron nitride (hexagonal, commercial grades)	BN	27–55
Borosilicate glass	$PbO \cdot B_2O_3 \cdot SiO_2$	0.8
Diamond	Crystalline	1,500–2,000
Diamond	Chemical vapor deposited (CVD)	800–1,300
Gallium arsenide	GaAs	43–50
Indium phosphide	InP	68
Quartz	Crystalline SiO_2	1.4
Sapphire	Crystalline Al_2O_3	32–35
Silicon carbide	SiC	270
Silicon dioxide	SiO_2	1.5–7
Silicon nitride	SIN	170

Note: values vary depending on the test conditions, test method, crystal structure and purity. Ranges are based on values from various sources.

2.3.2 Thermal conductivities of adhesives

Unfilled polymer adhesives are inherently thermal insulators. Their thermal conductivities range from 0.1 to 0.3 W/m K.[54,55] By formulating resins with metal, metal oxide, or inorganic fillers, thermal conductivities can be increased 10-fold or better. Commonly used metal fillers include silver, gold, nickel, and copper. The use of metal fillers, however, also renders the adhesive electrically conductive and is not suitable where electrical insulation or isolation is required. Commonly used inorganic electrically nonconductive fillers include alumina, silica, beryllia, aluminum nitride, and cubic boron nitride. Generally, as a minimum thermal conductivity of 1.5 W/m K is specified for electrically conductive adhesives and 0.15 W/m K for electrically insulative adhesives. Some recent epoxy formulations filled with diamond particles or with cubic boron nitride are reported to have thermal conductivities as high as 10 and 11.4 W/m K, respectively.[56] Furthermore, some silver-filled epoxies are reported to have thermal conductivities as high as 5.7 W/m K (Table 2.7).

In addition to the type of filler, thermal conductivity depends on many other factors including the amount of filler used, whether it is by weight or by volume, its form and size, the surface treatment of the filler particles, the completeness of cure, and the test conditions used for measurement. Test methods are described in MIL-STD-883, Method 1012.1 and in ASTM C518.[57,58]

2.3.3 Effect of fillers on thermal conductivity

The conductivities of adhesives can be increased by formulating them with the maximum amount of the highest thermal filler. Thermal conductivity increases modestly, up to a filler level of approximately 70% by weight, then increases significantly, suggesting a percolation threshold similar to that for electrically conductive adhesives. As such, viscosity also increases rapidly at about 70% filler loading (Fig. 2.14). With filler levels above 80% by weight, the adhesive becomes so viscous that it cannot be easily dispensed. Even with such high filler loadings, the thermal conductivities of the formulations are still far below those of the filler itself (Tables 2.5–2.7). Advancements in this area will be required to meet the thermal requirements for the next generation of high-density circuits.

The thermal conductivity (K_e) of a filled adhesive may be predicted from the Lewis and Nielsen model.[60]

$$K_e = K_c \frac{1 + AB\phi_2}{1 - B\phi\phi_2} \tag{2.14}$$

where K_c is the thermal conductivity of the continuous phase (the polymer matrix) (typically 0.2 W/m K), A is the shape parameter that increases with aspect ratio, ϕ_2 is the volume fraction of the filler, ϕ_m is the maximum filler level while maintaining a continuous matrix phase (packing factor), ϕ is calculated from ϕ_2 and the packing factor ϕ_m as in Eq. (2.15)

$$\phi = 1 + \left(\frac{1 - \phi_m}{\phi_m^2}\right)\phi_2 \tag{2.15}$$

Table 2.7 Thermal conductivities of adhesives and plastics commonly used in electronic devices and assemblies

Material	Filler	Commercial product	Supplier	Thermal conductivity (W/m K)
Thermoplastic adhesive (paste)	Unfilled	Staystik® 383	Cookson Electronics	0.25–0.3
	Silver	Staystik® 181		3.0–3.5
	Alumina	Staystik® 783		0.5–1.0
	Aluminum nitride	Staystik® 282		1.0–1.5
Thermoplastic adhesive (film)	Unfilled	Staystik® 482	Cookson Electronics	0.25–0.3
	Silver	Staystik® 581		3.0–3.5
	Alumina	Staystik® 872		0.5–1.0
	Boron nitride	Staystik® 1172		8.0–10.0
	Aluminum nitride	Staystik® 682		1.0–1.5
	Alumina	TP 8550	AI Technology	6.4
Epoxy film, thermally conductive	Glass fabric	Ablefilm® 506	Henkel	0.9
Tack-free epoxy film	Alumina	ESP 7455-W	AI Technology	>1.5
Epoxy paste adhesive	Silver	Ablebond® 84-1 LMI SR4	Henkel	2.5
	Silver	EG 8020	AI Technology	5.7
	Inorganic	ME 7155		>1.7
	Silver	Epo-Tek® 35-175MP	Epoxy Technology	1.5
	Gold	Ablebond® 85-1	Henkel	2.0
	Diamond	ME 7156	AI Technology	11.4
	Alumina	Ablebond® 84-3	Henkel	0.8
	Silica	LCA4-BA5	Bacon Industries	0.8
Epoxy-glass laminate	Glass fabric reinforced	FR-4, G-10	Various	0.2–0.3
Polyimide coating	Unfilled	PI-2611D	DuPont	0.1
Polyimide-glass laminate	Glass fabric reinforced	Various	Various	0.45
Polyurethane	Unknown	Flexobond 442	Bacon Industries	0.86
Silicone paste adhesive	Unknown	S-6605	Dow Corning	0.85
Silver-glass paste adhesive	Silver	QMI 3555R	Henkel	90

Note: values are those reported in suppliers' data sheets. Please contact the supplier for values to be used in specifications.

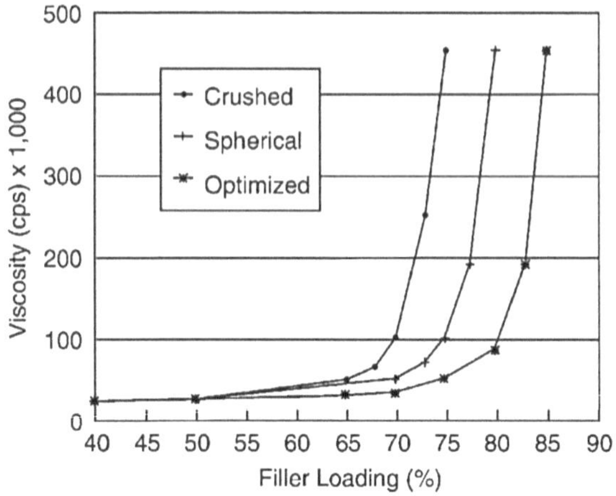

Figure 2.14 Percolation threshold of thermally conductive fillers.[59]

B is calculated from the ratio of thermal conductivities of the two phases and A as in Eq. (2.16).

$$B = \left\{ \frac{\left(\dfrac{K_f - 1}{K_c} \right)}{\left(\dfrac{K_f + A}{K_c} \right)} \right\} \tag{2.16}$$

where K_f is the thermal conductivity of the filler.

Graphs of the actual and predicted thermal conductivities for a silver-filled epoxy composition are shown in Fig. (2.15).

2.4 Stress dissipation

A function of adhesives that is becoming increasingly important is to absorb and dissipate stresses produced in electronic assemblies due to changes in environmental conditions or accelerated testing. The need for stress-absorbing and dissipating adhesives has increased with advancements in microelectronics. Some advancements requiring low-stress adhesives are:

- Introduction of large-chip devices (up to 1-in square).
- Increase in the number of I/Os per device (over 500 microbumps for flip-chips).
- Use of thinned silicon or GaAs die (as low as several mils).
- Greater use of fragile components (glass diodes, fiberoptics, MEMS).

In addition, there has been greater use of flexible circuitry and foldable flex circuits requiring a corresponding degree of flexibility in the adhesives used. For these applications, solders or other metallurgical attachment materials have often failed due to cracking, delamination, electrical opens, or changes in electrical parameters. Some

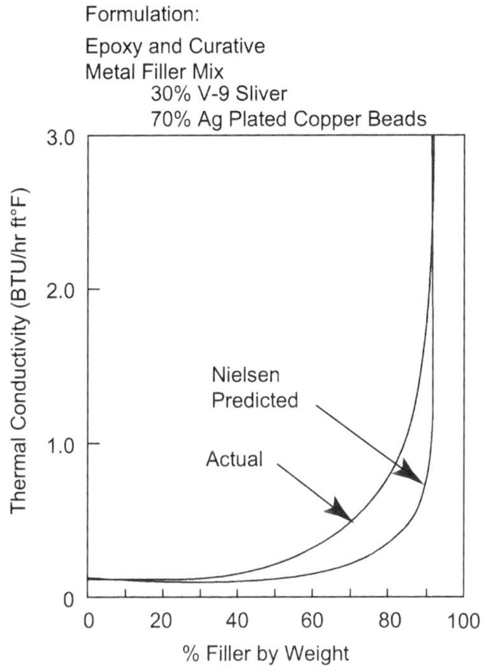

Figure 2.15 Predicted thermal conductivity using the Nielsen model compared with actual data. *Copyright:* Adhesives Age, 1989, reproduced with permission.

of these failures occur immediately after assembly while others are induced on aging, temperature cycling, vibration, thermal or mechanical shock, or exposure to moisture. A major condition for failures is a large mismatch in the CTEs of adherends bonded with high-modulus materials and subjected to large temperature extremes.

Adhesives formulated from flexible, elastomeric resins are now commercially available and effectively used to dissipate stresses. They are low-stress or "no-stress" adhesives that have a low modulus of elasticity. Several guidelines that can be followed to reduce stresses are the following:

- Selecting an adhesive that has a low shrinkage during cure (<3%).
- Selecting an adhesive that after curing matches the CTEs of the adherends as closely as possible.
- Selecting an adhesive that has a low modulus of elasticity.
- Moderating the cure conditions, for example, employing a step cure instead of a rapid high-temperature shock cure.
- Selecting an adhesive whose T_g lies below or above its expected operating or testing temperature.

2.4.1 Thermal expansion and stresses

The degree of linear or volume expansion of a material with increasing temperature is an important thermomechanical parameter in predicting and assessing stresses.

The coefficient of thermal expansion (CTE) may be based on either linear or volume expansion. Unless otherwise stated, CTE for this discussion is based on linear expansion and is the ratio of the change in length per °C to the length at 0 °C, as follows:

$$l_T = l_0(1 + \alpha T) \tag{2.17}$$

or

$$\alpha = (l_T - l_0)/l_0 T$$

where l_T is the length at temperature T, l_0 is the length at 0 °C, and α is the CTE. Coefficients of thermal expansion are reported as cm/cm/°C or, more generally, as unit/unit/°C or ppm/°C. CTEs vary with temperature and are usually reported for a temperature range. The coefficients of volume expansion are generally three times those for linear expansion.

A knowledge of the CTEs of materials being bonded and of the attachment materials is extremely important in predicting and avoiding stresses due to large mismatches in expansion during temperature changes. Use of a rigid, high-modulus of elasticity material to join members that have large differences in their CTEs invariably results in warpage, cracking, and delamination. Conversely, materials having closely matched CTEs such as small silicon die (CTE = 2.6–3.0 ppm/°C) can be attached reliably to Kovar® leadframes (CTE = 5.1–5.5 ppm/°C), even with high-modulus adhesives. With the introduction and wider use of copper leadframes (CTE = 16–17 ppm/°C), only the low-modulus, low-stress adhesives can be used without delamination at the die interface or sometimes cracking of the die.

Low-stress and stress-dissipating adhesives were also found to be essential as underfills for flip-chip bonded devices, chip-scale packages, and BGA packages. Underfill adhesives are designed to protect solder-ball interconnections by minimizing the stress and strain caused by mismatches in the CTEs of the adherends, especially during temperature cycling. In effect, the underfill bonds the lower-CTE, high-modulus silicon chip to the higher-CTE, lower modulus substrate (typically a plastic-laminate circuit board or a ceramic interconnect substrate).[61] The result is that the bonded surfaces assume a "composite" CTE, constraining the substrate modulus with the high stiffness of silicon. The underfill also protects the part during shock and vibration.[62] Ideally, one would select all materials with matched or closely matched CTEs, but this is seldom possible in practice. Thus, flexible adhesives having low moduli are used and, although they may have moderate to high CTEs, are able to absorb and dissipate stresses produced. The moduli for low-stress or no-stress adhesives range from approximately 15,000 psi to 30,000 psi at 25 °C. Coefficients of thermal expansion for some non-organic materials and organic materials commonly used in electronic assemblies are given in Tables 2.8 and 2.9.

2.4.2 Glass-transition temperature and stresses

In plotting the expansion of a polymeric material as a function of temperature, there is a temperature at which an abrupt or significant increase in expansion occurs. This temperature, known as the *glass-transition temperature* (T), is the temperature that

Table 2.8 Coefficients of thermal expansion for some non-organic materials used in electronic assemblies

Material	CTE (ppm/°C)
Alumina ceramic	5.6–6.7
Alumina, 92%	7
Alumina, 96%	7.7, 6.6
Aluminum	23.5
Aluminum nitride	3.8–4.4
Beryllia	4.7
Chromium	4.9, 6.3
Copper	16–17
Copper–tungsten (20% Cu)	6.5–8.3, 7.0
Copper–molybdenum (20% Cu)	5.8–8.3, 7.2
Diamond, single crystal	1.1
Diamond, CVD	0.4–1.5
Gallium arsenide	6.86, 5.9
Glass, borosilicate	3.24–4.5
Gold	14.1
Indium phosphide	4.75
Invar	0.64, 1.5
Kovar®	5.1–5.5
Lanxide (Al-SiC)	6.7–7.3
Lead	29
Molybdenum	4.8
Nickel	7.8, 13.3
Palladium	11.8
Platinum	8.9
Quartz (crystalline SiO_2)	0.3–0.5
Sapphire (crystalline Al_2O_3)	5.5
Silicon	2.6–3
Silicon carbide	2.8
Silicon–germanium	4
Silicon oxide (fused)	0.5
Silicon nitride	2.3
Silver	19.1
Tin	20
Tin–lead solder (60–40)	24.6
Tungsten	4.5

CTE values were collected from various sources. Ranges of values or different values are due to differences in test methods, test conditions or purity of the sample.

corresponds to the intersection of lines drawn tangent to the leading baseline and the steepest portion of the post transition curve. Figure 2.16 is a generalized thermal expansion curve showing the inflection point at which the T_g occurs.

At the T_g a material change occurs from a solid rigid glassy state to a soft amorphous state. The polymer molecules are transformed from a somewhat ordered

Table 2.9 Coefficients of thermal expansion and glass-transition temperatures for some adhesives and plastics used in electronic assemblies

Material	Product	Supplier	T_g (°C)	CTE, <T_g (ppm/°C)	CTE, >T_g, (ppm/°C)
Thermoplastic adhesive (paste)	Staystik® 383	Cookson Electronics	98	N/A	N/A
	Staystik® 181		98	N/A	N/A
	Staystik® 783		98	N/A	N/A
	Staystik® 282		98	N/A	N/A
Thermoplastic adhesive (film)	Staystik® 472	Cookson Electronics	25	N/A	N/A
	Staystik® 571		25	N/A	N/A
	Staystik® 872		25	N/A	N/A
Thermoplastic adhesive (film)	Staystik® 672	Cookson Electronics	25	N/A	N/A
	TP 8550	AI Technology	−70	N/A	65
Epoxy film, thermally conductive	Ablefilm® 506	Henkel	3	45	350
	Ablefilm® 550K		105	95	480
Tack-free epoxy film	ESP 7455	AI Technology	−60	N/A	110
Epoxy paste adhesives	Ablebond® 84-1 LMI SR4	Henkel	103	50	200
Epoxy paste adhesives	EG 8020	AI Technology	50	40	N/A
	ME 7155		−25	N/A	120
	Epo-Tek® 35-175MP	Epoxy Technology	>100	31	97
	Ablebond® 85-1	Henkel	100	50	200
	ME 7156	AI Technology	−25	N/A	120
	Ablebond® 84-3	Henkel	85	40	100
	LCA4-BA5	Bacon Industries	84	27	N/A
	Epo-Tek® 930-4	Epoxy Technology	≥90	27	136

Epoxy polyphenylene oxide – E glass[a,b]	Various	Various	180	12–18 (*x,y*)	150–170 (*z*)	–
Epoxy-glass laminate[a,c]	FR-4, E-glass, G-10	Various	140–200	13–18 (*x,y*)	55–100 (*z*)	–
Epoxy – Aramid, woven[a,b]	Various	Various	140–180[b]	6–8 (*x,y*)	90–150 (*z*)	–
Epoxy – Aramid, nonwoven[a–d]	Various	Various	140–180	7–9 (*x,y*)	80–110 (*z*)	–
Polyimide – E-glass laminate[b–e]	Arlon® 33N	Arlon	>250	13–17 (*x,y*)	45–75 (*z*)	–
Polyimide – quartz[d]	Various	Various	>250	6–8 (*x,y*)	34 (*z*)	–
Polyimide – Aramid[b,d]	Various	Various	>250	13–15 (*x,y*)	4–100 (*z*)	–
Polyimide – Aramid[a,c]	Various	Various	>250	7–9 (*x,y*)	75–95 (*z*)	–
Polyetherimide-glass[f]	ULTEM® 2300	GE	215	23 (*x,y*)	–	–
Cyanate ester – E-glass[b,g]	Various	Various	>230	9–10 (*x,y*)	45–55 (*z*)	–
Cyanate ester – S-glass[a,f,g]	Various	Various	230	8–10 (*x,y*)	40 (*z*)	–
PTFE – woven glass[h]	GX	Rogers	N/A	15 (*x,y*)	200 (*z*)	–

[a] Hu M. Choosing laminates. *Advanced Packaging*, pp. 16–17 (Fall 1993).
[b] www.istrep.com/Downloads/PWB%20Properties.doc.
[c] Non-woven Aramid Reinforcements: Controlled Thermal Expansion Prepreg and Laminate for Printed-Wiring Boards, *Nepcon West*, New Substrates, Sessions TS-01 and TS-00 (Mar. 27, 1995).
[d] Thermount Applications Guidelines, Arlon Materials for Electronics Division (undated).
[e] Arlon 33N Product Data Sheet, 0901-RI (2000).
[f] www.boedeker.com/ultem/p.html.
[g] Markstein H. MCM-L offers easiest path to high performance, *Electronic Packaging and Production*, pp. 48–51 (Dec. 1993).
[h] Comparing Circuit-Board Materials for DBS Applications, Rogers Corp. brochure (undated).

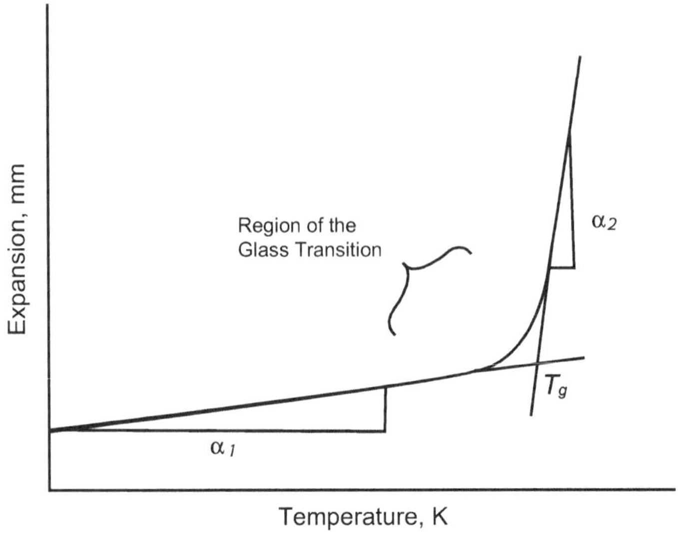

Figure 2.16 Generalized thermomechanical analysis curve for epoxy adhesive.

state to a random state of high molecular motion. The change in CTE at the T_g can be 2–5 times greater than its value before the T_g is reached and can induce major stresses. For this reason, adhesives suppliers should report the T_g for each adhesive and the CTEs for temperatures below and above the T_g. Generally, suppliers report two CTE values: α_1 from -55 °C to the T_g and α_2 from the T_g to 155 °C.

The glass transition is a second-order transition, that is, the second derivative of the free energy function. In comparison, a first-order transition, such as the melting of a material, is the first derivative of Gibbs free energy. The equation used to determine the glass transition[63] is:

$$\alpha = \left(\frac{1}{V}\right)\left(\frac{\partial V}{\partial T}\right)_p = \left(\frac{1}{V}\right)\left(\frac{\partial}{\partial T}\right)\left[\left(\frac{\partial G}{\partial P}\right)_T\right]_P \tag{2.18}$$

where α is the expansion coefficient, V is volume, P is pressure, T is temperature, and G is Gibbs free energy.

The glass transition occurs where α changes abruptly from a low to a high value. It is determined by monitoring either dimensional changes over a temperature range [for example, by thermomechanical analysis (TMA)] or by measuring heat flow using differential-scanning calorimetry (DSC).

2.4.3 Residual stresses

Residual stresses are the result of shrinkage of the adhesive during curing and polymerization and during the cool-down from the curing temperature to room temperature. Residual stresses may be quantified by measuring the warp that results in a substrate from curing the adhesive on the substrate. One method is to calculate the bending stress from the radius of curvature where bending stress, S_b, is defined as:

$$S_b = \frac{Eh}{2\gamma R} \tag{2.19}$$

where E is Young's modulus, γ is Poisson's ratio, h is the bondline thickness, and R is the radius of curvature.[64] A high radius of curvature indicates a low stress in the bondline. The radius of curvature is calculated by measuring the amount of warp. Radii of curvature of 2,000 mm or more are possible when bonding silicon ICs to silver-plated copper leadframes with stress-absorbing epoxy adhesives.[65]

A practical test to measure the effects of residual stress is to measure the out-of-plane deformation or warpage of a substrate upon which the adhesive is cured; the greater the warpage, the greater the stress, and the lower the radius of curvature. In one example, a 300-mil-square silicon die bonded with an electrically conductive epoxy to a copper leadframe, produced 10-μm warp after cure. Exposure to subsequent processing temperatures from wire bonding and over molding with epoxy further increased the warpage to 13 μm. The warp further increased to 32 μm when a larger die (500-mil square) was used.[66] The warp test has been used by formulators to develop and evaluate new low-stress adhesives.

A one-dimensional model of stress buildup in silicon devices over a temperature range is given by the following equation:[67]

$$\sigma_{Si} = \varepsilon_a \times E_a = (\alpha_a - \alpha_{Si}) \times E_a \times \Delta T \tag{2.20}$$

where σ_{Si} is the stress on the silicon, ε_a is the strain of the adhesive, E_a is the modulus of the adhesive, α_a is the CTE of the adhesive, α_{Si} is the CTE of silicon, and ΔT is the temperature difference.

Since the coefficients of thermal expansion below the T_g for most adhesives range from 25 to 100 ppm/°C and are greater than the CTEs of the device or substrate materials, stresses in bonded parts are compressive on the die and substrate and tensile on the adhesive.

Another equation for maximum stress in rectangular devices addresses differences in both coefficients of expansion and moduli as follows:[68]

$$S_{max} = K(\alpha_{sub} - \alpha_{Si})\,(T_o - T)\sqrt{\frac{E_a E_{sub} L}{x}} \tag{2.21}$$

where S_{max} is the maximum stress, K is a geometric constant related to shape and filleting, α_{sub} is the CTE for the substrate, α_{Si} is the CTE for the silicon die, E_a is the modulus for the adhesive, E_{sub} is the modulus for the substrate, L is the length of the die, x is the bondline thickness, T_o is the cure temperature, and T is the application temperature.

In summary, the stress (typically at the edge of the die) increases with die size, modulus of the adhesive, temperature, difference in expansion coefficients, and increasing bondline thickness.[44]

Finite-element analysis has also been applied to predict internal stress distributions and to help explain stress transfer mechanisms such as shrinkage during cure.[69]

2.4.4 *Effect of curing conditions on the glass-transition temperature*

Curing a polymer resin via polymerization involves several stages. For an epoxy resin cured with a hardener, viscosity increases as crosslinking of the monomer units occur. At some time–temperature condition, gelation occurs whereupon the mass becomes semisolid and rubbery. With further crosslinking and increasing molecular weight, the gel converts to a glassy solid stage (vitrification).

Gillham[70–72] developed a time–temperature-transformation (TTT) diagram (Fig. 2.17) to represent the curing process. The TTT diagram is a plot of the log of time on the *x*-axis versus the curing temperature on the *y*-axis. There are five distinct regions in the diagram:

- Liquid region
- Ungelled glassy region
- Gelled glassy state
- Gelled rubbery region
- Char region

The graph shows three critical temperatures:

$T_{g\infty}$: the maximum T_g of the fully cured material

Gel T_g: the temperature at which gelation and vitrification occur simultaneously

T_{g0}: the T_g of the mixed reactants, corresponding to a minimum cure

In isothermal cures, at temperatures between the gel T_g and T_{g0}, the resin undergoes gelation followed by vitrification. Since T_g is a function of both the degree of cure and the cross-link density, it increases to a point at which all polymer reaction sites have been consumed. At that point, the T_g reaches a plateau and does not increase further with temperature until the decomposition temperature where a char region occurs (Fig. 2.17). An example of the effect of cure conditions on T_g and CTE values was shown by Konarski[74] who varied the cure cycles of an anhydride-cured epoxy from 130 to 175 °C as shown in Table 2.10. The value of T_g continued to

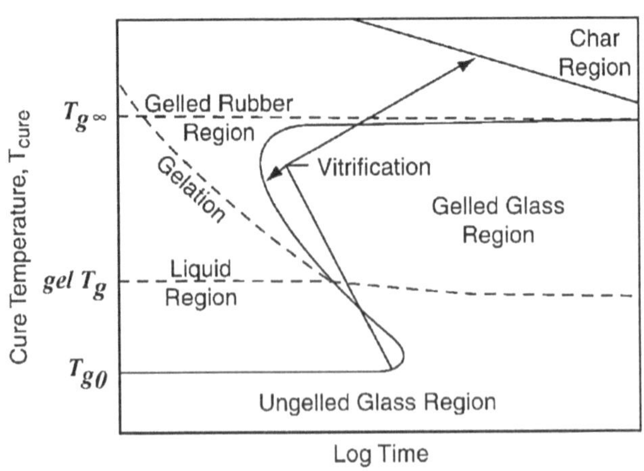

Figure 2.17 Time–temperature-transformation (TTT) diagram.[73]

Table 2.10 Effect of cure schedule on T_g and CTE for an epoxy adhesive[74]

Cure schedule (°C/hours)	T_g (°C, by TMA)	CTE (ppm/°C, 25–100 °C)
130/2	117	22.9
146/2	127	23.8
160/2	137	26.3
175/2	146	24.3

increase with increasing cure temperature while the expansion coefficient increased up to the 160 °C cure.

The dependence of curing temperature and time on the T_g and on the flexural properties of an experimental anhydride-cured epoxy containing 73% by weight silica filler was also demonstrated by differential-scanning calorimetry (DSC). Flexural strength and modulus were measured at 80 °C according to ASTM D790.[75] Figure 2.18 shows the increase in T_g with increasing cure temperature and time. The epoxy, cured at 140 °C for 50 minutes, had a T_g of 150 °C. However, increasing the cure conditions to 165 °C for 90 minutes raised the T_g to 185 °C. The benefit of increasing the cure schedule to improve the T_g reached a maximum at about 170 °C for this particular formulation. Overcuring, which results in excessive crosslinking and embrittlement, only slightly affected the T_g, but degraded both the flexural strength and modulus. Figures 2.19 and 2.20 show the effects of cure conditions on modulus and flexural strength.

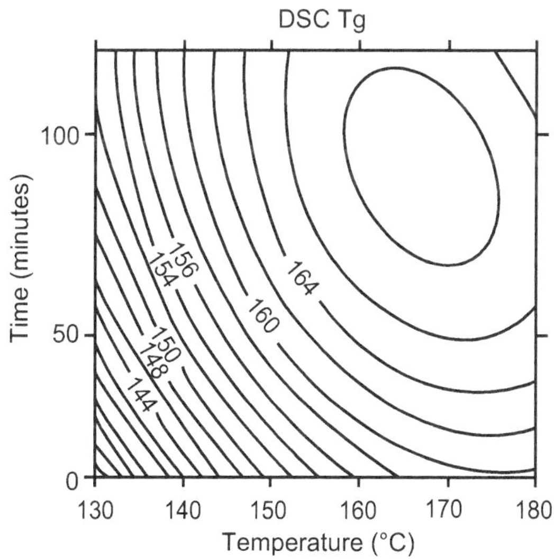

Figure 2.18 Plots of T_g (°C) versus cure conditions.
Source: Dexter Hysol, now Henkel.[75]

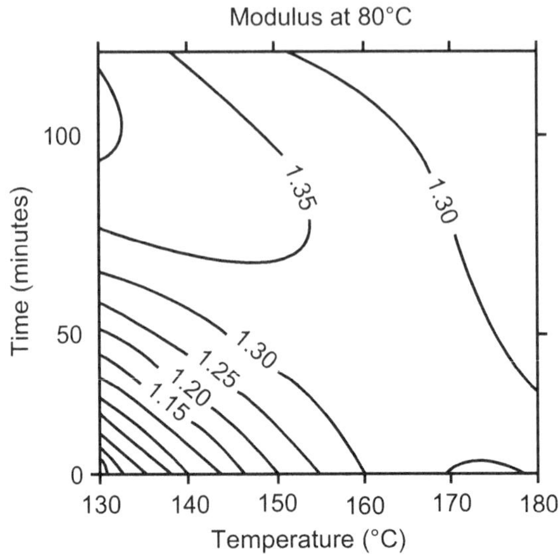

Figure 2.19 Plots of modulus (Mpsi) versus cure conditions.
Source: Dexter Hysol, now Henkel.[75]

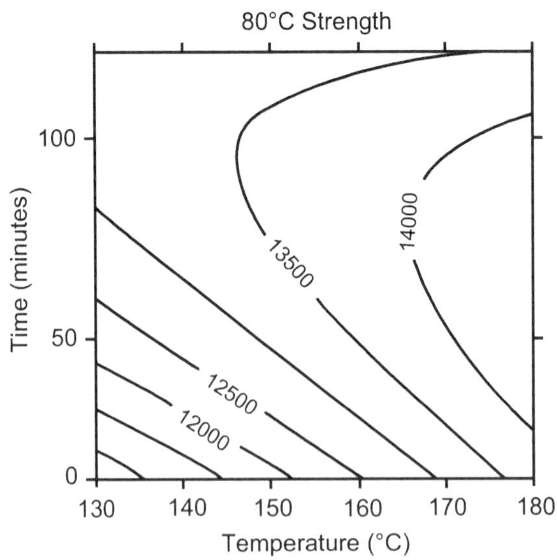

Figure 2.20 Plots of flexural strengths (psi) versus cure conditions.
Source: Dexter Hysol, now Henkel.[75]

2.4.5 Effects of gelation and curing conditions on stresses

In addition to the cure temperature and time, the manner by which an adhesive is cured affects its thermomechanical properties and stresses. Stresses are greatly reduced by step curing, that is, by allowing the adhesive to gel at a temperature significantly lower than the final cure temperature. Konarski[74] found that a pre-gel prior to final cure cycles of 130–175 °C lowered both the CTEs and stresses compared with samples that were not pre-gelled. The pre-gel step allows the polymer chains to "anneal" prior to full cure. Step curing is often used for flip-chip underfill adhesives where stress reduction is important in preventing solder ball interconnect failures, especially for large ICs.

References

1. Loctite Technical Data Sheet, Product 3607; Jul. 2003.
2. Lee H, Neville K. *Handbook of Epoxy Resins*. McGraw-Hill; 1967 [chapter 6].
3. Shukla RK, Mencinger NP. A critical review of VLSI die attachment in high reliability applications. *Solid State Technol*. Jul. 1985;67-74.
4. Tan N, Chin B, Lim KHK, Bourdillon AJ. Engineering surfaces in ceramic pin grid array packaging to inhibit epoxy bleeding. *Hewlett Packard J*. Aug. 1998.
5. Mittal KL, ed. *Surface contamination: an overview*, Surface Contamination: Genesis, Detection and Control. Plenum Press; 1979, 1:3-46.
6. Mittal KL. Adhesion Science and Technology Training Course, Appendices A and E, Center for Professional Excellence, San Mateo, CA (Sep. 1986)
7. Chung K, Ozguz V. Flexible conductive adhesive as solder replacement in flip chip interconnection. In: *Proc. Semicon. West*. 1998:127–131.
8. Bonding Solutions, 3M Co. www.3M.com/reclosables/html/fundamentals.jhtml; Aug. 26, 2002.
9. Sabreen S. Basics of surface wetting and pretreatment methods. *Plastics Technical Reference Library*. Jul. 16, 2001. The Sabreen Group.
10. *Technical Data Sheet for Ablebond 826-1DS*. Ablestik Laboratories; Oct. 2000.
11. Hannafin JJ, Pernice RF, Estes RH. Die Attach Adhesives: A Guide to Material Properties and Applications. Epoxy Technology, Technical Paper GB-48.
12. McCabe JD. Selecting the right die-attach adhesive. *High Density Interconnect*. Jul. 2001:32-6.
13. Technical Data Sheets for Epibond 7002 and 7200, Vantico; Feb. 2001.
14. Test Method Standard Microcircuits, MIL-STD-883D, Method 2019.5; 1993.
15. Ablebond 84-1 LMINB, Ablestik Laboratories Technical Data Sheet (Oct. 2000); NB1, revised (Jul. 2003).
16. Kraus HS. Chip adhesives update. *Solid State Technol*. Mar. 1987;33-4.
17. Hiraka S, Habashi M. The influence of selenium deposits on silver plating on adhesive strength. In: *Proc. 29th An. Int. Reliability Physics Symp*. Apr. 1991:8–11.
18. Summer A, Friesen D, Matthews M. *Viscosity and Its Effect in Silver-filled Adhesives*. Ablestik Laboratories Technical Paper; Aug. 1998.
19. Koh W. Encapsulants for Chip-on-board and Chip Scale Packaging. Dexter Technical Paper; Feb. 1996.
20. Loctite Technical Data Sheets: 3560 (Feb. 1997); 3561 (Nov. 1996); 3563 (Dec. 1999); and 3567 (Feb. 2000).

21. Fay HG. Rheology for the Bewildered: A Loctite Guide. Loctite Technical Paper; Aug. 1977.
22. Dixon D, Kazalski J, Murch F, Marongelli S. Practical issues concerning dispensing end factors. *Circuits Assembly* Aug. 1997;8(8):36.
23. Rasiah IJ, Ho PS, Manaharan M, Ng CL, Chau M. Rheological analysis of an underfill material. In: *Proc. 1998 IEEE/CPMT Electronics Packaging Conf.* 1998:354–358.
24. Billmeyer FW. *Textbook of Polymer Science.* New York: Interscience, Div. of John Wiley; 1962 [chapter 6].
25. McNeilly K, Pernice RF. Silver flake increases performance of conductive adhesive. *Adv Packaging.* Jun./Jul. 1999.
26. Schwiebert MK, Leong WH. Underfill flow as viscous flow between parallel plates driven by capillary action. *Proc. IEEE Trans Components Packages Mfg. Technol, Part C* Apr. 1996;19(3):133-7.
27. Gordon MH, Ni G, Schmidt WF, Silvan RP. A capillary driven underfill encapsulation process. *Adv Packaging.* Apr. 1999:34-7.
28. Bacher B. Flip-chip Process Guidelines and Considerations, Semicon West, San Jose, CA; Jul. 1999.
29. Private communication with P. Morganelli, Emerson and Cuming; Jul. 26, 2001.
30. Shi DM, McKeen K, Jensen B. Advances in flip-chip cure rates and reliability. In: *Proc. Fourth Int. Symp. and Exhibition on Advanced Packaging Materials, Processes, Properties, and Interfaces.* Braselton, GA: IMAPS; Mar. 1998.
31. Technical Data Sheets for JMI 8806 (Dec. 2, 1999); JM18802B (UF8802B) Honeywell (Oct. 2000); Technical Data Sheets for Ablefill JM8806 (Jan. 2002); Ablefill UF8802B, Ablestik Laboratories; Jul. 2001.
32. Newbold J, Lewis A. The effects of new trends in flip-chip packages on automating the underfill process. In: *Proc. NEPCON West*; Mar. 1998.
33. Gilleo K. The chemistry and physics of underfill. In: *Proc. NEPCON West*, Anaheim, CA; Mar. 3–5, 1998:280–292.
34. Anderson B, Bacher B, Gomez M. Development Methodology for a High Performance Snap-cure, Flip-chip Underfill. Dexter Electronic Materials Technical Paper; Mar. 1999.
35. Gilleo K., Blumel D. The great underfill race. In: *Proc. Int. Microelectronics and Packaging Soc.* 1998:701–706.
36. Private communication with M. Todd, Henkel-Loctite; Sep. 4, 2001.
37. Lewis A, Babiarz A, Ness CQ. Solving liquid encapsulation problems in IC packages. *Electron Engineer.* Apr. 1998.
38. Lau J, Wong CP, Prince JL, Nakayama W. *Electronic Packaging Design Materials, Processes, and Reliability.* McGraw-Hill; 1998.
39. Welsh DJ, Pearson RA. Fundamental study on adhesion improvement for underfill using adhesion promoter. In: *Proc. 51st Electronic Components and Technol. Conf.* Orlando, FL; May 2001.
40. Deleo JA, Gupta MR. *Proc. SPE, Second Int. Conf. Polyimides*, Ellenville, NY; 1985.
41. Pauling L. *The Nature of the Chemical Bond.* Cornell University Press; 1951.
42. Moghadan FK. Development of adhesive die attach technology in CerDip packages: materials issues. *Solid State Technol.* Jan. 1984.
43. Ying L. A novel approach—thermoplastic die attach adhesive. *Solid State Technol.* Sep. 1987.
44. Nguyen MN. Low stress silver-glass die attach material. *IEEE Trans Components Hybrids Mfg Technol.* Sep. 1990;13(3).
45. Trigwell S. Die attach materials and methods. *Solid State Technol.* Apr. 1995.

46. Li L, Lizzul C, Kim H, Sacolick L, Morris JE. Electrical, structural and processing properties of electrically conductive adhesives. *IEEE Trans Components Hybrids Mfg Technol.* Dec. 1993;16(8):843-51.
47. Lu D, Wong CP. Properties of conductive adhesives based on anhydride-cured epoxy systems. *J Electron Mfg.* Sep. 1999;9(3):223-32.
48. Lu D, Tong QK, Wong CP. Conductivity mechanisms of isotropic conductive adhesives. *IEEE Trans Electron Package Mfg.* 1999;22(3).
49. Leong WH. Developing an underfill process for dense flip-chip applications. In: *Proc. 1996 IEEE/CPMT Int. Electronics Mfg. Technol Symp.* Oct. 1996:10-17.
50. Jost EM, McNeilly K, Sexton P. Silver flakes for state of the art conductive adhesive applications. *Proc. Seventh Int. Microelectronics Conf.* 1992:442-8.
51. MD-200 Technical Data Sheet, Thermoset, Lord Chemical Products; Dec. 18, 1997.
52. Watanabe I, Takemura K. Anisotropic conductive adhesive films for flip-chip interconnection. In: Liu J, ed. *Conductive Adhesives for Electronic Packaging.* Electrochemical Publications Ltd; 1999 [chapter 10].
53. Minning C. Thermal management. In: Licari JJ, ed. *Multichip Module Design, Fabrication, and Testing.* McGraw-Hill; 1995 [chapter 5].
54. Bolger JC, Morano SL. Conductive adhesives: how and where they work. *Adhes Age.* Jun. 1984.
55. Cross R. Synergistic combinations of thermally conductive fillers in polymer matrices. *Proc. NEPCON West.* Feb. 1996;531-8. I.
56. Technical Data Sheet for ME-7156, AI Technologies.
57. Thermal Characteristics, MIL-STD-883, Method 1012.1; Nov. 1980.
58. Steady State Heat Flux Measurement and Thermal Transmission Properties by Measure of Heat Flow Meter Apparatus, ASTM C518, Am. Soc. Testing and Materials.
59. Goodrich B, Toyama B. Liquid encapsulation considerations. Dexter Technology Paper; Dec. 1994.
60. Hermansen RD, Tunick SA. Formulating custom-tailored thermal transfer adhesives. *Adhes Age.* Oct. 1989.
61. Gilleo K, Colterman B, Chen T. Molded underfill from flip chip in package. *High Density Interconnect.* Jun. 2000;28-31.
62. Yaeger E, Carson G. Beyond flip chip: underfills enhance CSP reliability. *Chip Scale Rev.* Mar. 2001.
63. Ng LV, Fick HJ, Bui M. How dielectric T_g affects substrate performance. *Adv Packaging.* Nov./Dec. 2000.
64. Loh KI. Low-stress adhesive. *Adv Packaging.* Winter 1993:18-22.
65. *Technical Data Sheet for Ablebond 965-IL.* Ablestik Laboratories; May 1991.
66. *Technical Data Sheet for Ablebond 826-IDS.* Ablestik Laboratories; Oct. 2000.
67. Chung KK, Dreier G, Avery E, et al. Tack-free flexible film adhesives. *Hybrid Circuit Technol.* May. 1990.
68. Mooney CT, Bolger J. Die attach in Hi-Rel PDIPS: polyimide or low-chloride epoxies. *Trans IEEE.* 1984.
69. Kelly G. *The Simulation of Thermomechanically Induced Stress in Plastic Encapsulated Packages.* Boston: Kluwer Academic Publishers; 1999.
70. Gillham JK. *Developments in Polymer Characterization-3, Applied Science*, In: Dawkins JV, ed. England 1982.
71. Enns JB, Gillham JK. *J Appl Polymer Sci.* 1983;28:2567.
72. Enns JB, Gillham JK. *Polymer Characterization: Spectroscopic, Chromatographic, and Physical Instrumental Methods*, In: Craver CD, ed. Am. Chem. Soc. 1983

73. Tess RW, Poehlein GW. *Applied Polymer Science, ACS Symp. Series* 1985;285:950.
74. Konarski MM. Effects of Tg and CTE on Semiconductor Encapsulants. Loctite Electronics Technical Paper.
75. Naito CM, Todd M. Effects of Processing Parameters on the Performance of Epoxy Encapsulant Materials. Dexter Technical Paper; Dec. 2000.

3 Chemistry, Formulation, and Properties of Adhesives

Chapter Outline

Adhesives Technology for Electronic Applications. DOI: 10.1016/B978-1-4377-7889-2.10003-8

Adhesives are used in a wide range of electronic applications and must withstand diverse environments from benign office conditions to severe long-term military, space, and automotive environments. A design engineer must, therefore, have a basic understanding of the chemistry and properties of adhesives in order to make the right selection. Knowledge of the polymer resins, formulations, and polymerization and curing mechanisms, and their bearing on the desired electrical, mechanical, and thermal properties assures the reliability of the final product.

Epoxy adhesives are the most widely used and most versatile of the polymer types. They combine numerous desirable properties including adhesive strength, thermal stability, reparability, dielectric properties, and retention of the properties under high thermomechanical stresses and environmental exposures such as humidity. Through the choice of hardeners, catalysts, fillers, and other additives and by controlling the degree and method of curing, properties can be altered and tailored for specific applications. Silicones meet the needs for very high and low temperature extremes and, because of their low moduli of elasticity and flexibility, are able to dissipate stresses and dampen vibration.

Besides epoxies and silicones, other polymer-based adhesives used for electronic applications include:

- Polyimides for die and substrate attachment and for multilayer circuit insulation where resistance to high temperatures and electrical stability are required.
- Polyurethanes for their thermoplastic properties, ease of rework, flexibility, vibration damping, and stress dissipation.
- Cyanate esters for single-chip attachment in single packages where cure temperatures of approximately 175 °C and postcure temperatures of approximately 200 °C result in high thermal stability and high glass-transition temperatures (T_g).
- Modified cyclo-olefin thermosets (MCOT) for chip attachments requiring very low moisture absorption.
- Acrylics and acrylates for low cost, rapid processing especially in UV-curing formulations.

Lastly, silver-glass adhesives are included in this discussion. Though not polymer types, silver-glass adhesives contain temporary polymer/organic binders that are burned off after bonding. However, very high processing temperatures, up to 400 °C, are required for silver-glass adhesives, thus limiting their use in many temperature-sensitive applications.

3.1 Chemistry

The basic chemical reactions, used to synthesize monomer and polymer resins and the chemistry involved in the use of curing agents to polymerize the resins, have been extensively studied and are well documented. This section serves as a summary only of those polymers that are primarily used in adhesives formulations for electronic applications. Among these polymers are the epoxies, silicones, polyurethanes, polyimides, acrylates, cyanate esters, and cyclo-olefins. Further technical detail for these polymers may be acquired through literature searches in the transactions of the American Chemical Society (Polymer Group), Society of Plastics Engineers (SPE), the Society for the Advancement of Materials and Process Engineers (SAMPE) and transactions of the Institute for Electrical and Electronics Engineers (IEEE) such as Electronic Packaging Manufacturing and Advanced Packaging.

3.1.1 Epoxies

Epoxy adhesives are widely used in electronics assembly and packaging.[1] They are formed by polymerizing epoxy resins having one or more terminal glycidyl ether (epoxy) groups with a variety of curing agents,* hardeners, or catalysts.

*Curing agent is often used interchangeably with hardener and catalyst. Strictly speaking, *curing agent* is the broader term encompassing both hardeners and catalysts. A *hardener* is a curing agent that reacts with a resin and becomes part of the polymer through cross-linking while a *catalyst* is also a curing agent that initiates polymerization of the resin, but does not become part of the final polymer structure.

3.1.1.1 Epoxy resins

Most commercially available epoxy resins are based on the diglycidylethers of bisphenol-A, bisphenol-F or other phenolic compounds. A second family of epoxy resins, also used in adhesive formulations, is based on epoxidized olefin compounds such as epoxidized polybutadiene or cycloaliphatic olefins (Figure 3.1).[2–4] Epoxy resins based on the diglycidylether of bisphenol-A (DGEBA) having viscosities between 8,000 and 20,000 cP are most commonly used in formulations for adhesives, coatings, laminates, and encapsulants. The DGEBA resins are formed when an excess of epichlorohydrin reacts with bisphenol-A in the presence of an alkali such as sodium hydroxide (Figure 3.2a). The mechanism of epoxy resin synthesis involves a step-growth polymerization. Initially, the hydroxyl anion from sodium hydroxide abstracts a hydrogen atom from the bisphenol-A (Figure 3.2b). The intermediate

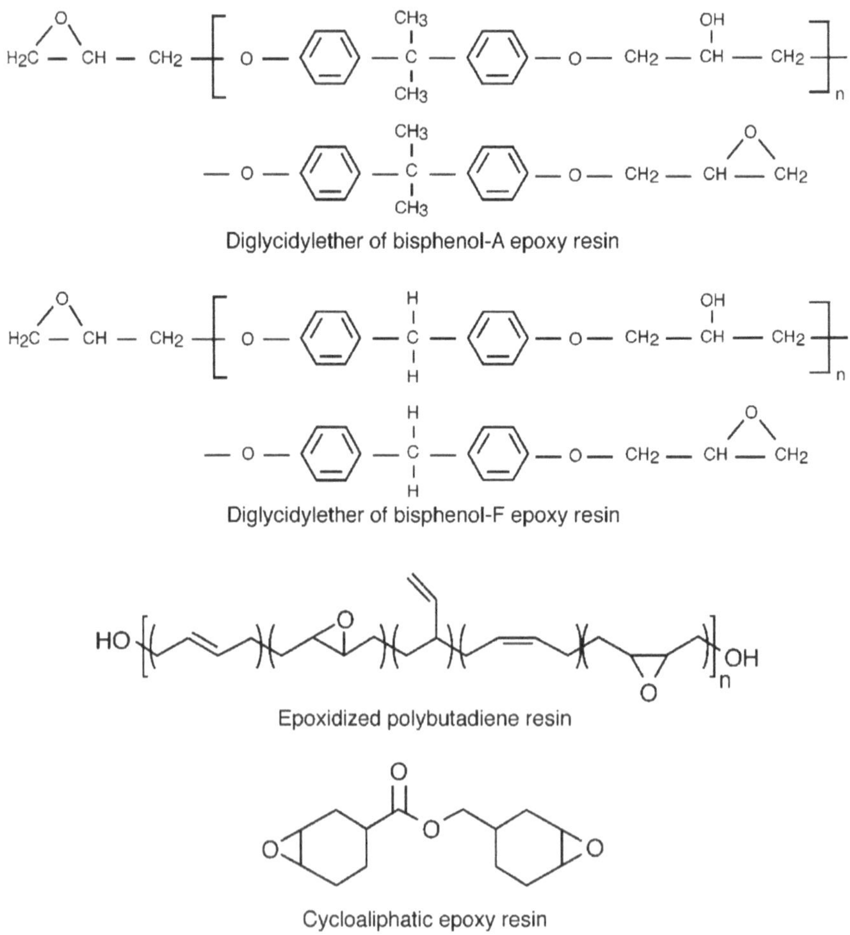

Figure 3.1 Main epoxy resin structures.

anion next displaces the chlorine atom from the epichlorohydrin to produce the bisphenol-A intermediate (Figure 3.2c). The reaction may terminate there with DGEBA being produced or may continue to produce higher molecular weight polymers, depending on the ratio of bisphenol-A to epichlorohydrin used (Figure 3.2d). Finally, an excess of epichlorohydrin terminates the reaction, again by a nucleophilic displacement (Figure 3.2e). Because of the retention of sodium chloride, a by-product of the synthesis, the resin must be purified either by molecular distillation or by repeated extractions with deionized water.

The extent of reaction, ρ, is determined from the initial amount of monomeric material, N_o, and the amount remaining, N_t, after time, t, where $N_0 - N_t$ is the amount of monomer converted.

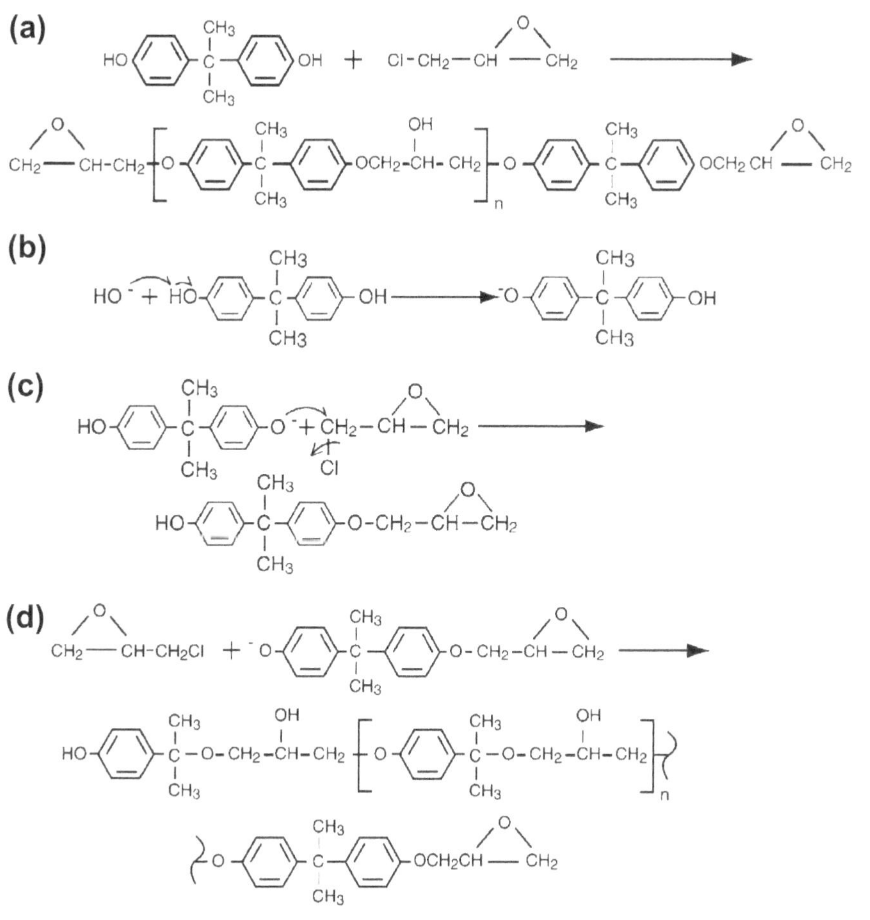

Figure 3.2 (a) Epoxy resin synthesis: generalized reaction; (b) reaction of hydroxyl group of NaOH with proton; (c) displacement of the chlorine atom by the hydroxyl group; (d) chain propagation; (e) termination step.

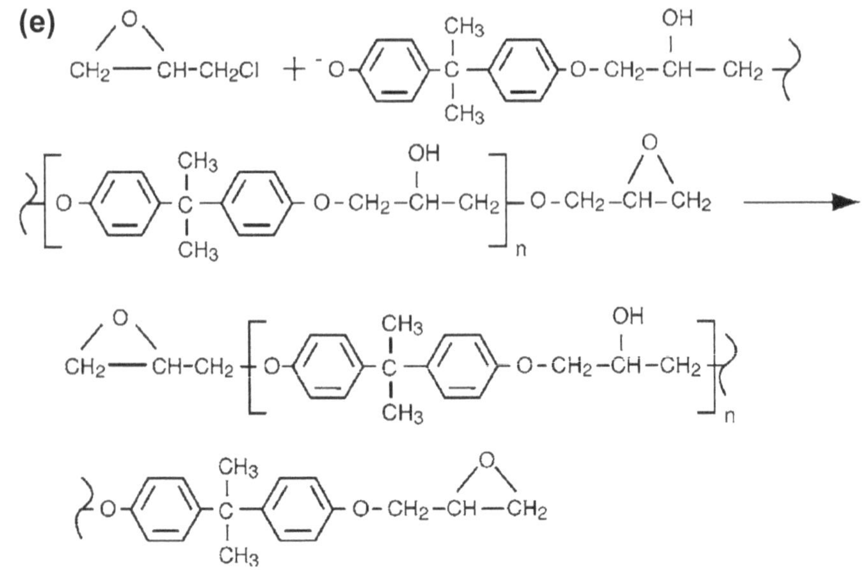

Figure 3.2 (*Continued*).

$$\rho = \frac{N_o - N_t}{N_o} \tag{3.1}$$

Since molecular weight is a function of the number of repeating monomer units, x_n, the average molecular weight, M_n, is given by:

$$M_n = M_o x_n \tag{3.2}$$

where M_o is the monomer weight. Since $x_n = N_o/N_t$, the degree of polymerization and the number of repeating units are related by:

$$x_n = \frac{1}{(1-\rho)} \tag{3.3}$$

Based on the generalized formula (Figure 3.2a), pure diglycidylether of bisphenol-A has a repeating unit value of $n = 0$. Other epoxy resins have n values ranging from less than 1–18.[5] Epoxy resins having n values from 0 to 1 are typically liquids, while those with values greater than 1 are solids. The Dow Chemical Company "300 series" epoxy resins have n values less than 1; for example, the value of n for DER 331 is 0.15.[6] Examples of the main epoxy resins used in adhesive formulations are given in Table 3.1.[7]

3.1.1.2 Curing of epoxy resins

Epoxy resins are characterized by their *epoxy-equivalent weights* (EEW), that is, the weight of resin per epoxy group. Thus, DGEBA, which has a molecular weight of 340 grams and two epoxy groups, would have an epoxy equivalent weight of

Table 3.1 Typical properties of some epoxy resins used in adhesive formulations

Epoxy Resin Type	Examples: Resin/Supplier	Viscosity (poise @ 25 °C, (unless indicated)	Epoxy Equivalent Weight (gm)
High purity diglycidyl ether of bisphenol-A	EPON® 825/Hexion[a]	50–65	175–180
	D.E.R.™ 332/Dow Chemical Co.	4,000–6,000 mPa s	171–175
Diglycidyl ether of bisphenol-A (standard, undiluted resin, general purpose)	EPON®828/Hexion	110–150	185–192
	D.E.R.™ 331/Dow Chemical Co.	11,000–14,000 mPa s	182–192
Reaction product of epichlorohydrin and bisphenol A to provide low viscosity without the use of a diluent	D.E.R.™ 330/Dow Chemical Co.	7,000–10,000 mPa s	176–185
Diglycidyl ether of bisphenol-F	D.E.R.™ 354/Dow Chemical Co.	3,400–4,200 mPa s	167–174
Reaction product of epichlorohydrin with bisphenol-F and bisphenol-A (achieves performance properties of bisphenol A combined with low viscosity of bisphenol F)	EPON® 862/Hexion[a]	25–45	165–173
	D.E.R.™ 351/Dow Chemical Co.	4,500–6,500 mPa s	169–181
Elastomer-modified epoxy functional adduct formed by reacting DGEBA with 40% carboxyl- terminated butadiene-acrylonitrile elastomer	EPON® 58005/Hexion[a]	3,000–8,000	325–375

[a] EPON is a registered tradename of Hexion Specialty Chemicals. Epon epoxy resins were originally developed by Shell Chemical Co.

170 grams. Epoxy-equivalent weights are important in calculating the stoichiometric (theoretical) amounts of hardeners or curing agents needed to achieve optimum polymerization. With amine-curing agents, it is customary to allow one epoxy group for each active hydrogen of the amine compound. The calculated amount should then produce nearly optimum properties in the polymerized epoxy. Non-stoichiometric amounts, for example in using excess amine, tend to stop chain building and produce low-molecular-weight polymers that tend to be brittle. On the other hand, too little amine results in incomplete cure.[2,8] The optimum ratios of commercially available two-part epoxy adhesives have generally been calculated and experimentally verified by the supplier so that the user needs only to follow the given directions for mixing and curing.

Epoxy resins may be cured with a wide variety of curing agents, in general, any compound that has a labile hydrogen atom or hydroxyl group. These curing agents may be amines, polyamides, carboxylic acids, anhydrides, imidazoles, or amide-amines. In general, almost any compound that has a reactive hydrogen atom can react with the oxirane (epoxy) rings and initiate the polymerization process. Compounds that can donate electrons such as Lewis bases or accept electrons such as Lewis acids may also be used as catalysts. Examples of typical epoxy-curing agents are given in Table 3.2. Advantages and disadvantages of various epoxy-curing agents are compared in Table 3.3.

3.1.1.3 Amine curing agents

Amines may be aliphatic (straight chain or branched) or aromatic (cyclic benzene structure) and, in turn, may be primary (two free hydrogens), secondary (one free hydrogen), or tertiary (no hydrogens). Primary amines react rapidly with epoxy resins, even at room temperature, by opening up the epoxy ring and forming a secondary amine structure (Figure 3.3a). Each primary amine compound has two reactive hydrogens and, theoretically, can react with two epoxy functional groups. After a stoichiometric amount of amine has been consumed, the resulting secondary amine reacts further with the remaining epoxy group (Figure 3.3b) producing a tertiary amine and generating more hydroxyl groups that can further react and cross-link with other epoxy molecules (Figure 3.3c). The theoretical ratio of epoxy to active hydrogen groups is 1, but some adhesives suppliers recommend using slightly less curing agent for improved properties. Amine-cured epoxies are characterized by moderate-to-high exotherms, fast curing at low temperatures, and short pot lives. Often the pot lives are too short to be used in production processes. Both aromatic and heterocyclic amines are also used as curing agents for epoxy resins. Aromatic and heterocyclic curing agents impart higher temperature stabilities than their aliphatic amine counterparts.

3.1.1.4 Polyamides

Polyamide resins are characterized by the presence of the repeating amide group [–C(=O)–NH–] along the polymer chain. The polyamides are synthesized by the condensation polymerization of diamino compounds with dicarboxylic acids or

Table 3.2 Typical epoxy curing agents

Curing Agent	Application	Types (Suppliers)	Formulation and Gel Times
Amines	Room temperature or low-temperature cure	Versamine® (Cognis), Ancamine® (Air Products), Aradur® (Huntsman), and Epi-Cure® (Hexion Specialty Chemicals)	1–15 parts per hundred resin (phr), fast gel times
Polyamides	Room temperature or low-temperature cure	Versamid® (Cognis), Ancamide® (Air Products), Aradur® (Huntsman), and Epi-Cure® (Hexion Specialty Chemicals)	50–70 phr, intermediate gel times
Anhydrides	Heat cure	HHPA, MHHPA, and MTHPA (China Aroma Chemical Co., Dixie Chemical)	0.5–0.75 parts anhydride to 1 part epoxy, slow gel time
Lewis acids	Catalytic cure (homopolymerization)	Modified amine complex, Anchor® 1040 (Air Products)	5–12 phr, fast gel time
Imidazoles	Accelerator for anhydride systems	Curezol® 2E4MZ (Shikoku)	1–4 phr, fast gel time
Dicy	Accelerator and adhesion promoter for epoxy systems	Dicy-100 (Suzhou Fun Chemical Co.)	2–5 phr (P-104) and 4–6 phr (P-106), fast gel time

Table 3.3 Advantages and disadvantages of epoxy curing agents

Curing Agents	Advantages	Disadvantages
Aliphatic amines	Rapid cure	Critical mix ratios
	Room-temperature cure	Strong skin irritant
	Low viscosity	High vapor pressure
	Low cost	Blushes
		Often exothermic
Polyamides	Convenience	Somewhat higher cost than amines
	Room-temperature cure	High viscosity
	Low toxicity	Low heat resistance
	Good flexibility or resilience	Low vapor pressure
	Good toughness	Tendency to outgas
Aromatic amines	Moderate heat resistance	Solids at room temperature, difficult to mix with epoxy resins
	Good chemical resistance	Long, elevated-temperature cure schedules
Lewis acid or base complexes (boron trifluoride amine complex)	Long pot life	Tendency to exotherm
	High heat resistance	Poor moisture resistance, corrosion risk
Anhydrides	High heat resistance	Long, elevated-temperature cure schedules
	Good chemical resistance	Critical mix ratios
	Low moisture absorption	Moderate moisture absorption

esters. Naturally occurring fatty acids such as linoleic acid, or their derivatives, are often reacted with diamines to yield a series of liquid resins. Polyamides are thermoplastics used primarily as hot-melt adhesives that bond well to porous surfaces and have excellent flexibility at low temperatures.

Polyamides that contain amino or hydroxyl groups along their chains or carboxyl groups that terminate a chain can be used to cure epoxy resins, imparting their inherent flexibility.[9] The Versamids® are examples of polyamide curing agents.[10]

3.1.1.5 *Amide-amine curing agents (Dicy)*

The most notable and extensively used amide-amine-curing agent is Dicy, (dicyandiamide or l-cyanoguanidine). Dicy is used as a curing agent for film epoxy adhesives and for B-staged epoxy prepregs. Unlike free amines, Dicy will not react with epoxy resins at room temperature or low temperatures. Dicy may be considered a latent curing agent since temperatures of 145–165 °C are required to initiate polymerization. At those temperatures, Dicy decomposes releasing the very active free amines. This is a benefit in producing one-part adhesives and films that have a long pot life and shelf life.

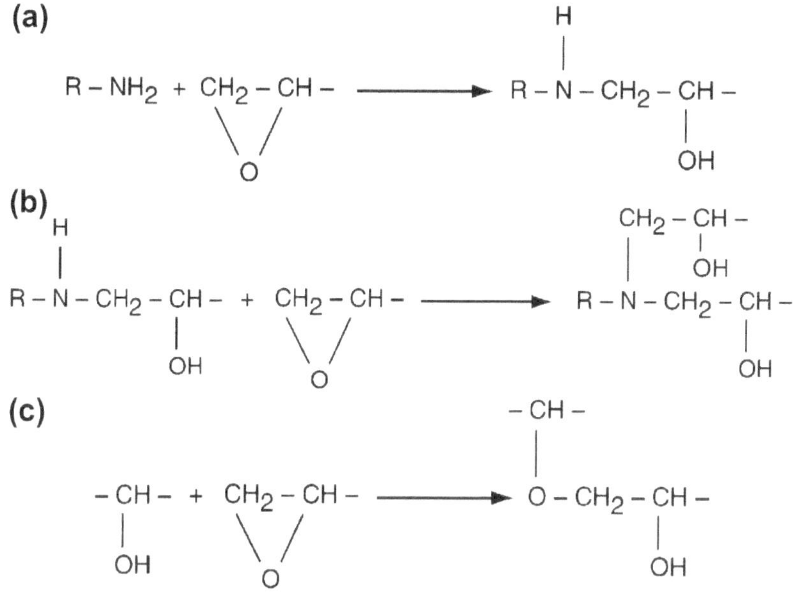

Figure 3.3 Amine curing mechanism for epoxies: (a) initial step, formation of a secondary amine and more hydroxyl groups; (b) formation of tertiary amine; (c) continued cross-linking through reaction of hydroxyl groups.

Several problems that have been encountered during Dicy's long use in electronics should be mentioned. Because of its low solubility in the epoxy resins, Dicy is often first dissolved in organic solvents. Traces of these solvents are retained in the cured material. Because Dicy is hygroscopic, it also absorbs and retains small amounts of moisture and, if the Dicy has not been completely reacted or if an excess of Dicy was used, ammonia evolves as a decomposition product. These outgassed products have been detected in sealed packages through residual gas analysis and may or may not be serious depending on their amounts and their interactions with the electronic devices and materials. Extended vacuum baking prior to sealing has been effective in removing volatile constituents and reducing the outgassing.

3.1.1.6 Anhydride curing agents

Besides amines, liquid and solid anhydrides are extensively used to cure epoxy resins. Examples of typical anhydride-curing agents are given in Table 3.2. In contrast to amines, the reaction rates of anhydrides with epoxy resins are slow, requiring temperatures of 150–170 °C for 2–8 hours. Adding small amounts of catalysts such as a tertiary amine (0.5–3% by weight) or cobalt (II) acetylacetonate, however, can accelerate curing,[11] Benzyldimethylamine, in catalytic amounts, is also widely used as an accelerator. Solid anhydrides include phthalic anhydride (PA), hexahydrophthalic anhydride (HHPA), maleic anhydride, and pyromellitic dianhydride (PMDA). Among popularly used liquid anhydrides are methyl nadic anhydride (MNA) and dodecenyl

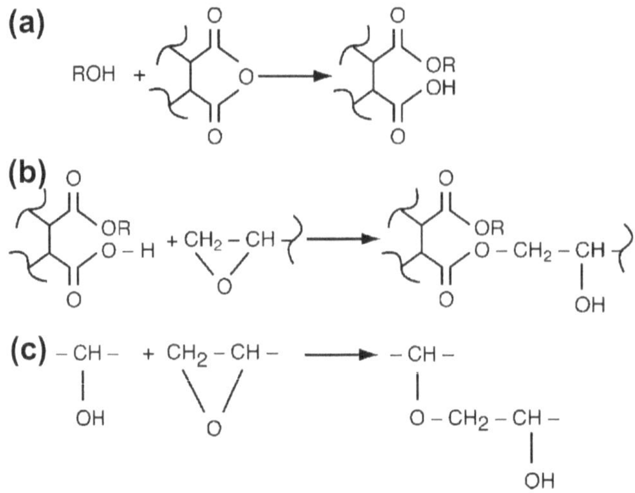

Figure 3.4 Anhydride-curing mechanism for epoxies: (a) reaction of hydroxyl group (from epoxy resin) with anhydride; (b) reaction of carboxylic acid group with epoxy; (c) reaction of hydroxyl group directly with epoxy.

succinic dianhydride. Epoxy resins cured with the solid anhydrides, especially the aromatic types such as PMDA, display high-thermal stabilities, high heat distortion temperatures, high radiation resistance, low weight loss after 500 hours at 200 °C,[2] and long pot lives.[6] However, because of their low solubilities and high melting temperatures, solid anhydrides are more difficult to mix with epoxy resins than liquid anhydrides; often, elevated temperatures and solvents must be used. The liquid anhydrides are much easier to process, have lower viscosities, and are easier to blend with epoxy resins. Because all anhydrides are hygroscopic to some extent, humidity, and exposure to moisture should be controlled prior to and during cure.

Epoxy/anhydride reactions vary significantly depending on the presence of a catalyst.[12] In uncatalyzed systems, the anhydride reacts with hydroxyl groups of the epoxy resin to produce monoesters and generates a free carboxyl group (COOH) (Figure 3.4a). The free carboxylic acid group then reacts with the epoxy group forming a second ester linkage and a free hydroxyl group (Figure 3.4b). Lastly, the hydroxyl group can react directly with another epoxy group (Figure 3.4c). Without a catalyst, the chemical reaction shown in Figure 3.4c occurs to a more significant extent, even though those in Figures 3.4a and 3.4b are the principal reactions. In the presence of a catalyst, formation of polyesters dominates (Figure 3.4b). Maximum polymerization occurs with an anhydride/epoxy ratio of 0.85:1, although, in practice, ratios up to 1:1 are used and yield useful properties.[12]

3.1.1.7 *Catalytic-curing agents (Lewis bases and Lewis acids)*

Both Lewis bases and Lewis acids in catalytic quantities (2–4 phr) cure epoxy resins rapidly without requiring heat. Lewis bases are compounds that have a free pair of

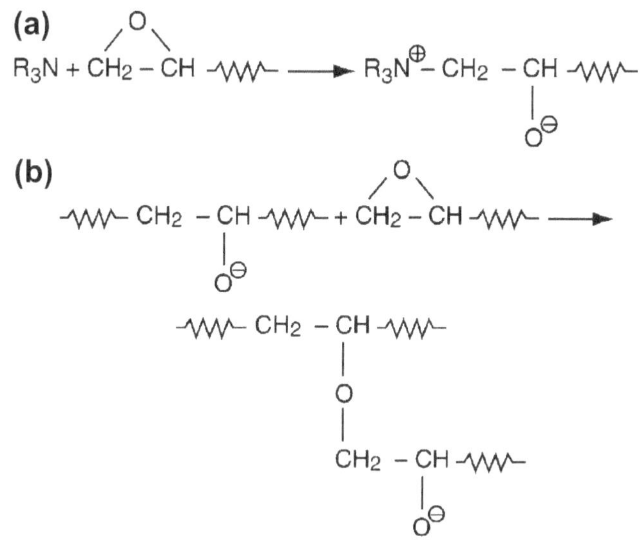

Figure 3.5 Polymerization of epoxy with a Lewis base catalyst: (a) initial ring opening; (b) chain propagation and cross-linking.

electrons such as tertiary amines while Lewis acids are compounds that can accept a pair of electrons such as boron trifluoride. Lewis acids such as boron trifluoride are so reactive that, in addition to being highly volatile and toxic, their pot lives are too short and exotherms too high to be useful. Although quite reactive separately, Lewis acids can form complexes with many amines neutralizing their polarities and reducing their reactivities. As complexes, Lewis acids and bases are latent catalysts that are fairly stable when mixed with epoxies until heated to a temperature at which the complexes dissociate into the reactive species. For example, the monoethylamine complex of boron trifluoride (BF_3MEA) has negligible reactivity with epoxy resins at room temperature, but becomes reactive when heated to approximately 80 °C. Latent catalyst complexes are easier to handle, provide longer pot lives, and are lower in toxicity and vapor pressure than their individual constituents. Triethylamine boron trifluoride, monoethylamine boron trifluoride, piperidine boron trifluoride, and aniline boron trifluoride are examples of latent-curing catalysts for epoxy resins.

Once the Lewis base or Lewis acid is released from the complex, the polymerization reactions involved are similar to those for curing an epoxy resin with a tertiary amine (Figure 3.5).

3.1.1.8 Levels of curing agents used

Optimum properties are achieved when the amount of curing agent used is *stoichiometric*, that is, the number of active hydrogen atoms from the curing agent and number of epoxy groups are equal. For amine-curing agents, the *amine-equivalent weight* (AEW) is the molecular weight of the amine divided by the number of active hydrogen atoms. Using the epoxy-equivalent weight (EEW), the theoretical amount of curing agent, A_c, can be calculated as in Eq. (3.4).

$$A_c = \frac{\text{AEW}}{\text{EEW}} \times 100 \tag{3.4}$$

where A_c is the curing agent required to cure 100 parts of resin (phr).

Epoxy resins are frequently formulated with reactive diluents, solvents, and fillers. These additives affect the calculation so that the EEW needs to be modified as in Eq. (3.5).[6]

$$\text{EEW} = \frac{w_t}{\left(\frac{w_r}{\text{EEW}_r} + \frac{w_a}{\text{EEW}_a} + \frac{w_b}{\text{EEW}_b}\right)} \tag{3.5}$$

where w_t is total weight, w_r is weight of resin, w_a is weight of reactive diluent, w_b is the weight of any other additive having a reactive epoxy group.

One possible adhesive formulation contains 20 pbw (parts by weight) of epoxy resin, 70 pbw of filler, and 10 pbw of a reactive diluent and an aromatic amine curing agent such as diamino diphenylsulfone (DADS) whose amine-equivalent weight is 62. If the resin is Epon®* 825 or DER®* 332, having an epoxy equivalent of 174, the modified EEW will be:

$$\text{EEW} = \frac{20 + 70 + 10}{\left(\frac{20}{174} + \frac{10}{130}\right)} = 521.4 \tag{3.6}$$

Thus, the amount of DADS required would be 62 × 100/521.4 or 11.9 pbw, about 12% of the total composition.

3.1.2 Silicones

Silicones are a unique class of polymers due to their semi-organic molecular structure. Instead of the normal carbon-to-carbon backbone structure of most polymers, silicones have a silicon-to-oxygen structure that gives them advantages of very high thermal stabilities (up to 300 °C, in some cases), flexibility at subzero temperatures (−80 °C), and excellent electrical properties under both extreme conditions. Silicone adhesives, coatings, and encapsulants have been used from the inception of electronics and their formulations have improved steadily with each new generation of microelectronic assemblies.

The –Si–O–Si–O– backbone of silicones is referred to as siloxane. The silicon atoms may be linked to a wide variety of aliphatic or aromatic groups, as shown in Figure 3.6 where the R groups are commonly methyl ($-CH_3$), phenyl (C_6H_5-), allyl ($-CH_2-CH{=}CH_2$), or vinyl ($-CH{=}CH_2$). Silicones used in electronic assembly and

*(Epon is a registered tradename of Hexion Specialty Chemicals. Epon epoxy resins were originally developed by Shell Chemical Co.). (DER is a registered trademark of the Dow Chemical Co.)

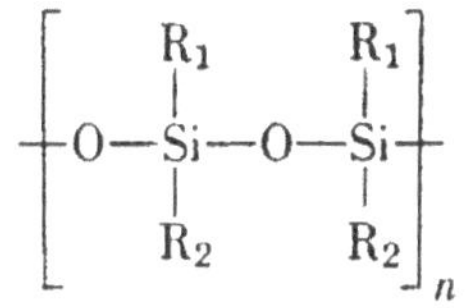

Figure 3.6 General structure for a linear silicone polymer.

Table 3.4 Polymer resins, cross-linkers, and catalysts used as raw materials for silicones

Silicone Type	Polymer Resin	Cross-linker	Catalysts
One-part RTVs	Silanol-terminated polymers	Vinyl, methyl, ethyl, or phenyl-substituted silanes	Titanate catalyst (0.5 phr), tin compounds (~50 ppm), and other metal organics
Two-part RTVs	Silanol fluids (~35%) and silicone fluids (~25%), and silica powder (~35%)	Ethyl silicates polydiethoxy silanes (~1%)	Dibutyl tin dilaurate catalysts (~2%)
Addition-cured silicones	Vinyl-terminated polydimethyl siloxanes (~45%)	Methylhydrosiloxane – dimethyl siloxane copolymer (~50%)	Platinum-complex catalysts (<1%)

Source: Compiled from suppliers' technical data sheets.

packaging may be either room-temperature vulcanizing (RTV) that cure by condensation polymerization or vinyl types that cure by addition polymerization initiated by a catalyst. Table 3.4 lists some formulations for each type.

3.1.2.1 Condensation-cured silicones

Both the heat-cured and RTV silicones are formed by condensation polymerization and are sometimes referred to as alkoxy- or acetoxy-cured systems.[13] The basic hydroxysilane starting materials are produced by the hydrolysis of mixtures of mono-, di-, and trichlorosilanes as depicted in Figure 3.7.[14] Condensation with elimination of water produces hydroxyl-terminated alkylpolysiloxanes. These resins can be further polymerized in the presence of alkoxysilanes such as propylorthosilicates to yield the RTV silicones (Figure 3.8). Catalysts such as organotitanates or dibutyltindilaurate[11] may accelerate the curing process.

3.1.2.2 Addition-cured silicones

A second class of silicones cures by addition polymerization initiated by a catalyst.[15] Polymerization occurs by a free-radical mechanism involving a vinyl, allyl, or other unsaturated group of a silicone monomer (Figure 3.9). Homopolymers can be formed

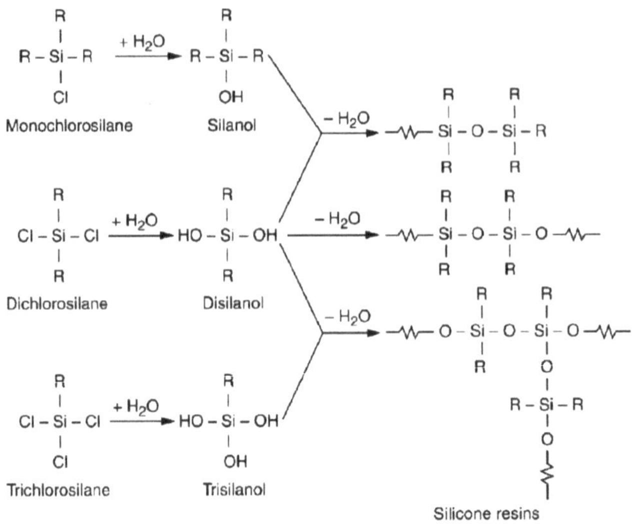

Figure 3.7 Synthesis of RTV silicone resins.[14]

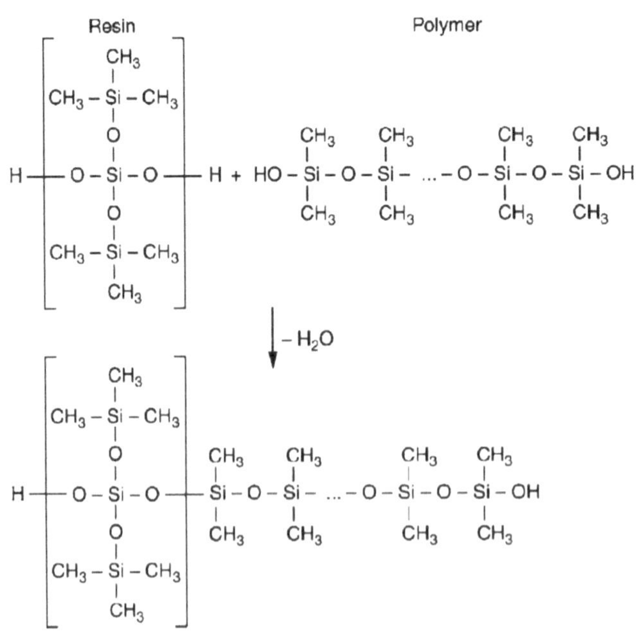

Figure 3.8 Step-growth polymerization mechanism for silicones.

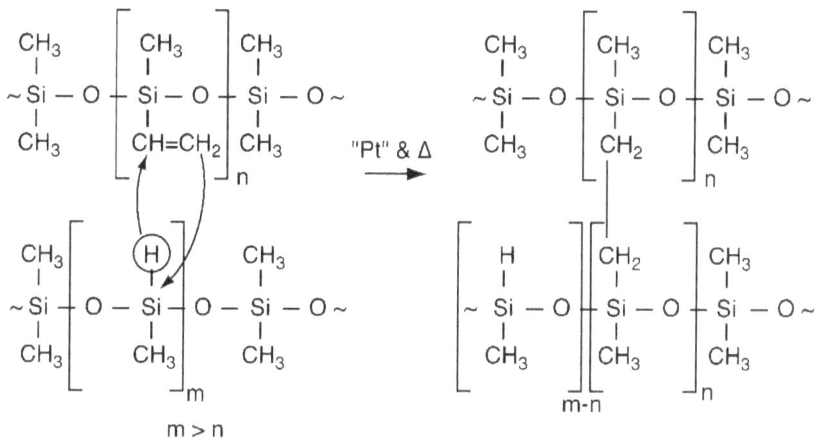

Figure 3.9 Addition polymerization of vinyl silicone.

by polymerization of the same monomer while copolymers are formed by reactions among different monomers. The first step involves thermally decomposing the catalyst to produce active species. Addition-cured silicones are often called platinum silicones since catalysts such as platinum-divinyltetramethyl complex and platinum-divinyltetramethyl disiloxane are widely used. Addition-cured silicone adhesives are either one-part or two-part, heat-cured systems and have an advantage over RTV types in not producing by-products during curing.

3.1.2.3 Moisture-cured silicones

Many of the silicone adhesives used in electronics are one-part, moisture-curing types, generally cured at room temperature in a range of 30–80% RH. Full physical and electrical properties are attained in 24–72 hours. Moisture-cured silicones consist of the normal hydroxyl-terminated silicones in which the hydroxyl groups have been blocked, rendering them inactive. On exposure to ambient moisture, the hydroxyl group is regenerated and can again enter into the polymerization reaction. Silicones may be blocked with either acetoxy or alkoxy groups, but the acetoxy types are not recommended for electronics due to the release of acetic acid during the cure with moisture. The alkoxy-blocked silicones are preferred since an alcohol is released that is generally noncorrosive. Moisture-cured silicones are used in many applications including sealing lids and housings, assembling components onto circuit boards, RF gasketing, tacking and sealing around wires, and also for chip-scale spacers and die attachment.

3.1.3 Polyimides

Polyimides are formed from polyamic acid or polyamic ester precursors by heating to temperatures as high as 400 °C. Polyimide precursors are synthesized by reacting

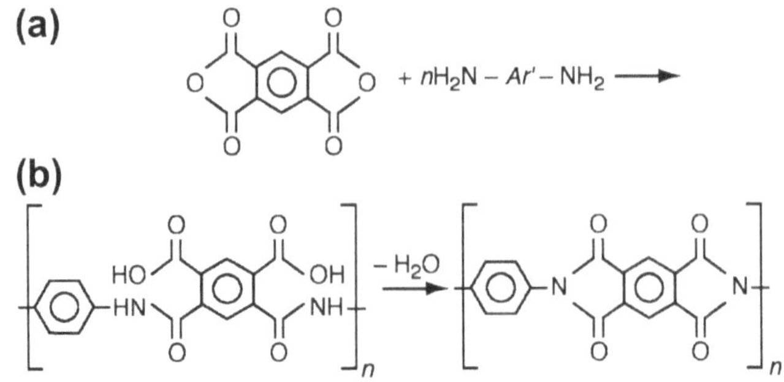

Figure 3.10 Polyimide reactions (a) preparation of polyimide precursors; (b) curing of polyimides by imidization.

equimolar amounts of aromatic diamines with aromatic dianhydrides forming polyamic acids or polyamic esters as in Figure 3.10a. Widely used aromatic amines and aromatic anhydrides are shown in Table 3.5. Of these, the prepolymer from pyromellitic dianydride (PMDA) and 4,4′-oxydianiline (ODA) has been the most widely studied and used.[16] The precursors are available as solvent solutions (generally, N-methyl-2-pyrrolidone) that can then be formulated as adhesives, coatings, or preforms by adding fillers and other additives. Curing is not the traditional cross-linking or chain propagation using curing agents, but occurs by heating to eliminate water and close the imide rings along the chain forming the very stable polyimide structure (Figure 3.10b). The process of ring closure is known as *imidization* and requires step curing, which, for silver-filled paste adhesives, is typically 60 minutes at 150 °C, followed by 90 minutes at 285 °C.[17] In some cases, however, step cures upto 400 °C are necessary to achieve complete imidization, an example of which is: 15 minutes at 135° C, followed by 30 minutes at 300 °C, and finally 10 minutes at 400 °C. The initial temperature exposure of 135–150 °C assures the removal of the N-methyl-2-pyrrolidone (NMP), a solvent that is present in almost all polyimide precursor formulations. For some adhesives, a prebake of 30 minutes at 80 °C (maximum) may be required to drive off water and volatile organic solvents.[17]

A second type of polyimide resin is based on a prepolymer that is completely imidized, but has acetylenic end-groups (triple bonds) which can then cure by addition polymerization (Figure 3.11). These resins do not require high curing temperatures and do not evolve water during cure as do the polyamic acid prepolymers.[18] The *bismaleimides* are yet another variation of polyimides. They are used as adhesives in surface mounting components onto printed-wiring boards and in attaching die in packages or onto interconnect substrates. Bismaleimides are known for their low moisture absorption and hydrophobic properties important in preventing the popcorn effect during solder reflowing (see also Chapter 6 on Reliability).

Lastly, resins are commercially available where siloxane structures have been incorporated into the polyimide chains. These *polysiloxaneimides* are noted for their flexibility and low moisture absorption. They are thermoplastics that are processed as hot-melt adhesives.[19]

Table 3.5 Typical diamine and dianhydride reactants for polyimides

Diamines	Supplier	Dianhydrides	Supplier
m-Phenylenediamine (m-PDA)	DuPont	Pyromellitic dianhydride (PMDA)	Lonza, BEYO Chemical Co.
p-Phenylenediamine (p-PDA)	Shanghai Amino-Chem Co., DuPont	Benzophenone tetracarboxylic dianhydride (BTDA)	Akron Polymer Systems, Chriskev, Lianyungang Darui Chemical Co.
4,4′-Oxydianiline (ODA)	BEYO Chemical Co., Shanghai Logic-chem Import and Export Co.	4,4′-Oxydiphthalic anhydride (ODPA)	Shanghai Research Institute of Synthetic Resins
Methylene dianiline (MDA)	BEYO Chemical Co.	Hexafluoroisopropylidene Bisphthalic Dianhydride (6FDA)	DuPont
4,4′–Diaminodiphenyl sulfone (4,4′-DDS)	BEYO Chemical Co.	3,3′4,4′-Biphenyl tetracarboxylic dianhydride (BPDA)	Shanghai Logic-Chem Import and Export Co., Chriskev
4,4′-Bisaminophenoxy biphenyl (BAPB)	AK Scientific, Inc.		
4,4′-Diamino-benzanilide (DABA)	Hangzhou Dayangchem Co.		
1,3′-Bis(4-aminophenoxy)benzene (134APB) 1,4′-Bis(4-aminophenoxy) benzene (144APB)	Chriskev (TP-Q, TP-R)		
3,3′-Diamino benzophenone (DABP)	ABCR GmbH & Co. KG.		

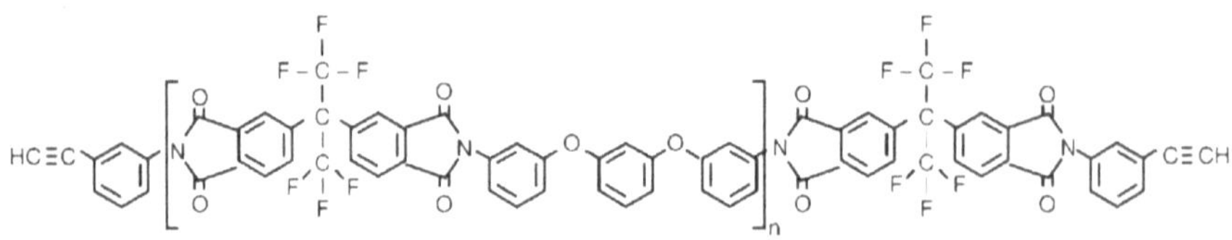

Figure 3.11 Acetylene-terminated polyimide resin structure.

Polyimide adhesives are commercially available as electrically conductive or insulative pastes used for die attachment. They are also available as thermoplastic preforms for chip-on-lead and lead-on-chip attachments.

3.1.4 Polyurethanes

Polyurethane adhesives are used primarily in the assembly of printed-wiring boards to attach large components such as capacitors, magnetics, heat sinks, and connectors. Their key benefits include flexibility, absorption of stresses, vibration damping, and thermal transfer. Being thermoplastic, polyurethanes are also easier to rework than epoxies.

Polyurethanes are distinguished from other polymer types in containing the repeating urethane group throughout its structure:

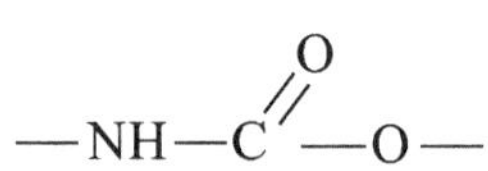

A typical polyurethane adhesive may contain, in addition to the urethane linkages, aliphatic and aromatic hydrocarbons, esters, ethers, amides, urea, and allophanate groups. Polyurethanes are formed by the addition reaction of diisocyanates or polyisocyanates with polyols (Figure 3.12) through a step-growth polymerization

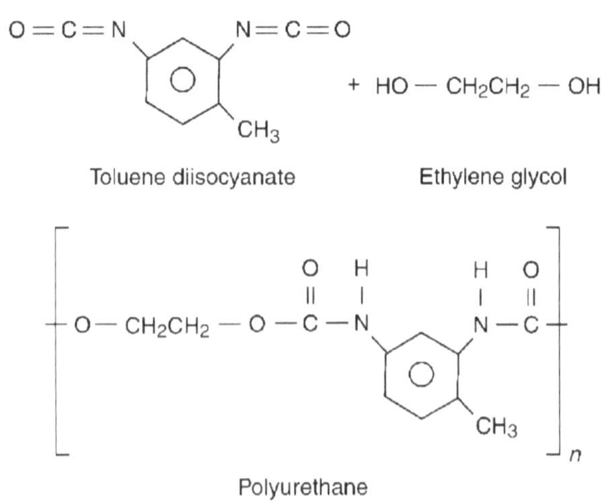

Figure 3.12 A basic polyurethane polymerization reaction.

mechanism. Polyurethanes may be either thermoplastic or thermoset, depending on the functionality of the monomers. Increasing the functional groups to three or more results in the formation of three-dimensional, highly cross-linked thermoset polymers. The most widely used diisocyanate starting materials are toluene diisocyanate (TDI) consisting of an 80:20 mixture of the 2,4- and 2,6-isomers, methylene-4,4′-di(phenylisocyanate) (MDI), and polymeric MDI (Table 3.6).[20,21] Also used are 2,2′,3- and 4,4′,4″-triphenylmethane triisocyanate (TTI). Widely used polyols include castor oil, hydroxyl-terminated polyesters, and polyether polyols. Bayer Chemicals is a major producer of both the isocyanates and polyol starting materials for the production of polyurethane resins. Among Bayer's products are the Mondur® series of aromatic diisocyanates, the Desmodur® series of aliphatic diisocyanates and polyisocyanates, the Desmophen® series of polyester polyols, and the Multranol® series of polyetherpolyols.

An example of a polyurethane adhesive used for surface mounting components consists of Mondur M cured with castor oil or with other polyols.[22,23] The molecular weight and properties of polyurethanes are optimized when one equivalent weight of the isocyanate reacts with one equivalent weight of the polyol. The equivalent weights are based on the percent of hydroxide of the polyol and the percent of isocyanate of the polyisocyanate[21] as follows: Eq. (3.7)

$$\text{Equivalent weight (isocyanate)} = \frac{(42 \times 100)}{\%\text{NCO}} \tag{3.7}$$

$$\text{Equivalent weight (polyol)} = \frac{17 \times 100}{\%\text{hydroxide}} \tag{3.8}$$

As an example, the amounts of Desmodur N-100 required to react with 300 grams of Desmophen R-221-75 polyol can be calculated (assuming a 1:1 ratio). Desmodur N-100 has a 22% isocyanate content and Desmophen R-221-75 has a 3.3% hydroxide content. Thus, the equivalent weights are:

$$\text{Equivalent weight of N-100} = \frac{42 \times 100}{22} = 191 \tag{3.9}$$

$$\text{Equivalent weight of R-221-75} = \frac{17 \times 100}{3.3} = 515 \tag{3.10}$$

The number of equivalents of the polyol is 0.58, so the amount of the isocyanate required to react in a 1:1 ratio will be 111 grams.[21] The more flexible polyurethanes used in printed-wiring board applications generally employ polyols that have lower hydroxyl contents.

3.1.4.1 Moisture-curing polyurethanes

Urethane polymer resins that still contain unreacted isocyanate groups react with atmospheric moisture and continue curing. The reaction is a two-step process. First, water reacts with some of the free isocyanate groups producing an amine and

Table 3.6 Polyisocyanate and polyol starting materials for polyurethanes

Polyisocyanate/ Supplier[a]	Description	Equivalent Weight	% NCO	Polyol/ Supplier[a]	Description	Hydroxyl Content	Equivalent Weight
Mondur® TD 80/Bayer Materials Science	Toluene diisocyanate (TDI) (mixture of 98% 2,4 and 2% 2,6 isomers)	N/A	N/A	Desmophen® 1700/Bayer Materials Science	Linear hydroxyl-bearing medium viscosity polyester resin (solvent-free)	3 ± 0.09	~1,310
Mondur® TDS/ Bayer Materials Science	Toluene diisocyanate (TDI) (mixture of 98% 2,4 and 2% 2,6 isomers)	N/A	N/A	Desmophen® 1800/Bayer Materials Science	Slightly branched polyester polyol (solvent-free)	82 + 0.09	N/A
Mondur® MR/ Bayer Materials Science	Aromatic polymeric isocyanate based on diphenylmethane-diisocyanate (MDI)	≥31	133	Desmophen® 2000/Bayer Materials Science	Polyethylene adipate diol	52–58	N/A
Mondur® CD/ Bayer Materials Science	4,4, diphenyl methane diisocyanate	N/A	29–30	Multranol® 9133/Bayer Materials Science	Poly propylene oxide-based triol	1,035–1,055	N/A
Desmodur® IL BA/Bayer Materials Science	Aromatic polyisocyanate based on TDI (dissolved in n-butyl acetate)	525	7.8–8.2	Multranol® 9144/Bayer Materials Science	Polyether polyol	140–160	N/A
Desmodur® N 75 BA/X/ Bayer Materials Science	Aliphatic polyisocyanate based on Hexamethylene diisocyanate (HDI) (dissolved in n-butyl acetate and xylene) [1:1]	255	16.2–16.8	T41/Eagle Specialty Products	Urethane grade castor oil (maximum purity and minimum moisture)	160–168	N/A

Desmodur® N 100/Bayer Materials Science	Aliphatic polyisocyanate based hexamethylene diisocyanate (solvent free)	191	21.8–22.2	T31/Eagle Specialty Products	Urethane grade castor oil (maximum purity and minimum moisture)	160	N/A (362 Equivalent weight)
Desmodur® N 3200/Bayer Materials Science	Aliphatic polyisocyanate HDI trimer (solvent free)	181	22.5–23.5				
Desmodur® N 3300/Bayer Materials Science	Aliphatic polyisocyanate HDI trimer (solvent free)	~193	21.5–22.1				
Desmodur® N 3300 BA/Bayer Materials Science	Aliphatic polyisocyanate HDI trimer (90% in n-butyl acetate)	~214	19.3–19.9				
Desmodur® W/Bayer Materials Science	Monomeric cycloaliphatic diisocyanate	131	≥31.8				
Desmodur® H/Bayer Materials Science	Monomeric aliphatic diisocyanate	84	≥49.7				
Desmodur® HL BA/Bayer Materials Science	Aliphatic/aromatic polyisocyanate copolymer based on TDI and HDI (dissolved in n-butyl acetate)	400	10–11				

–Mondur, Desmodur, Desmophen, and Multranol are tradenames and products of Bayer Materials Science.
(*Source:* Compiled from suppliers' technical data sheets.

releasing carbon dioxide as a by-product; then, the amine further reacts with the remaining isocyanate groups forming urea linkages and fully cured polyurethane urea. Moisture-curing polyurethanes can penetrate into pores and tight spaces where moisture is usually present to form strong chemical bonds.

Moisture-curing polyurethanes can be formulated to produce a wide range of physical properties, from elastomeric and flexible to hard and rigid materials. Among some noteworthy properties are high abrasion resistance and high impact resistance.

3.1.5 Cyanate Esters

Cyanate-ester adhesives are produced from resins that contain cyanate groups

$$\left(-O-C\equiv N\right)$$

which on heating, and in the presence of a catalyst, polymerize to form highly aromatic triazine ring structures.[24–26] Monomers that contain two or more cyanate groups polymerize to form a highly cross-linked triazine network, a process called *cyclotrimerization* (Figure 3.13).[27,28] Examples of some commercially available cyanate monomers include the dicyanate of bisphenol-A (Ciba Geigy's AroCy B-10®), the polycyanate of phenolformaldehyde (Allied Signal's Primaset®), the dicyanate of hexafluorobisphenol-A (Ciba Geigy's AroCy F®), and a number of other dicyanates of substituted bisphenols.[29]

Catalysts may be the soluble salts or complexes of transition metals, for example, those of tin, copper, iron, or cobalt, or compounds that have an active hydrogen atom such as mercaptans, phenols, or amines.

The triazine structures, like the imide structures of the polyimides, impart high thermal stability to the cured adhesive. Some silver-filled cyanate-ester adhesives have glass-transition temperatures as high as 245 °C and can withstand temperatures of 200 °C or more. However, to achieve these properties, cure temperatures of 175 °C for 1–2 hours followed by a postcure above 200 °C are required, making them more suitable for attaching single die in metallurgically-sealed ceramic packages than for attaching die on plastic laminates or in hybrid or multichip modules. An example of a commercially available cyanate-ester adhesive is Henkel's Hysol® QMI-301.

3.1.6 Modified Cyclo-olefin Thermosets (MCOT)

Modified cyclo-olefin thermoset (MCOT) resins are liquids that cure when heated at 150 °C for a minimum of 30 minutes to form the general polymer structure shown in Figure 3.14. The monomers cure by an addition polymerization and, as such, no water or other by-products are evolved. The molecular structures are essentially nonpolar which impart very low moisture absorption properties. For two formulations developed by Johnson–Matthey, water absorptions of 0.05% were recorded when exposed to 85 °C/85% RH conditions for 168 hours. Only 0.1% water absorption was measured even after pressure cooker testing at 150 °C, 85% RH, for 168 hours.[30] These low moisture absorption values make MCOT adhesives and molding

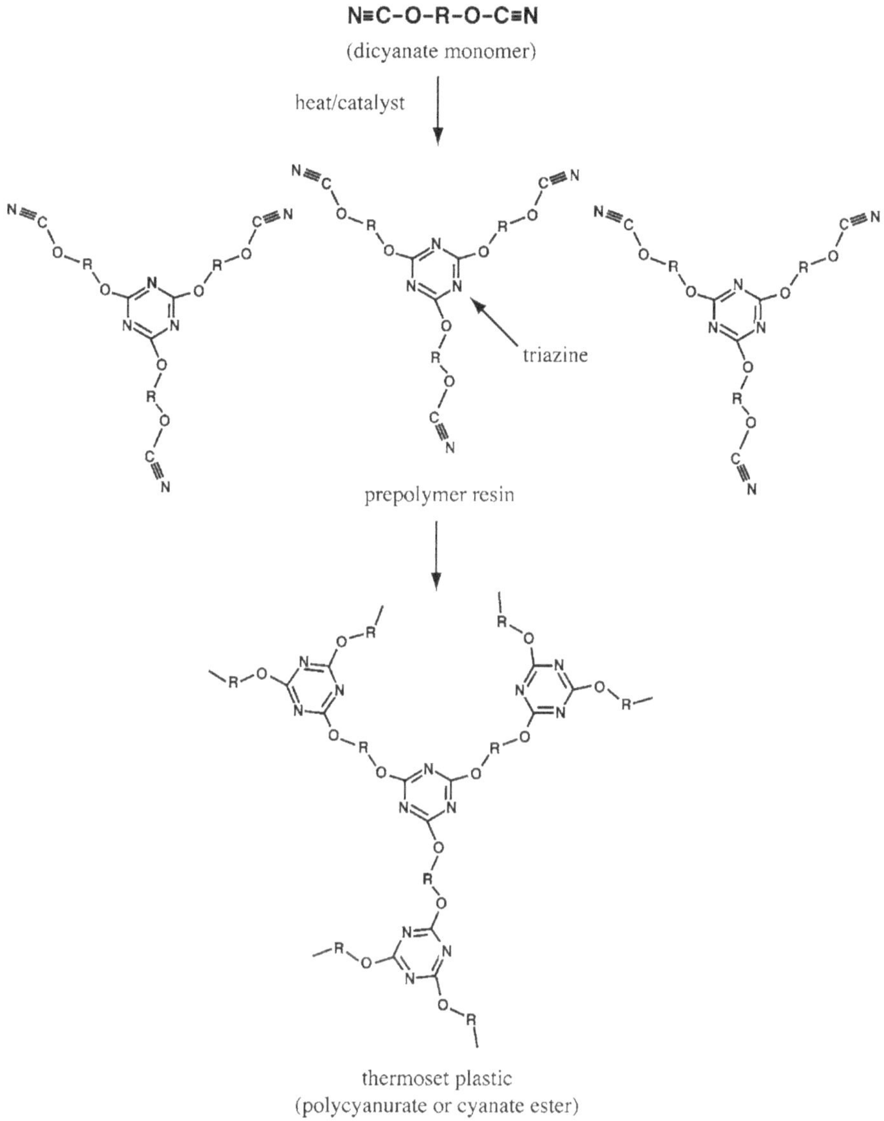

Figure 3.13 Cyanate-ester polymerization.

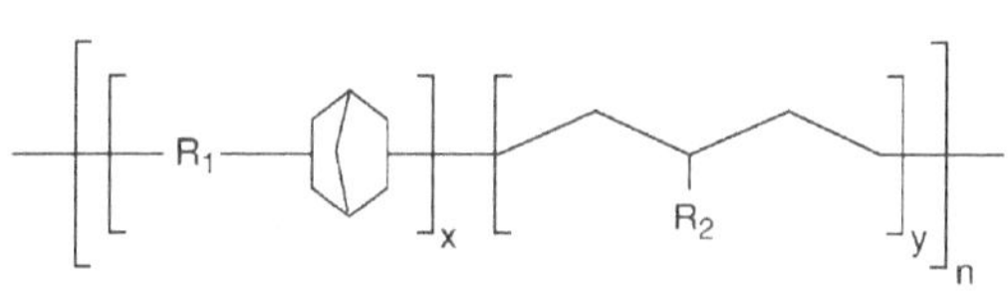

Figure 3.14 Modified cyclo-olefin thermoset (MCOT) polymer structure.

compounds particularly suitable for assemblies that must subsequently be solder reflowed where the "popcorn" reliability problem arises (see Chapter 6). Plastic materials having water retentions of greater than 0.1% undergo rapid evaporation of the water at the 230–250 °C solder reflow conditions inducing stresses in the adhesive and adhesive interfaces resulting in delamination and cracking. MCOT adhesives are block copolymers.[31,32] The cyclic olefin block is a rigid segment while the linear olefin is a soft (flexible) block. In Figure 3.14, R_1 and R_2 comprise the nonpolar and polar portions, respectively, that control the amount of moisture absorption. The optimized value for R_1 is 18% and for R_2, 36%. A molecular weight of 3,262 is reported to provide the best combination of adhesion, moisture resistance, and viscosity properties.[33]

3.1.7 Acrylics and Acrylates

Acrylic adhesives were developed in the late 1960s in Germany as an outgrowth of poly(methylmethacrylate) chemistry. Because of their low cost and rapid processing, the largest market for acrylic polymers is in commercial/industrial applications such as the bonding of wood, steel, and plastics. However, some unique properties, such as their ability to be cured with UV/visible light, their transparency, and rapid low-temperature cures, are finding a niche in optoelectronics and medical applications.

3.1.7.1 Basic acrylate chemistry

The basic acrylic monomers or oligomers contain unsaturated double bonds (vinyl groups), and consequently cure by addition polymerization involving a free-radical reaction. Free-radical-producing compounds such as peroxides, peracetic acids, and sulfones are added to acrylic resins to initiate polymerization. Free-radical polymerization of acrylics may also be induced by exposure to UV or visible light. These UV-curing adhesives, most of which are based on acrylic or modified acrylic resins, are of increasing commercial importance today for the rapid bonding of electronic devices, fiber optics, and liquid crystal flat-panel displays. A key feature of acrylics is their transparency and refractive index making them useful for the bonding of optical parts.

Reactive methacrylate monomers include methylmethacrylate, diethyleneglycol dimethacrylate, and butyl methacrylate. The polymerization reaction for the basic monomer is depicted in Figure 3.15.

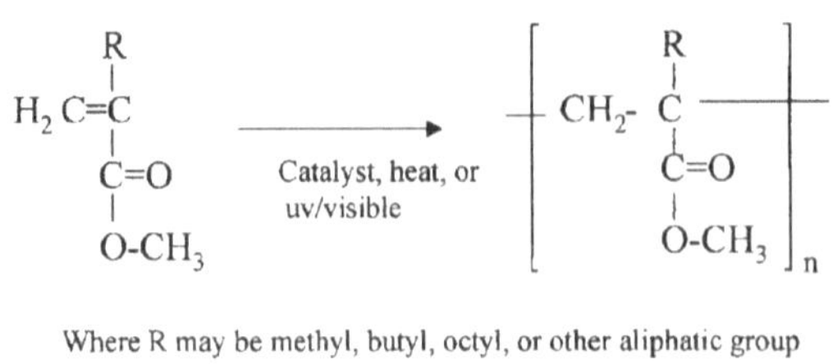

Figure 3.15 Basic acrylic polymerization reaction.

3.1.7.2 Cyanoacrylates

Cyanoacrylates are one-part, highly polar thermoplastic polymers. The resin monomers cure in seconds when in contact with a weak base such as the moisture that is present on most surfaces. Many cyanoacrylate-adhesive formulations are commercially available, but not widely used in electronics assembly because of their poor resistance to solvents and moisture at elevated temperatures (>70 °C). Cyanoacrylates have relatively low impact and peel strengths and may be brittle unless toughened by the addition of elastomeric resins.

3.1.8 Silver-glass adhesives

Although silver-glass adhesives, strictly speaking, are not polymer adhesives, a polymer binder is used in the paste to provide ease of dispensing, but is subsequently burned off during processing. Silver-glass adhesives contain 80–85% by weight of silver-flake filler dispersed in a lead borate glass, a polymer binder, and an organic solvent. After automated dispensing or screen-printing onto a substrate and attaching the die, the adhesive is fired in an oven or furnace. The processing is similar to that for thick-film conductor pastes. The solvent is first evaporated at low temperatures of 50–200 °C, followed by burning out the organic binder at 300 °C; at approximately 340 °C the glass begins to soften and sintering of the silver and glass begins to occur. Completion of the sintering and bonding to the substrate occurs during the last step of 10 minutes at 410–430 °C.[34] Of course, at these high temperatures, silver-glass adhesives can only be used on ceramic, metal, or silicon substrates or packages.

3.2 Formulation additives

Besides the resin and hardener portions, adhesives are formulated with a number of additives to enhance or modify their basic properties. Fillers are extensively used to improve flow properties, reduce thermal expansion, reduce shrinkage during cure, increase thermal conductivity, or impart electrical conductivity. Reactive diluents are used to reduce viscosity and avoid the use of volatile organic solvents. Elastomeric polymers may be added to impart toughness and flexibility. Catalysts and accelerators in small amounts may also be used to initiate sluggish polymerizations. Hundreds of formulations from dozens of suppliers are available and literally tens of thousands are possible. Many commercially available formulations are based on the same or similar resins and hardeners, yet differ in the amount of ingredients used, the number and types of additives, and the techniques used in processing, giving rise to proprietary formulations that have unique or application-specific properties. In fact, the number and nature of the additives and the manner in which they are processed is as much an art as a science and constitutes the most proprietary aspect of a supplier's formulation.

3.2.1 Fillers

The most important additive in an adhesive formulation is the filler. Fillers such as metal particles are used to render adhesives both electrically and thermally conductive while electrically insulative fillers such as metal oxides are used to provide thermal conductivity alone. Regardless of whether they are electrically conductive or not, fillers provide numerous simultaneous benefits, including the following:

- Increased thermal and electrical conductivity
- Increased thermal stability
- Reduced CTE
- Reduced shrinkage and stresses during cure
- Improved bond strengths
- Improved flow properties
- Extended pot life
- Reduced cost.

3.2.1.1 Electrical and thermal conductivity

To meet requirements for electrical and thermal conduction, polymer resins must be highly filled with metal particles. Silver is the most widely used filler because of its high electrical conductivity and retention of conductivity on exposure to air, heat, and humidity. Although a surface oxide is formed, silver oxide, unlike the oxides of other metals, is still electrically conductive. Copper, although also highly electrically conductive, forms an oxide that is electrically resistive so that inhibitors or plating with silver or gold must be used to stabilize its properties. Gold, nickel, and carbon are also used to a lesser extent either because of high cost, in the case of gold, or poorer electrical conductivities in the case of nickel and carbon. Gold and silver-palladium fillers are used in adhesives where silver migration may be of concern. Plated particles such as gold-plated nickel are widely used as fillers for anisotropic adhesives. Silver, nickel, or gold-plated nonconductive core particles such as glass or plastic spheres are also widely used in anisotropic film and paste formulations.

Resins can be filled to a maximum of approximately 85–90% by weight (40–50% by volume, depending on resin density) after which their viscosity increases to an extent that the material cannot easily be dispensed. The effect of increasing levels of silver filler and temperature on the thermal conductivity of an epoxy adhesive can be seen in Figure 3.16.[35] To achieve high thermal conductivity and still preserve good electrical insulation properties, resins must be filled with ceramic or inorganic fillers such as alumina, silica, beryllia, aluminum nitride, boron nitride, and even diamond. Examples of some fillers used in adhesive formulations and their properties are given in Table 3.7.[36–42]

The processing and surface preparation of silver particles is critical to their performance as a filler. The first step in manufacturing silver flakes is to produce silver powder. Powdered silver may be produced by chemical precipitation, electrolytic precipitation, or melt atomization. Most commercially available silver powders are precipitated by reducing silver from a silver nitrate solution. Particle sizes of powdered silver range from submicron to 5 μm.[43] Silver powders are selected based on size and

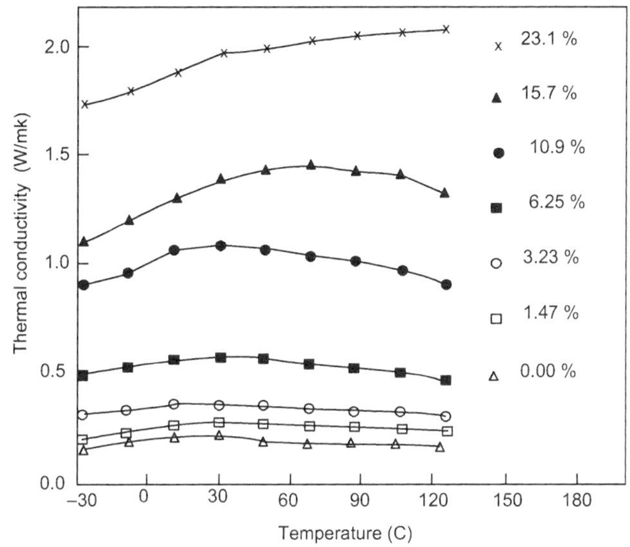

Figure 3.16 Effect of increasing amounts of silver filler and of temperature on the thermal conductivity of an epoxy adhesive. The concentrations are by volume (23% by volume corresponds to 80% by weight).[35]

morphology then milled by conventional means or, more often, in Attritor® grinding mills[44] to yield flakes of the desired size. Ball milling the powders produces a mixture of approximately 80% flat flakes, 5–10 μm in size, and irregularly shaped particles, 1–5 μm size.[43,45–47] Most silver powder feedstocks are in the 0.5–8 μm range and typically produce silver flakes of 1–10 μm. In practice, a broad distribution of particle sizes is better than a narrow distribution in optimizing electrical conductance. Bimodal distribution also appears to enhance conduction.[48,49] Besides particle-size distributions, fillers are also characterized by surface area, tap density, and percent weight loss when heated in air or in a nitrogen/hydrogen atmosphere. The weight loss at 110 °C is a measure of the amount of water and solvents retained by the filler, while the weight loss at 538 °C is a measure of the amount of lubricant used in coating the particles. Both tests are used for quality control. Table 3.8 lists some silver fillers used in epoxy formulations. The morphology of the silver flakes is influenced by numerous processing conditions including the following:

- Length of milling time
- Type and size of the milling media (for example, 3/16 in. steel balls or 6 mm glass balls)
- Rotational speed of the milling machine
- Ratio of silver-to-media
- Type of lubricant used
- Amount of solvent.

Lubricants and solvents are added to the filler during milling to prevent agglomerates from forming due to cold welding of the malleable filler particles.[43]

Table 3.7 Electrical and thermal properties of fillers and filled adhesives

Material	Specific Gravity (g/cm$^{3)}$	Volume Resistivity (ohm-cm)	Thermal Conductivity (W/m K)
Silver	10.5	1.6×10^{-6}	429
Silver palladium	–	2.3×10^{-6}	150
Copper	8.9	1.8×10^{-6}	401
Gold	19.3	2.3×10^{-6}	318
Aluminum	2.7	2.9×10^{-6}	237
Nickel	8.9	10×10^{-6}	91
Platinum	21.5	21.5×10^{-6}	72
Tin/Lead solder (60/40)	–	$20\text{–}30 \times 10^{-6}$	53
Tin/Lead solder (40/60)	–	$20\text{–}30 \times 10^{-6}$	42
Carbon	–	3×10^{-3}	–
Aluminum oxide	3.45	1×10^{14}	17–40[a]
Fused silica	2.2	1×10^{14}	1.4
Aluminum nitride	3.26	10^{14}	170–260[a]
Boron nitride	2.25	1×10^{15}	130–300+[a]
Diamond	3.51	1×10^{13}	800–2,000[a]
Unfilled epoxy adhesives	1.1	10^{14} to 10^{15}	0.2–0.3
Typical silver-filled epoxies	–	1×10^{-4}	2.4–8
Gold-filled epoxies	5.2	8×10^{-4}	2–3
Silver-palladium filled epoxies	2.1	4×10^{-3}	1.3–2.0
Silver-glass	–	$\leq 1.5 \times 10^{-5}$	60- >80
Oxide filled epoxy adhesives	1.5–2.5	$10^{14}\text{–}10^{15}$	1–2
Boron-nitride-filled epoxy adhesives	–	$10^{14}\text{–}10^{15}$	3–4
Diamond-filled adhesives	–	$10^{14}\text{–}10^{15}$	10–12
Silver-filled polyimide	–	3×10^{-4}	2
Boron nitride-filled bismaleimide	1.25	$>10^{15}$	0.9
Boron nitride and silica filled modified cyclo-olefin thermoset	–	–	~1.0
Unfilled silicone	1.08	2.9×10^{13}	0.19
Silicone rubber adhesive (thermally conductive)	2.1	1.6×10^{13}	0.88

[a] Depending on purity level. Values are for comparative purposes only since values for specific compositions, processing conditions, and temperatures will vary.

Lubricants are fatty acids or salts of fatty acids such as stearic or oleic acids in ethanol solvent.[50] The lubricant coats the silver particles during the milling process and is retained during subsequent processing because it becomes chemisorbed on the particle surfaces. Untreated particles appear as aggregates of submicron particles, typically 100 μm in size, presumably due to the high surface tension of the

Table 3.8 Silver fillers used in epoxy adhesives

Product/Supplier	Description	Application	Tap Density (gm/cm^3)	Surface Area (m^2/gm)	Weight Loss (%)		Particle Size Distribution (μm)		
					110 °C (1 hr in air)	538 °C (1/2 hr in air or H_2)	90% less than	50% less than	10% less than
Silver Flake #15/Ferro	General purpose, fine silver flake with mid-range density	Die attach adhesives with good rheological properties	3.0–3.9	0.60–1.20	≤0.10	≤0.60	13.0	5.5	1.5
Silver Flake #78/Ferro	High-purity silver flake with unique rheological properties	Die-attach adhesives (low viscosity, high thixotropy)	4.5–5.6	0.15–0.30	≤0.10	≤0.50	20.0	10.0	5.0
Silver Flake #118/Ferro	High-purity silver flake with unique rheological properties	Die attach adhesives	3.5–4.6	0.60–0.75	≤0.10	≤0.50	20.0	8.0	4.0
Silflake® 135/ Technic Inc.	Mechanically flattened silver powders	Conductive adhesives	2.0 –3.0	0.5–1.1	0.10	0.70	N/A	N/A	N/A
Silflake® 235/ Technic Inc.	Mechanically flattened silver powders (cleaner version of 237)	Conductive inks and adhesives	2.5–4.0	0.6–1.2	0.1	0.7	N/A	N/A	N/A
Silflake® 237/ Technic Inc.	Mechanically flattened silver powders	Conductive inks and adhesives	2.5–4.0	0.6–1.2	0.1	0.7	N/A	N/A	N/A

(Continued)

Table 3.8 Silver fillers used in epoxy adhesives—*cont'd*

					Weight Loss (%)		Particle Size Distribution (μm)		
Product/Supplier	**Description**	**Application**	**Tap Density (gm/cm^3)**	**Surface Area (m^2/gm)**	**110 °C (1 hr in air)**	**538 °C (1/2 hr in air or H_2)**	**90% less than**	**50% less than**	**10% less than**
Silflake® 360/ Technic Inc.	Mechanically flattened silver powders	Dispensible, conductive die-attach adhesives	3.1–4.1	0.4–0.9	0.1	1.2	N/A	N/A	N/A
EA-3106/Metalor	N/A	N/A	3.7	0.55	N/A	0.45	15.5	5.9	1.6
EA-0298/Metalor	N/A	N/A	3.6	0.85	0.02	0.47	10.4	3.4	1.0
EA-0015/Metalor	N/A	N/A	4.1	0.52	0.01	0.35	13.5	5.7	1.8
AA-4703/Metalor	N/A	N/A	3.6	0.99	0.01	0.52	11.4	3.9	1.1
AA-0101/Metalor	N/A	N/A	5.0	0.13	N/A	0.13	18.3	7.7	2.8
AA-0005/Metalor	N/A	N/A	3.5	0.73	N/A	0.37	7.6	3.0	1.0
SA-2831/Metalor	N/A	N/A	4.7	0.82	0.01	0.34	6.1	2.4	0.8
AA-192N/Metalor	N/A	N/A	3.8	0.75	0.02	0.30	18.7	9.2	4.6

N/A – Not available.
(*Source:* Compiled from suppliers' technical data sheets).

untreated metal.[45] Lubricant properties necessary in processing silver filler are as follows:

- Providing lubricity to avoid agglomeration of particles.
- Allowing adequate wetting of the filler.
- Being compatible with the polymer resin and not affect its shelf life or other properties.
- Preventing degradation of the electrical or thermal conductivity of the filler or the filled cured adhesive.

The type of lubricant used may have a major effect on the volume resistivity of a cured epoxy. In one example, a conductive epoxy formulated with silver flakes using a thiol lubricant in a proprietary solvent showed consistently higher volume resistivity (4–5 orders of magnitude) than a similar composition where the silver flakes were coated with a fatty acid lubricant.[51]

3.2.1.2 Increased thermal stability

Weight loss is reduced by the high-temperature stabilities of the inorganic or metallic fillers compared to the resin. Thus, the more filler formulated into an adhesive, the lower the weight loss at elevated temperatures. This is due, not only to the reduced weight of resin by adding filler, but also to reduced shrinkage and increased conductivity that result from the filler.[52]

3.2.1.3 Reduced coefficients of expansion

Expansion coefficients are reduced by adding fillers that have a lower coefficient of thermal expansion (CTE) than the organic resin. In effect, the concentration of the higher-CTE organic phase is lowered by the presence of filler.[53] This effect is illustrated by the rule of mixtures used to predict the CTE of an adhesive (α_{adhs}) which is:

$$\alpha_{adhs} = (\alpha_{resin} \times \text{vol.\%}_{resin}) + (\alpha_{filler} \times \text{vol.\%}_{filler}) \tag{3.11}$$

By lowering the volume percent of resin and increasing the volume percent of the filler, the composite CTE is reduced. This is best illustrated in Figure 3.17[52] where an epoxy resin formulated with 15 phr of *m*-phenylenediamine has a CTE of 50 ppm/°C without filler. The addition of about 60% by weight of alumina, for example, lowered the CTE to about 25 ppm/°C.

3.2.1.4 Reduced shrinkage

Fillers have a pronounced effect on reducing shrinkage during cure. Early data reported by Lee and Neville,[2] although for casting resins, are also applicable to adhesives. Unfilled epoxies may shrink 1% to 8% by volume, but adding 20% filler can reduce this shrinkage to below 1%. The 7.17% shrinkage of an unfilled DGEBA resin cured with 12 phr of diethanolamine was reduced to 4.85% when filled with 100 phr of mica. In another example, an unfilled DGEBA resin cured with a primary amine had a shrinkage of 0.91%, but was reduced to 0.67% when filled with 20% of quartz and to 0.77% when filled with 20% silica.[54]

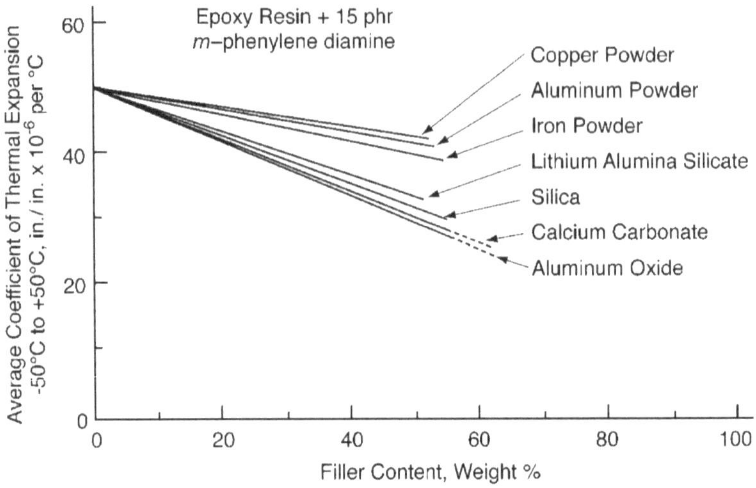

Figure 3.17 Effect of filler content on the expansion coefficient of an epoxy resin.[52]

Another measure of shrinkage, besides volume, is linear shrinkage. An unfilled epoxy has a linear shrinkage of 3.5×10^{-3} in/in that decreases to below 2.0×10^{-3} in/in when 60% by weight of filler is added.[52]

3.2.1.5 *Improved bond strengths*

The tensile-shear strength of an epoxy adhesive can increase from 1,000 psi (unfilled) to 3,500 psi with approximately 50 phr alumina filler.[2] The improvement in shear strength was also shown by Wiles.[55] An epoxy formulation with no filler had a shear strength of 2,760 psi at 25 °C and 450 psi at 105 °C. By adding 100 phr of alumina filler, these strengths increased to 3,750 psi at 25 °C and to 1,650 psi at 105 °C.

Another example of the effect of fillers on improving bond strength is reported for a silicone RTV-1 sealant formulated with Cabot Corporation's Cab-O-Sil® LM-130 and TS-720 materials. Cab-O-Sil® fumed silicas are very effective reinforcing agents due to the very-large surface area available for polymer/filler interactions. The very low, 45 psi, tensile strength of the unfilled silicone increased to 210 psi with 14 phr of LM-130 fumed silica and to 150 psi with the same level of TS-720 fumed silica.[56]

3.2.1.6 *Improved flow properties*

The flow properties of an adhesive are largely controlled by its inherent viscosity and by the alteration of its viscosity through the addition of diluents and fillers. Flow properties are important for screen-printing, automatic dispensing, and capillary flow, and can be tailored by the type of filler, amount of filler, and the particle-size distribution.

The surface area of the filler has a pronounced effect on viscosity. As the surface area of silver-flake filler increases, the frictional forces between particles increase

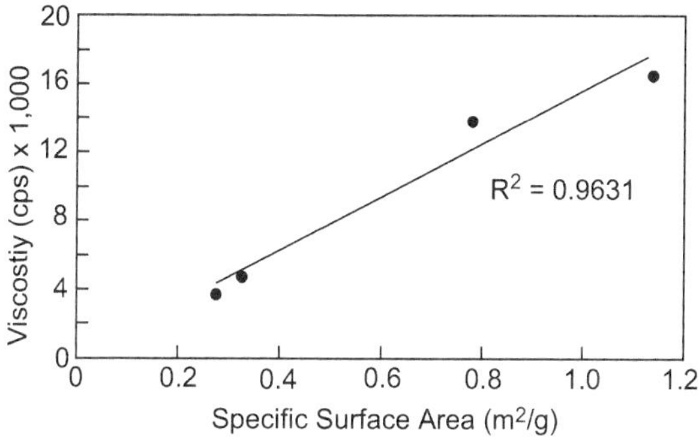

Figure 3.18 Viscosity as a function of surface area (fixed flake size).[50,57]

resulting in higher viscosity. Epoxy formulations filled with constant-size silver flakes of various surfaces areas show this effect when viscosity is plotted as a function of the filler surface areas (Figure 3.18).[50,57]

Cab-O-Sil® also enhances the flow properties as follows:

- Improves flow control, including thixotropy.
- Minimizes or eliminates settling during storage.
- Allows extrusion control during dispensing.
- Prevents sag or slump during cure.

Surface treatments for Cab-O-Sil® influence the viscosity and thixotropic properties of an adhesive as shown in Figure 3.19 where MS-5 is untreated Cab-O-Sil®, TS-720 is treated with polydimethylsiloxane, and TS-610 and TS-530 are treated with dimethyldichlorosilanes.[56] The silica filler treated with polydimethylsiloxane (TS-720) produces greater van der Waals forces of attraction and thus higher viscosity compared to TS-610 and TS-530.[56]

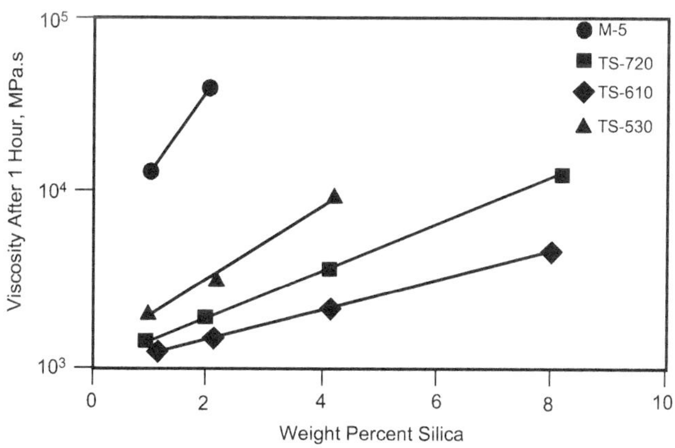

Figure 3.19 Viscosity increase with increasing Cab–O–Sil® filler and effects of surface treatments.[56]

3.2.1.7 Extended pot life

In general, pot life is increased through formulation with fillers. Fillers lower the exotherm occurring during curing by controlling the concentration of reactants in a given mass. The exotherm heat is dissipated over the sample size, especially in the case of fillers having high thermal conductivities.[53]

3.2.2 Nanofillers and Nanotechnology

Nanotechnology is a subject of intensive research in studying and developing new materials having dimensions less than 100 nm (submicron), even down to the molecular level. Although nanotechnology is an emerging technology for electronics, medical and other fields, it really began in 1917 with the experiments of Langmuir in which he demonstrated that organic molecules having a highly polar unit on one end and a non-polar unit on the other oriented and aligned themselves on a trough of water as a monolayer with all the polar ends attracted to the highly polar water surface. This approach is now referred to as "bottom up," that is, it starts at the molecular level and builds up various material structures. A second approach, "bottom down," consists in taking current macromaterials and breaking them down into nano dimensions. While the reduction in dimensions of electronic circuits continues with the features of many devices already in the submicron range,[58] interconnections and packaging compatible with these devices has not kept up. Nanotechnology now offers the opportunity to develop submicron connections to fill this gap.

Properties of nanofillers recently developed nano materials are reported to display greater mechanical strength, greater thermal conductivity and improved electrical performance when compared to materials of normal particle sizes. Nano dimensional materials are being studied as fillers in polymer matrices in a variety of formulations for electrically conductive adhesives, thermally conductive adhesives, encapsulants, printed circuit boards, coatings, catalysts, underfills for flip-chip-attached devices and wafer-level connections.[59]

Of particular interest to adhesives formulators are nanofillers such as carbon nanotubes (CNT), silica, alumina, magnesium oxide, titanium dioxide, zirconium oxide (ZnO_2), silver, copper, and nickel). Of these, carbon nanotubes are the most widely studied for electrically conductive adhesives to attach microdevices, to interconnect microcircuits and to increase I/O densities at the device level.[60]

A key finding of many investigators has been the improved mechanical and adhesion strength properties of epoxy and other polymer materials when filled with nanomaterials. The tensile strength, shear strength, and Young's modulus of epoxy adhesives filled with single-wall CNTs (30–60 nm diameter and approx. 200 nm long) or with alumina nanofibers (2–4 nm diameter and approx. 200 nm long) improved with increasing percentage of the fillers upto 8–10% by weight. Above 10%, however, the strength properties began to deteriorate.[61] These findings were corroborated by others using nano alumina- filled epoxy film adhesives [62] and epoxies filled with homogeneously dispersed CNTs and alumina.[63] Nanoparticles (20 nm) of ZrO_2 as a filler for epoxy resins were studied and reported to improve the lap-shear strengths for aluminum panels in loadings 10–20 wt %.[64] Improved cohesive strength of an adhesive formulation containing up to 10% by wt. of nano silica particles also was demonstrated.[65] A dry adhesive formed from vertically-aligned multi-walled CNTs was reported to display anisotropic properties whereby the adhesive had strong adhesion to a vertical surface from a shear force but could still be removed by pulling away in the normal direction.[66]

In addition to improving strength, increasing electrical conductivity and reducing interconnect dimensions, formulations containing nanofillers have been reported to improve flame retardant and humidity resistance properties of polymer resins.[67]

Nanoparticles of silver and carbon are also being studied as replacements for copper or gold plating or for copper filling of microvias in interconnect substrates.[68]

Nanofillers have also been used to improve the optical properties of adhesives. Magnesium oxide nanofillers in cyclic olefin or polymethylmethacrylate resin formulations have resulted in improved temperature-stable optically transparent adhesives. Filler sizes well below the wavelength of visible light (400–700 nm) were reported to greatly reduce scattered light of filled polymers and to provide optical stability and stable refractive indices over a range of temperatures.[69]

3.2.2.1 Preparation of nanoparticles

Nanoparticles can be produced by several methods including mechanical, electric arc discharge, laser ablation, chemical vapor deposition (CVD), plasma-enhanced CVD (PECVD) and wet chemical reactions.

Nanoparticles formed by mechanical methods (bottom down) such as ball milling or planetary grinding of larger particles are best suited to large-scale production utilizing low-cost equipment but suffer from particles agglomerating as they are formed, upon storage, or when admixed with a liquid polymer resin. Mechanical processes are best suited to forming nanoparticles of ceramic, mineral, or metal oxides. Nanoparticles may also be formed (bottom up) by "growing" them in place using well-known thin-film deposition processes such as CVD, PECVD, sputtering, arc discharge and laser ablation.[70,71]

Carbon nanotubes are probably the mostly widely studied nanomaterial currently produced. Interest lies in eventually incorporating CNTs into scalable integrated circuits.[72]

Multi-wall and single-wall CNTs have been produced either by arc discharge or laser ablation. In arc discharge, the negative electrode consists of graphite and, during electrical discharge at 100 amps and high-discharge temperatures, produced up to 30% by weight of carbon nanotubes with lengths up to 50 μm.

In laser ablation, a pulsed laser beam vaporizes a graphite target in a high-temperature reactor while an inert gas such as argon is flowed into the chamber. Nanotubes form on the cooler surfaces of the reactor as the vaporized carbon condenses.[73] Later, a composite target of graphite with cobalt and nickel catalyst was used and found to increase the yield of single-wall CNT to 70%.[74] Both vacuum- and laser-deposited nanofilms are of great interest in electronics because they can be formed as aligned monolayers in features compatible with the nano dimensions of current semiconductor devices and integrated circuits.

For hundreds of years, nano particles have also been produced by wet chemical processes such as colloid dispersion and precipitation from chemical solutions. The exposure of silver salt solutions to normal light or UV-light, precipitating metallic silver particles, used in black-and-white photographic film is a well-known practical example. Nickel nanoparticles have been produced by hydrolysis of Mg_2Ni in water resulting in free nickel, magnesium hydroxide and hydrogen.[75] Wet chemical techniques are reported to produce better quality nanoparticles than mechanical methods. The particles are already formed as a dispersion thus minimizing or avoiding agglomeration and better controlling the chemical composition and morphology of the particles.

Most high-quality nanoparticles are still produced in experimental quantities at universities or research laboratories and thus are still quite expensive and not widely available for use as fillers in adhesives.

3.2.2.2 Safety issues

Concurrent with the increasing development and uses of nanomaterials, there is an equal effort in studying their potential toxicity. Titanium dioxide, for example, is widely used in food products and cosmetics and considered safe, but studies of nano titanium dioxide particles which are small enough to enter the body in several ways may be toxic according to some studies.[76] The expanded use of nanofillers in consumer products has generated safety concerns that have resulted in proposals for detailed studies by the Food and Drug Administration.[77]

In summary, nanotechnology is an emerging technology with a promise for numerous applications in electronics, medical and other fields but there are still many unknowns including its toxicity and safety.

3.2.3 Thixotropic Agents

Thixotropy is a very important property of adhesives in managing their flow properties during and after dispensing. For rapid dispensing, low viscosity and high flow are required while, immediately after dispensing, reversion to low flow is important so that the adhesive is precisely placed and cured without any lateral or vertical movement. This occurs because of the non-Newtonian behavior (thixotropy) of many filled adhesives whereby the shear force of screen printing or forcing the adhesive through apertures reduces its viscosity but upon removal of the force the material regains its original higher viscosity, preventing it from sagging or creeping until it can be fully cured (see also Chapter 2). Thixotropic adhesives also allow for thicker more controlled bond lines.

In many cases, highly filled paste adhesives are inherently thixotropic due to the inorganic fillers used such as alumina, silica, other metal oxides, or metals. However, small amounts (1–5%) of thixotropic agents can be added to further control the flow properties. A popular and widely used thixotropic agent is fumed silica, also known as amorphous fumed silica or colloidal silica and sold under the tradenames of Cab-O-Sil® (Cabot Corp.) and Aerosil® (Evonik Degussa). Other thixotropic materials include microparticles of alumina, aluminum nitride, carbon black, and nanocarbons.

3.2.4 Coupling Agents

In blending inorganic fillers with polymer resins, the adhesion of filler particles with the resin may require enhancement by treating the fillers with a coupling agent that bridges the filler to the resin molecules (see also Chapter 2). Adhesion is strengthened by the formation of covalent bonds or by van der Waals forces of attraction between the two materials. Among the most widely used coupling agents are the organosilanes in which two different reactive groups are part of the organosilicon molecule. One group (such as hydroxy, methoxy, or ethoxy) bonds to the inorganic filler (such as silica, alumina, or metals) while the other reactive group (such as vinyl, epoxy, methacryl, or amino) bonds to the resin polymer. Silane coupling agents may be used in both thermosetting and thermoplastic adhesive formulations resulting in improved mechanical and electrical properties.[78]

3.2.5 Diluents

Diluents are low-molecular-weight, low-viscosity compounds that are used to reduce the viscosity or enhance the solubility of a resin and/or hardener. Diluents may be either reactive or non-reactive. However, the reactive types are more desirable since they combine chemically with the main resin during cure and are not free to outgas or leach, especially during thermal-vacuum exposure. Epoxy diluents, for example, are formulated together with epoxy resins to reduce the viscosity and still remain 100% solids. Normally, solvents would be used to reduce viscosity, but most organic solvents are now being disallowed because of their high volatile emissions. Some long-chain aliphatic reactive diluents can also improve the flexural strength of epoxy adhesives.

Examples of reactive diluents for epoxy resins include: phenylglycidyl ether, butylglycidyl ether, allylglycidyl ether, butanediol diglycidyl ether and glycerol-based epoxy resins. Diluents should be used sparingly in amounts of 5–15 phr, enough to reduce the viscosity of the adhesive, yet not degrade the properties of the cured material.

3.2.6 Flexibilizers

Flexibilizers improve peel and impact strengths of adhesives by allowing them to deform under stress and reduce mechanical damage by lowering the modulus of elasticity.

Adding small amounts of reactive flexibilizers to an epoxy resin can reduce the modulus of elasticity, glass-transition temperature (T_g), and CTE of an otherwise hard and brittle material. Rubber-like materials such as a carboxyl-terminated

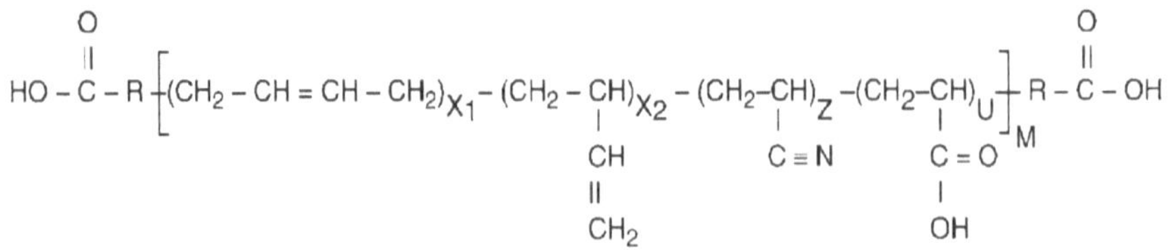

Figure 3.20 Acrylonitrile-butadiene copolymer.

acrylonitrile–butadiene copolymer (Figure 3.20) are used in epoxy formulations to reduce stresses and prevent cracking. The carboxyl group reacts with the epoxy resin to produce an epoxy-terminated rubbery polymer. Small amounts of liquid rubber domains become interspersed within the epoxy matrix and mitigate crack propagation.[79] Masterbond® 10HT, an example of a rubber-toughened, stress-absorbing epoxy, was qualified for bonding ceramic components to printed-wiring Boards.[80,81]

Polyamides containing amino groups are also excellent flexibilizers for epoxies since they co-react with the epoxy resin and introduce their inherent thermoplastic properties.

A second approach to reducing modulus and flexibilizing an adhesive is to add a plasticizer to the formulation. Since most plasticizers are of relatively low-molecular weight and do not co-react with the resin, there is a risk of separation, leaching, or outgassing during thermal or thermal-vacuum environments.

In general, flexibilizers improve the low-temperature bond strengths at −50 °C or lower by reducing stresses, but often at the expense of reducing high-temperature strengths.

3.2.7 Flame Retardants

With the ever increasing use of adhesives in electronics manufacturing, flammability becomes a safety issue especially where faulty devices overheat, hot spots develop on a circuit board, or arcing occurs. All adhesives based on organic polymers are flammable at some elevated temperature in the presence of oxygen. However, depending on the chemical structure and degree of polymerization or cross-linking, there are wide variations in flammability as determined by Underwriters Laboratory UL-94.[82] For example, adhesives based on polyimides, silicones, fluorocarbons and highly cross-linked epoxies are among the most stable while polyurethanes and acrylics are less stable. Resins or hardeners that have a high aromatic structure (benzene rings, phenyl groups) or a nitrogen heterocyclic structure also improve the flame-resistance properties.

Besides the danger of propagating a fire, the decomposition products are generally noxious or toxic. Polyurethanes, during burning, are reported to produce dense toxic fumes containing hydrogen cyanide, free isocyanates, carbon monoxide, carbon dioxide and nitrogen oxides[83] while acrylics and epoxies produce nitrogen oxides, carbon monoxide, water, and carbon dioxide, among other gaseous fragments.

There are two basic and widely used approaches to reduce the flammability of adhesives and other polymeric materials:

- Incorporate halogen flame-retardant groups such as bromo or chloro groups as part of the main polymer resin or hardener molecules.
- Blend into the formulation inorganic compounds of known flame retardancy, generally inorganic salts or hydrates of inorganic compounds.

On initial burning, the halogenated flame retardants decompose evolving bromine, chlorine and low-molecular-weight halogenated compounds that, as heavy gases, settle over the flame and retard its further propagation by excluding oxygen.

For many years, flame-retardant epoxies have been formulated using halogen-containing epoxy resins or hardeners notably the chlorinated or brominated versions of DGEBA. Antimony compounds such as antimony trioxide and antimony pentoxide have been shown to provide a synergistic effect when used with the halogenated flame retardants.[84] Although effective, halide-containing adhesives are being restricted in their use and phased out because of the toxicity and enviromental hazards of the halides particularly polybrominated bisphenol A, brominated biphenyl ethers and other bromine-containing organic compounds. Numerous studies have shown the presence and accumulation of these compounds in streams, water systems, and food chain.[85] The phaseout of brominated flame retardants presents a major challenge to manufacturers in finding effective non-toxic alternates that are compatible with the resin systems and are not deleterious to the final physical and electrical properties of the cured adhesives.

Non-halogenated flame retardants such as metal hydrates, metal hydroxides, hydrated metal salts, antimony oxides, borates and phosphorus-containing compounds such as organophosphates are now being addressed as substitutes for the halogenated flame retardants.[86] Organophosphorus compounds, especially, are being widely investigated as reactive monomers (Figure 3.21) for epoxy and

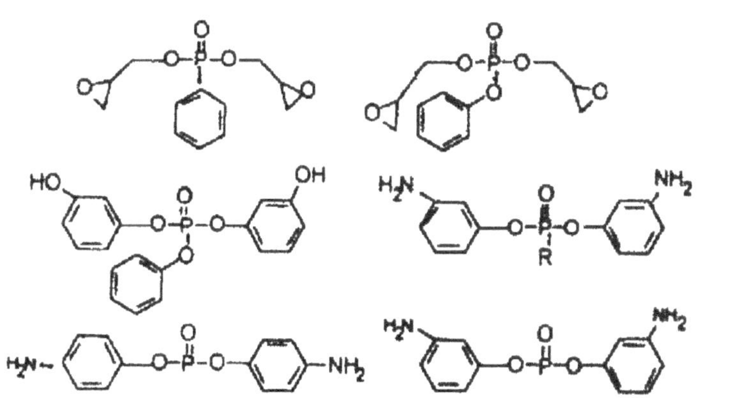

Figure 3.21 Organophosphorus monomers for producing flame-retardant adhesives.[87,88]

polyester adhesives.[87,88] As such, organophosphorus groups can be introduced into resins such as an epoxies, polyesters, and polyurethanes thus providing flame-retardant properties yet maintaining the physical and electrical properties of the resin.[87,88]

3.2.8 Stabilizers

Some polymer formulations, both before and after curing, are susceptible to degradation at elevated temperatures in air, to UV light, to humidity or to combined weathering conditions. Degradation includes oxidation, free-radical formation, or hydrolysis generally resulting in cleavage of the polymer chains into smaller fragments with changes in electrical and physical properties. Among adhesives most likely to be affected are those based on polyesters, polyurethanes, acrylics, polyolefins and polyamides. In such cases, antioxidants and UV absorbers are added to the formulations to stabilize them.

Ultraviolet-light stabilizers are of two types: those that preferentially absorb UV, thus shielding the polymer structure from UV light and those that protect by quickly scavenging the free radicals as they form. The latter belong to a group of compounds known as hindered amine light stabilizers (HALS).

Ultraviolet-light-absorbing stabilizers include benzene-based highly aromatic compounds and nitrogen heterocyclic compounds such as benzophenones, benzotriazoles, and phenyltriazenes. HALS stabilizers may be piperidine or phenolic compounds in which the nitrogen or carbon within the ring structure is hindered by carbon groups that are linked to the adjacent carbon atoms. Often both types are used to provide a synergistic effect. UV-absorbers are used in amounts of 0.1–1%, depending on the composition of the particular polymer system.[89]

A series of HALS and UV-absorbers trademarked Chimassorb® and Tinuvin® are available from and registered trademarks of Ciba (now BASF).

Antioxidants are added to some formulations to protect them from degradation caused by atmospheric oxygen especially at elevated temperatures. Partial oxidation resulting in free-radical formation without molecular scission usually results in yellowing and minor changes in mechanical or electrical properties. Excessive oxidation, however, results in chain scission into small free-radical fragments which continue to propagate or terminate by recombining with other free radicals. Antioxidants interrupt the degradation process by reacting with the free radicals eliminating them from continuing the propagation process. Antioxidants also react with hydroperoxides to yield non-reactive species. Antioxidants are especially useful in formulations of polyesters, polyolefins, ethylenic polymers and polyamides. Hot-melt adhesives, because they are kept for long periods of time at 150 °C in air during application, generally require the addition of one or more antioxidant. Examples of some antioxidants include tris-nonylphenylphosphite (TNPP), 2,6-ditertiarybutyl p-cresol, and dilaurylthiopropionate.[90]

Fortunately, many adhesives used in electronics such as epoxies, acrylics, and silicones are highly resistant to oxidation under their normal operating temperatures and do not require antioxidants.[91]

3.2.9 Pigments and indicators

Pigments and fluorescent indicators in small amounts of 0.1 to 1% by weight may be added to adhesives formulations to identify a particular manufacturer or for quality inspection. Also, in two-part adhesives, a pigment such as carbon black is often added to one part so that after mixing the degree of mixing can be assessed visually when the color is uniform.

Colored pigments or fluorescent UV indicators may be used to inspect for bleedout from adhesives during and after cure. They may also be used to inspect for particles of cured adhesives that may have detached, migrated and contaminated other parts of an electronic assembly. Particles of electrically conductive or nonconductive adhesives can cause failures in microelectronic assemblies, gyros, accelerometers, or other electromechanical systems and are especially serious in space electronics. Rhodamine B which is pink to red in both the visible and UV regions and 2,6-distyrylpyridine which is visually colorless but intensely blue in the UV region are effective indicators in some adhesives formulations.

In using organic or inorganic pigments or colorants, considerations should be given to their influence on the final performance properties, especially electrical properties and any interactions with other additives of the formulation.

3.3 Formulation processes

In general, specific processes for formulating adhesives are proprietary with the adhesives manufacturer. However, some basic methods and considerations for formulating paste, film, and frozen adhesives are given here.

3.3.1 Paste adhesives

Paste adhesive formulations are produced by mixing the base resin with low-molecular-weight diluents or solvents, if necessary, to reduce viscosity. Other prepolymer resins may be added and agitated until dissolved. Viscosity and temperature are continuously monitored and more diluents or resin added as required. Finally, one or more types of fillers are added and mixed and the viscosity of the mixture is adjusted. In incorporating the filler into the resin, best results are obtained when the filler is preheated to drive off moisture and adsorbed gases. Thorough mixing in a three-roll mill minimizes the occlusion of air during mixing. The final mixture is usually filtered and degassed prior to being packaged as Part A of a two-part system. Frequently, the resin batch is filtered through a nylon mesh. Where the curing agent provides a long pot life or shelf life, or where the curing agent is of the latent type, it may mixed with the Part A resin, degassed, packaged, and stored either at room temperature or at −40 °C, depending on the reactivity of the mixture. In the case of one-part frozen adhesives, the resin and hardener components are rapidly mixed, degassed, inserted in syringes, and quick frozen (Figure 3.22).

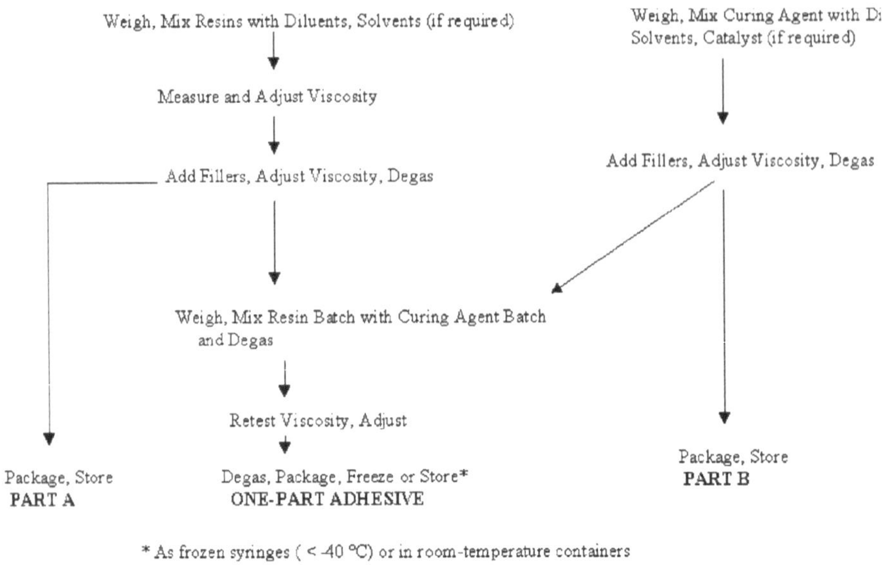

Figure 3.22 Manufacturing steps in formulating one-part and two-part paste adhesives.

The hardener portion (Part B) may also be mixed with fillers, diluents, solvents, and other additives. Fillers may be mixed in the resin portion alone or partly in the resin and partly in the hardener portion. Adhesion promoters such as Dow Corning A-180 silane may also be added to enhance the adhesion of the filler particles to the resin or hardener. Examples include several electrically conductive epoxy formulations that were used to evaluate their electrical stability on non-noble metal surfaces.[47] The base epoxy resin was the diglycidylether of bisphenol-F (DGEBF, equivalent wt. = 170 g). Methylhexahydrophthalic anhydride, hexahydrophthalic anhydride, and 4-methylphthalic anhydride were evaluated as curing agents with 4-methylimidazole as a catalyst. The manufacturing procedure was:

- Add curing agent to a weighed amount of epoxy resin (0.85:1 anhydride/epoxy ratio).
- Stir with heating until homogeneous.
- Cool to room temperature then add 0.5% (by weight of resin) of 2-ethyl 4-methylimidazole (Shikoku Chemical's Curezol® 2E4MZ) and mix. Curezol is a registered tradename of Air Products and Chemical Company, Inc.
- Add silver-flake filler (Ferro's 26LV) to the resin mixture at 80% by weight and blend in a three-roll mill until homogeneous.

Examples of formulations for specific applications are given in Table 3.9.

3.3.2 Film or preform adhesives

Film or preform adhesives are produced by casting the premixed resin and hardener onto release liners. Single or multiple master batches of the resin mixed with the

Table 3.9 Examples of adhesives formulations[a]

Adhesive Type	General Formulation	Examples
Surface-mount adhesives	35–50% epoxy resin, 1–10% amine or polyamine curing agent, and 10–30% mineral filler and thixotrope	Eccobond 125F (Henkel), Epo-Tek® 70E-4 (Epoxy Technology), Epi-bond® 7275 (Cookson Electronics)
	Mondur MC/castor oil and castor oil polyols (>60%), filler (<5%)	Flexobond 442 (Bacon Industries)
Dispensible electrically conductive adhesive	25–40% epoxy resin, 5–10% epoxy resin modifiers and diluents, 2–10% amine curing agents and catalysts, 50–85% silver flake	Ablebond® 84-1 LMI and 967-1 series (Henkel); Epo-Tek 3001 and Epo-Tek H20 (Epoxy Technology)
Dispensible electrically insulative adhesive	20–40% epoxy resin, 10–20% epoxy resin modifier, 1–5% amine, and >50% silica filler	Ablebond 84-3 (Henkel)
High-T_g die-attach adhesive	15–40% proprietary cyanate ester admixture, 60–85% silver filler	ME 8863 (AI Technology)
Underfill adhesive (rapid cure)	30–60% Bisphenol F epoxy reaction product, 1–5% imidazole catalyst, 30–60% silica filler	Loctite® 3563 (Henkel)
No-flow, fluxing type underfill	Epoxy resin, substituted phthalic anhydride (81-3) and methylhexahydrophthalic anhydride (2078E), phenolic compound, and liquid rubber component	STAYCHIP™ 2078E and STAYCHIP™ 81-3 (Cookson Electronics)
No-flow underfill	40–80% epoxy resin, 10–50% aliphatic amine complex, 1–5% aromatic aminederivative	UF-3667 (3 M)
Substrate-attach preform	35–50% rubber-modified epoxy, 10–20% epoxy resin, 1–5% organosilane, and 1–5% other (proprietary)	Ablefilm™ 550 (Henkel)
Silicone RTV (low tear strength)	5–10% methyltrimethoxy silane, 1–5% diisopropoxy di(ethoxyacetoacetyl) titanate	Dow Corning™ 838 (Dow Corning)
Silicone RTV (high tensile strength)	3–7% dimethyl methylhydrogen siloxane, 1–6% methylvinyl siloxane, dimethoxy (glycidoxypropyl)-terminated, and <1% tetra(trimethylsiloxy) silane	Dow Corning SE 1750 (Dow Corning)
Heat-cured silicone	3–7% dimethyl methylhydrogen siloxane with methylsilsequioxane, 1–6% glycidoxypropyl trimethylsiloxy silane	Dow Corning EA-6052 (Dow Corning)

[a] Excludes primers, stabilizers, and other additives.

hardener are first weighed, mixed, and milled, similar to the paste adhesives. The mixture may be screened through a nylon mesh to remove gelled particles and degassed. Adhesive is applied in a two-step process. First, adhesive is coated onto a release liner and laminated to a support reinforcement such as glass fabric. Primer, such as Dow Corning Z-6040, may be applied to the fabric prior to applying the adhesive. Adhesive may also be applied to the second side of the preform or reinforcement in a "second pass" lamination. The final product is accumulated between release liners (Figure 3.23).

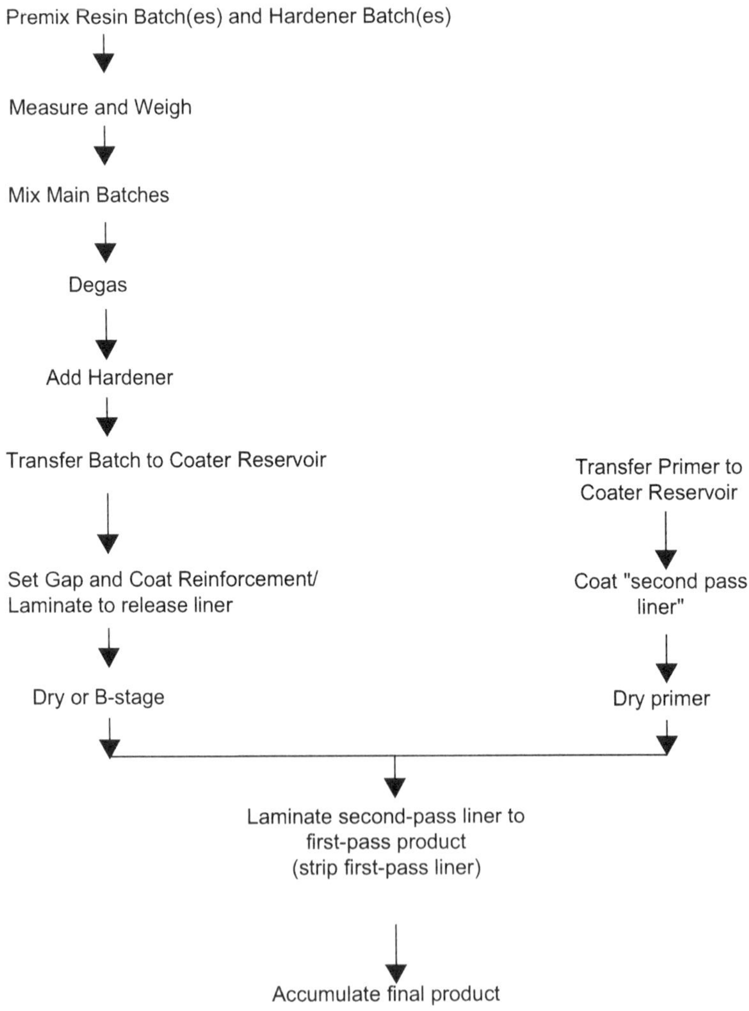

Figure 3.23 Manufacturing steps in formulating film/preform adhesives.

3.3.3 Premixed frozen adhesives

Two-part adhesives may be mixed, deaerated, packaged in tubes or syringes and frozen. The processing of frozen adhesives is best done by the adhesives supplier who is experienced and equipped to perform these steps.[92] The user thus avoids errors that can be made by production workers in accurately weighing, mixing, and vacuum degassing the adhesive prior to use. The use of frozen adhesives also minimizes handling and reduces exposure of workers to toxic or irritant components of the adhesives, especially curing agents such as amines used with epoxies.

Frozen adhesives are transported in containers with dry ice at −80 °C and must be stored at −40 °C until ready for use, at which time only the amount needed, is thawed and dispensed. The adhesive should be thawed to room temperature in its original closed container since thawing while exposed to air causes moisture condensation and absorption. After dispensing, any unused adhesive should not be refrozen since a portion of its pot life has been expended. Adhesives that are shipped in dry ice or stored at −40 °C may contain a temperature indicator, a cube of red material that melts if the temperature exceeds −40 °C.

3.4 Properties

The properties of cured adhesives are a function primarily of their chemical structures, formulation ingredients, and curing conditions. To a large extent, properties such as thermal stability, moisture resistance, and electrical performance can be predicted based on the molecular structures of the basic resins and curing agents used and the type and degree of polymerization that occur. Thus, polymers that are highly aromatic or heterocyclic (imides and triazines) and those that are highly cross-linked are more thermally stable and radiation resistant than their aliphatic linear counterparts. In Table 3.10, the general properties of the major polymer types used in adhesive formulations are compared. These should be considered general guidelines because of the numerous possible variations in the chemical structures of any one type of polymer resin, variations in the type of curing agent that can be used, and the large number of additives that can be formulated. The properties of a specific adhesive should always be experimentally measured.

3.4.1 Properties of epoxy adhesives

Of the many polymer types, epoxies have the best combination of properties that are desirable for electronics packaging and consequently are widely used. Among these properties are their high adhesion strength to almost all surfaces, good electrical insulating properties even under elevated temperature and humidity conditions, ease of curing and processing, low cost, and ease of reworking. Some limitations, which can be obviated, are their moderate-to-high absorption of water due to their highly polar structures, high amounts of chloride and other ions if not well purified, and risk of outgassing if not cured well or not vacuum baked.

Table 3.10 General comparison of adhesives for electronic applications

Material	Advantages	Disadvantages
Epoxies	High-temperature use. Good moisture and chemical resistance. High purity. Low outgassing.	Long cure cycles with anhydride hardeners. Degassing required for two-part systems. Exotherms in large quantities for amine-curing agents.
Silicones	Highest purity. Stress absorbing. High and low temp. stability.	Migrate to other circuit elements. Low surface energy. Swelled by nonpolar solvents.
Polyurethanes	Good flexibility at low temperatures. Stress absorbing. Highly versatile chemistry.	Lower thermal stability and service temperature than epoxies (150 °C–163 °C). Average bond strength unless primer is used.
Polyimides	Higher temperature stability compared to epoxies. High ionic purity. Reduced bleedout.	Trapped solvent can produce voids under large ICs. Multi-step curing required to volatilize solvent. High-stress materials. May absorb moisture in cured condition. Cannot be B-staged.
Silver glass	High-temperature stability. Low outgassing. Close CTE match to silicon.	High-firing temperatures. High-stress adhesive. Voiding due to solvent.
Cyanate esters	High adhesion strength. High thermal stability. High T_g, low CTE.	High-moisture absorption. Popcorn susceptibility.
Modified cyclo-olefin thermoset (MCOT) technology	Low moisture absorption. Reduced popcorning. Excellent adhesion.	Marginal adhesion to some substrates.

3.4.1.1 Purity

Major improvements have been made in reducing the amounts of chloride, potassium, and sodium ions and in the total ionic content of adhesives that are sold for electronics applications. Most electronic-grade epoxy adhesives contain only 35–60 ppm of chloride ion, 15–20 ppm of sodium ion, and 1–5 ppm of potassium ion. In fact, some formulations are available that have less than 5 ppm total for all three ions. These low ion levels have been achieved through distillation of the resins, water extractions, or by using epoxidized olefin resins instead of bisphenol-A types where there are no sodium chloride by-products generated. The outgassing of ammonia and low-molecular-weight amines has also been reduced or eliminated by avoiding the use of excess amine-curing agents, postcuring or curing at higher temperatures to assure complete reaction, or by using more stable amine-curing agents such as imidazoles instead of Dicy.

3.4.1.2 Adhesion

Adhesion of epoxies is among the highest of all polymers. As discussed in Chapter 2, this is due to the highly polar structures of epoxies and their wettability to most surfaces. Lap-shear strengths of aluminum-to-aluminum bonded specimens are typically 2,000 psi, but may run as high as 6,000 psi for some formulations and cure conditions. Die-shear strengths for electrically conductive and electrically insulative epoxy adhesives range from 5,000 psi to 7,000 psi, with snap-cured adhesives being at the lower end of the range. Integrated-circuit die, as large as 500 mils (12.7 mm) square, bonded with silver-filled epoxies are capable of surviving severe accelerated tests such as constant acceleration in the Y_1 direction of 5,000 g as defined in MIL-STD-883, Method 5011.[93] Epoxy-attached components are also able to pass other MIL-STD-883 and MIL-PRF-38534[94] thermal cycling, heat aging, vibration, and mechanical shock tests. Failures have only occurred when the bonding surfaces were not well cleaned or when the adhesive was not processed according to the manufacturer's instructions.

3.4.1.3 Thermal and thermomechanical properties

The thermal conductivities of unfilled epoxies, as with all other unfilled polymers, are quite low, typically 0.1–0.2 W/m K. When filled with metal or thermally conductive nonmetal fillers up to 80–85% by weight, the thermal conductivities increase a minimum of tenfold. Some silver-filled epoxies are reported to have thermal conductivities as high as 6.0 to approximately 8 W/m K (AI Technology ESP 8450 W and ESP 8456-00, respectively).

Thermal stabilities of adhesives may be measured by conducting a thermogravimetric analysis (TGA) test, that is, measuring weight loss as a function of increasing temperature at a constant rate, usually 10 °C per minute. Thermal stabilities are often reported as weight loss at 300 °C for comparative purposes. TGAs may also be performed at constant temperature (isothermal weight loss) as a function of time. Epoxy adhesives should actually be compared at 200 °C or, at most, 250 °C because, above 250 °C, rapid decomposition begins to occur and also because adhesives used in electronic applications seldom require exposures above 250 °C. Figure 3.24 compares the TGA curve for an epoxy film adhesive cured at 150 °C for 2 hours with that of a PEEK® film. In this example, the epoxy shows a weight loss of slightly over 1% at 300 °C, but a sharp increase above this temperature, indicating the onset of rapid decomposition. In comparison, the PEEK film was thermally resistant above 300 °C.[95] The best silver-filled epoxy adhesives show losses between 0.2 and 0.5% at 300 °C.

3.4.1.4 Thermal expansion properties

The CTEs of most filled epoxy adhesives range from 25 to 50 ppm/°C but, as discussed in Chapter 2, can increase three- to four-fold above their glass-transition temperatures.

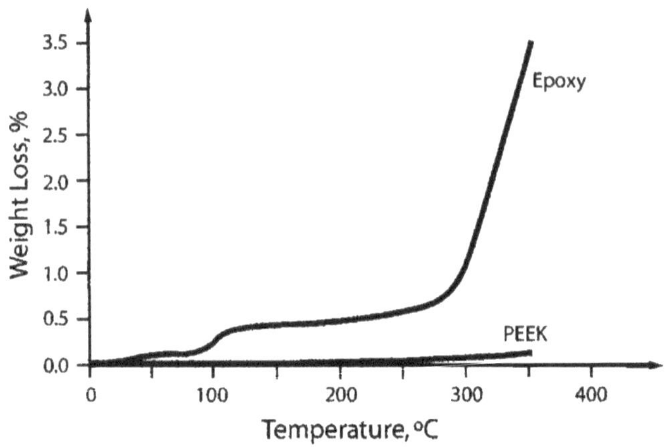

Figure 3.24 Thermal stability of an epoxy compared with PEEK® film.[95]

3.4.1.5 Electrical properties

The initial electrical properties of fully cured, electrically insulative epoxy adhesives, regardless of the curing agent used, are excellent under normal ambient conditions. Volume resistivities range from 10^{12} ohm-cm to above 10^{15} ohm-cm, dielectric constants from 2.8 to 4.8, and dissipation factors from 0.003 to 0.04 when measured at 25 °C and 1 MHz. Dielectric strengths range from 420 volt/mil to above 1,300 volt/mil. Although these initial electrical values are excellent, they can degrade on exposure to moisture and/or thermal aging. Both dielectric constants and dissipation factors increase with absorbed moisture (Table 3.11). Volume resistivities can drop several orders of magnitude to 10^{10} ohm-cm or below on temperature-humidity exposure such as the widely used 85 °C/85% RH test or the 10-day temperature-humidity cycle test per MIL-STD-202.[96] The high polarity of epoxy polymers that imparts such strong adhesion properties unfortunately also renders them hydrophilic,

Table 3.11 Electrical properties of an anhydride-cured bisphenol A epoxy[a]

Property	Initial Value	Water Immersion (24 hour)
Dielectric constant		
60 Hz	4.12	4.19
1 kHz	4.07	4.15
1 MHz	3.55	3.61
Dissipation factor		
60 Hz	0.0035	0.0043
1 kHz	0.015	0.016
1 MHz	0.032	0.032
Volume resistivity (ohm-cm)	1.81×10^{15}	2.31×10^{14}

[a] Cured with MNA (methyl nadic anhydride) 16 hours @ room temperature plus 4.5 hours @ 150 °C.

absorbing moisture to various degrees depending on the particular epoxy and its degree of cure. If after exposure to humidity and temperature, the adhesive has not decomposed or changed chemically, drying or vacuum baking usually restores the original electrical values.

For electrically conductive metal-filled adhesives, dielectric constants and dissipation factors are not meaningful parameters, but volume resistivities and contact resistances are significant. The volume resistivities for the best silver-filled epoxies range from 1×10^{-4} to 8×10^{-4} ohm-cm.

3.4.2 Properties of silicone adhesives

Among the major properties of silicones, distinguishing them from other polymer types, are their wide range of operating temperatures (from −80 °C to over 300 °C), their inertness and chemical stability, and their excellent electrical insulation and dielectric properties. Other unique features include high purity, low stress, low moisture absorption, resistance to oxidation and chemicals, and inherent flame retardancy. Among the limitations of silicones are their high moisture transmission rates, low tear and tensile strengths (for some formulations), and migration of low molecular weight species to adjacent surfaces during cure.

Depending on their molecular weights and degree of cure, silicones may exist as liquids, gels, or solids; may be solvent based or without solvent (100% solids); and may be either one-part or two-part systems. For adhesives applications, the one-part, 100% solids, room-temperature curing types are preferable. One-part silicones are generally the RTV and moisture-curing types. They are user-friendly requiring no weighing, mixing, or deaerating and cure on exposure to air moisture in a few hours. To assure complete cure, however, adhesive-bonded parts should cure for 24 to 72 hours. The cure time can be shortened by exposure to higher humidity than is present in air or by a short heat exposure. Some two-part silicones can be heat cured at 105–110 °C in only 10–15 minutes.

3.4.2.1 Thermal and thermomechanical properties

Silicones, as a class, are rated among the highest temperature stable polymers. They can withstand temperatures of 200 °C, almost continuously, without degradation of physical or electrical properties and have been used at temperatures as high as 300 °C. Because of their high thermal stabilities, they are used as adhesives and encapsulants for electronic modules that are expected to perform in extreme temperature environments, such as near automotive engines and in deep-well sensors. Because of their low moduli of elasticity, silicones also fare well at very low temperatures. They are rated for continuous use at −80 °C, but may be used at even lower temperatures.

Associated with their low moduli and elastomeric properties is their ability to dissipate stresses and to act as stress buffers between harder, less flexible materials such as epoxies. Because of their stress-dissipating properties, soft silicones are often used to bond or encapsulate fragile components such as glass diodes, laser diodes, and electro-optical devices. However, the CTEs of elastomeric silicones are generally

Table 3.12 Thermal and electrical properties of silicone adhesives[a]

Silicone Type	Thermal Conductivity, W/m K	CTE, ppm/°C	Dielectric Strength, V/mil	Dielectric Constant @ 100 kHz	Dissipation Factor @ 100 kHz
One-part, moisture-cure RTV	0.18	250–380	480–510	2.38–2.83	<0.001
Two-part, room-temperature cure	0.31	285	510	3.28	<0.002
Heat cure	0.32–0.35	256–300	350–577	2.78–3.02	<0.001–<0.0002

[a] Values compiled from a number of Dow Corning silicone adhesives and represent ranges for each category. Dow Corning should be contacted for specific formulations and values for specification purposes.

high (about 200 ppm/°C or higher) which, to an extent, offsets their stress-relieving properties. Other thermal properties are given in Table 3.12.

3.4.2.2 Humidity resistance

The behavior of silicones towards water differs considerably from that of other polymer types. Generally, a polymer that has high water absorption will also have a high moisture-vapor permeability. Silicones are unique in that they have low moisture absorption, yet high moisture-vapor transmission rates (MVTRs). Silicones are also hydrophobic and often used as water repellants. However, even their high MVTRs are not of great concern for electronics. Because of their high purity, especially the semiconductor-grade silicones, moisture has little effect on their dielectric properties.

3.4.2.3 Electrical properties

The electrical properties, especially for the high-purity semiconductor-grade silicones, are excellent even at temperature extremes of −80 °C and 200 °C and in high humidity. Silicone adhesives are available as electrically conductive and as electrically insulative types. Typical electrical properties for moisture- and heat-cured silicone adhesives are given in Table 3.12.

3.4.2.4 Physical properties

Among some limitations of silicones are their poor tear resistances and low tensile strengths. Young's modulus for some highly flexible silicones range from 0.5 MPa to 20 MPa, depending on the formulation and cure. Lap-shear strengths are also relatively low compared with epoxy adhesives. They generally range from 140 psi to 875 psi depending on the formulation. The heat-cured silicones have better mechanical properties than the moisture-cured RTV types. Other physical and mechanical properties are given in Table 3.13.

Table 3.13 Physical and mechanical properties of silicone adhesives[a]

Silicone Type	Tensile Strength, psi	Elongation, %	Tack-free Time, minutes. @r.t.	Unprimed Lap-shear Strength, psi	Unprimed Peel Strength, ppi
One-part, moisture-cure RTV	170–250	185–660	4–70	140–200	30–75
Two-part, room-temperature cure	250	225	–	230	10
Heat cure	769–950	150–265	–	550–875	–

[a] Values compiled from a number of Dow Corning silicone adhesives and represent ranges for each category. Consult Dow Coining for specific formulations and values for specification purposes.

3.4.3 Properties of polyimide adhesives

Polyimides are noted for their excellent thermal and radiation-resistance properties resulting directly from their highly aromatic polymer structures. Polyimide adhesives are stable in continuous operation at 250 °C and for shorter durations up to 400 °C. The glass-transition temperatures, typically 180–240 °C, are also higher than epoxies and most other polymer types. Thermal weight losses are also much lower than for some of the best epoxies. Figure 3.25 compares the weight loss at 250 °C of an anhydride-cured epoxy with a polyimide as measured by thermogravimetric analysis (TGA).

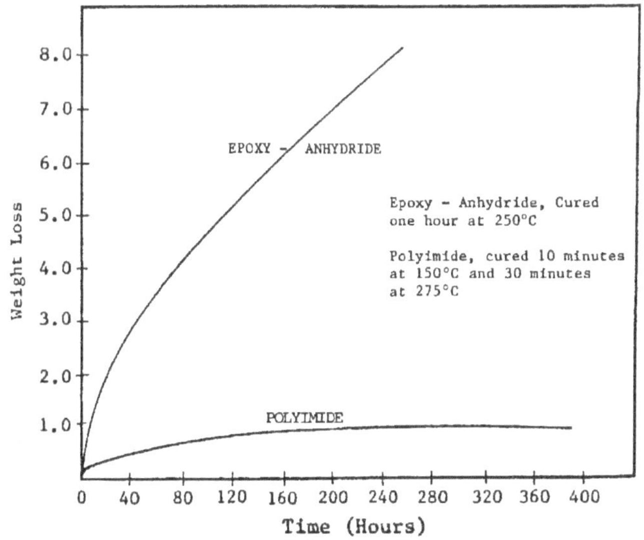

Figure 3.25 Weight loss of polyimide vs epoxy at 250 °C in nitrogen.

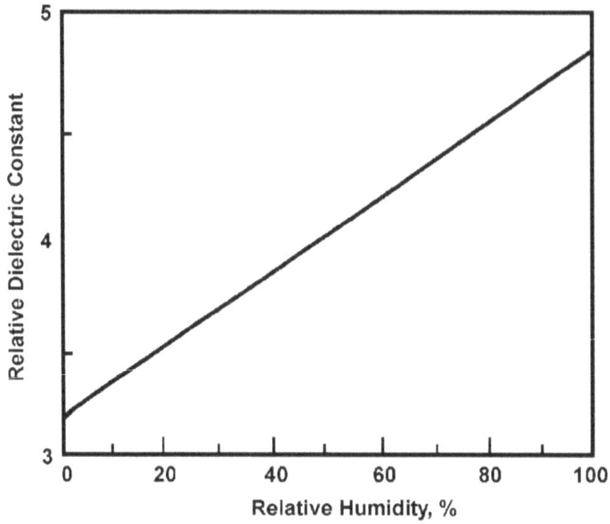

Figure 3.26 Effect of relative humidity on the dielectric constant of polyimide.[97,98]

Along with the advantages just cited, polyimides suffer some weaknesses, for example, they are much more difficult to process and cure than epoxies and other polymer types. The precursors must be dissolved in a highly polar, high-boiling solvent such as NMP (N-methyl-2-pyrrolidone) in order to process them as film or paste adhesives. To completely remove these solvents or traces of remaining solvent requires temperatures of 150–200 °C. The actual curing (imidization) requires step curing to temperatures as high as 400 °C although modified polyimides will cure at lower temperatures. Lastly, like epoxies, cured polyimides tend to absorb moderate to large amounts of water in humid environments. Water absorption may run as high as 2% at saturation, which degrades their electrical properties. Figure 3.26 shows a direct increase in dielectric constant as a function of exposure to increasing relative humidities. Figure 3.27 shows the increase in capacitance as a function of time exposed to 85% RH and 85 °C.

3.4.4 Properties of polyurethane adhesives

Polyurethanes, formulated as adhesives, are mostly two-part systems and must be mixed and deaerated prior to use. Like epoxies, they are highly polar, thus, have good adhesion to a large variety of surfaces, but their highly polar structures also result in a high affinity for moisture. On prolonged exposure to humidity-temperature environments, many of the early formulations reverted to the monomers. Reversion was caused by hydrolysis of the urethane and ester groups, breaking up the solid polymer into a sticky semisolid or fluid material. Some key properties of polyurethane adhesives are given in Table 3.14. Their advantages include the following:

- High flexibility and ability to dissipate stresses.
- Excellent adhesion to a wide variety of substrates.

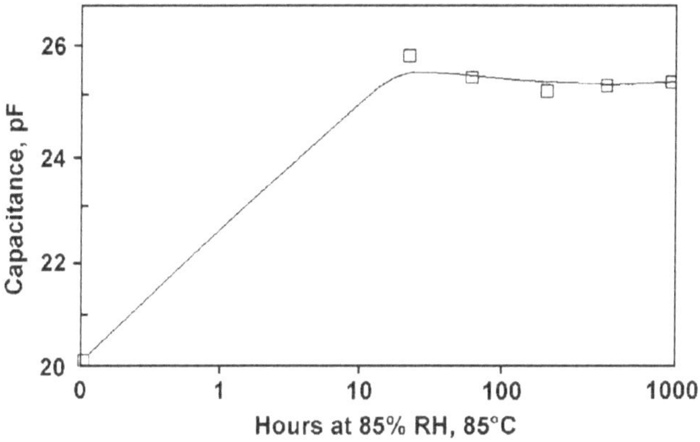

Figure 3.27 Increase in capacitance for polyimide on temperature/humidity exposure.

- Resistance to solvents commonly used in cleaning SMT assemblies.
- Ease of reworking by melting with a hot solder iron.
- Flexibility even at cryogenic temperatures.

Some limitations are the following:

- Less thermally stable than epoxies.
- Hardening and embrittlement on exposure to ozone and UV radiation.
- Difficulty in processing due to short pot lives.
- Extra safety precautions due to handling and exposure to toxic isocyanates.

3.4.5 Properties of cyanate esters

Cyanate-ester adhesives are available as either pastes or films. They are often blended or toughened with epoxy resins or other polymers and widely used for die attachment, underfill, and encapsulants. Cure schedules are generally 165–175 °C for 1–2 hours, but snap-cure formulations are available that cure in 10 seconds at 220 °C. The thermal properties of cyanate-ester adhesives are superior to those of epoxies. They are capable of withstanding temperatures of 200 °C when postcured and have glass-transition temperatures ranging from 140–245 °C, depending on the curing temperature. Curing at 165–175 °C for 1–1.5 hours generally results in T_g between 140 °C and 147 °C, while curing or postcuring at 300 °C for 30 minutes increases the T_g to 200 °C–250 °C.[99] Coefficients of expansion below the T_g range from 30–45 ppm/°C. The polymer derived from the basic dicyanate of bisphenol-A (Ciba Geigy's AroCy B®) is reported to have a T_g of 289 °C.[29] Cyanate esters have been selected for use in many space programs because of their low outgassing in thermal-vacuum environments and their high radiation resistance.[25] Because of their unique chemistry, cyanate-ester adhesives are reported to have low moisture outgassing (less than 1,000 ppm in a sealed cavity package) and less than 1% weight loss at 300 °C when measured by thermogravimetric analysis (TGA). This, coupled with their high purity and low ionic content (<20 ppm Na^+ or K^+ and <10 ppm Cl^-),[99] renders cyanate esters highly desirable as die-attach and underfill adhesives for space applications.

Table 3.14 Typical Properties of some polyurethane adhesives

Product Name/ Supplier	Description	Viscosity (cP)	Shelf life (−40 °C)/ Pot Life (25 °C)	Cure Schedule hour @°C	Hardness (Shore A)	Outgassing Test Results (ASTM E 595)	Lap-Shear Strength (psi)	CTE (ppm/°C) α_1	CTE (ppm/°C) α_2	T_g (°C)	Volume Resistivity (ohm-cm)	Thermal Conductivity (W/m K)
Arathane 7760/ Specialty Polymers and Services, Inc.	One-part thermal-transfer adhesive (CTE mismatched parts)	240,000–700,000	3 mos//>2 hours	0.5 @ 100 °C	70	0.471% total mass loss (TML), 0.011 collectible volatile condensible materials (CVCM)	200	29	88	−63	2.9×10^{14}	0.69
Flexobond 442/Bacon Industries	Thermally conductive, flexible system	Paste	3 mos/4 hours	2 @ 82°C	78	0.471% TML, 0.011% CVCM	500	N/A	N/A	N/A	1×110^{12}	0.864
DIS-A-PASTE 2000-A/B/ Aptek Labs.	Soft, two-part, mineral-filled electrically insulating	115,000 (Part A), 45 (Part B)	6 mos (25 °C)[a]/ 40 min	4 @ 100 °C	83	0.15% TML, 0.004% CVCM	425	31	147	−60	7×10^{14}	0.74
DIS-A-PASTE 2001 PMF Aptek Labs.	One-part, soft mineral-filled	140,000	6 mos (25 °C)[a]/ >2 hours	6 @ 85 °C	83	0.27% TML, 0.005% CVCM	425	31	137	−60	2×10^{14}	0.74
DIS-A-PASTE 2003-A/B/ Aptek Labs.	Two-part, mineral-filled	150,000 (A), 2,700 (B) (ASTM D-1824)	6 mos (25 °C)[a]/ >2 hours	6 @ 85 °C	83	0.27% TML, 0.005% CVCM	425	31	137	−60	2×10^{14}	0.74

[a] Sealed Containers.
Source: Compiled from suppliers' technical data sheets (not to be used for specification purposes).

Another property of cyanate esters important in reducing or avoiding popcorning during solder reflowing is their relatively low moisture absorptions. Moisture absorption at saturation in boiling water ranged from 0.7% to 2.5% depending on the starting cyanate monomer.

The electrical properties of cyanate esters are also quite good, even at high frequencies. Because of their low dielectric constants and dissipation factors extending into the gigahertz range, there has been considerable interest in cyanate esters for high-speed circuits. One formulation is reported to have a dielectric constant of 2.8 and a dissipation factor of 0.003 at 10 GHz.[100] Others range from 2.6 to 2.9 at 1 MHz. The bisphenol-A derivative is reported to have a dielectric constant of 2.9 at 1 MHz and 2.79 at 1 GHz. Its dissipation factor is 0.005 at 1 MHz and 0.006 at 1 GHz. The hexafluorobisphenol-A (Ciba Geigy's AroCy F®) exhibits even better electrical properties: dielectric constant of 2.66 and 2.54 at 1 MHz and 1GHz, respectively, and DF of 0.005 at both 1 MHz and 1GHz.[29] These electrical values are much better than those for epoxies and are attributed to the highly symmetrical macrostructures and absence of strong dipoles in the cured cyanate-ester polymers.

In spite of their low polarities, cyanate esters adhere well to metal conductors and maintain good adhesion up to 250 °C. They are thus compatible with solder-reflow temperatures. The adhesion strengths of toughened cyanate-ester adhesives, as measured from aluminum-to-aluminum lap-shear specimens are reported to be 5,900 psi at 25 °C, 2,572 psi at −67 °C, and 4,000 psi at 82 °C.

3.4.6 Properties of MCOT adhesives

In many respects, the properties of the modified cyclo-olefin thermoset (MCOT) adhesives are similar to those of the cyanate esters. Both types were developed to solve manufacturing problems in the die attachment and encapsulation for BGA and chip-scale packages where the high moisture absorptions of other polymer types, such as epoxies and polyimides, resulted in rapid moisture expulsion popcorn effect during subsequent solder-reflow steps. MCOT adhesives when cured at 150 °C for 30 minutes or less absorbed less than 0.05% water at 85 °C/85 RH conditions after 168 hours and 0.1% even after HAST* testing at 150 °C, 85% RH, 168 hours.[30] As with cyanate esters, MCOT adhesives are cured by addition polymerization, release no by-products, and contain very low traces of ionic impurities. Properties for two commercially available MCOT adhesives are listed in Table 3.15.

3.4.7 Properties of acrylates and UV-curing adhesives

Acrylate adhesives differ from other polymer types in their ability to be cured rapidly by exposure to UV and/or visible light. Their main advantage over heat-curing adhesives is in bonding heat-sensitive electronic devices in 5–30 seconds at room

*Highly Accelerated Stress Test per JEDEC-STD-22-A110.[101]

Table 3.15 Properties of MCOT adhesives[30]

Property	Thermally Conductive Paste	Electrically Conductive Paste
Filler type & percent	Boron nitride (40%)	Silver (80%)
Viscosity, Pa·s	10	10
Cure temperature	100–150 °C	100–150 °C
Cure time	<30 min	<30 min
Pot life @ 25 °C	16 hours	16 hours
Shelf life@ −40 °C	>6 months	>6 months
Extractable ions	<10 ppm	<10 ppm
% weight loss (up to 300 °C)	<1	<1
Cured Properties		
CTE (20–200 °C)	85 ppm/°C	78 ppm/°C
Modulus @ 25 °C	1,000 MPa	1,000 MPa
Modulus @ 200 °C	150 MPa	165 MPa
Glass-transition temp. (T_g)	65 °C	65 °C
Bulk thermal conductivity (W/m K)	1.0	2.3
Volume resistivity (ohm-cm)	NA	1.5×10^{-4}

temperature. Ultraviolet/visible light curing also allows time to position a device, a wire, or an optical fiber prior to cure (curing on demand). Acrylate adhesives are used for wire tacking, lens bonding, sealing, and fiber-optic circuit assembly. Adhesives as thick as 0.5 inches can be cured with UV light while depths of 1–3 inches can be cured with the visible and visible/UV types.[102] Formulations are available having a wide range of viscosities approximately 500 cP to 44,000 cP for 3M' s LC 1100 and 1200 series of Light-Cure® adhesives and 200,000 cP for ULTRA LIGHT-WELD® 9-202-69 Wire and Component Tacking Adhesive to 800,000 cP for Dymax 997 Multi-cure®.[103,104]

The mechanical and electrical properties of acrylates are similar to those for many other polymer types. Tensile strengths range from approximately 1,300 psi to 4,500 psi and shear strengths of PCB lap samples range from 424 to 566 psi. Aluminum-to-aluminum lap-shear strengths are reported at approximately 3,000 psi.[105] Electrical volume resistivities under room-ambient conditions ranged from 1×10^{14} ohm-cm to 1.4×10^{16} ohm-cm, dielectric constants from 3.15 to 3.43 at 10 MHz and dissipation factors from 0.017 to 0.024 at 10 MHz. Dielectric strengths ranged from 1,320 to 3,790 volts/mil.[102] Adhesives may range from semi-rigid to very flexible depending on the formulation.

Other properties of UV-cured adhesives of benefit to the electronics industry are: one-part systems, 100% solids (no solvents), and can be stored at room temperature.

3.4.8 Properties of film adhesives

Almost all film adhesives used in electronics assembly are either epoxies or polyimides on a glass fabric carrier. Film adhesives may be thermoplastic or thermosetting and are generally B-staged or partially cured. As with paste adhesives, film adhesives may also be electrically conductive or electrically insulative. Total film thicknesses (including the carrier) range from 3–10 mils with carrier thicknesses of 1–4 mils.[106,107] Heat curing is performed while a slight pressure is applied to the part being attached in order to equalize the flow throughout the surface and generate a small fillet at the edges.

Thermally conductive films are used to bond heat dissipating components to heat sinks. Electrically conductive silver-filled epoxy films when used to attach substrates also serve as ground planes and provide RF/EMI shielding. They are also used in the high production automated attachment of leadframes to chips and to bond chips vertically in chip stacks.

Most electrically conductive films are not as conductive as pastes, thus not suited for high frequency microwave circuits. Resistivities are generally an order of magnitude lower for conductive pastes (approx. 5×10^{-4} ohm-cm) than for their film counterparts (approx. 5×10^{-3} ohm-cm).

Electrically conductive film adhesives, if filled with silver particles, are also highly thermally conductive. Thermal conductivities range from 0.9 to 7 W/m K, depending on the formulation. Aluminum-to-aluminum lap shear strengths range from 2200 psi to 5100 psi. Films are commercially available in thicknesses from 2 to 6 mils.

In general, film adhesives are best suited as insulation to isolate circuitry or to attach large devices or substrates. Insulative films have volume resistivities from 10^{12} to 10^{14} ohm-cm, thermal conductivities from 0.8 to 1 W/m K, and lap shear strengths ranging from 1200 to 3000 psi. They are available in thicknesses from 2 to 8 mils.[108]

Properties of some film adhesives for die attachment and for substrate attachment are given in Tables 3.16 and 3.17.

3.4.9 Properties of silver-glass adhesives

The main advantages of silver-glass adhesives are their very high thermal conductivities and thermal stabilities compared with filled polymer resin formulations. Thermal conductivities of 65 W/m K to over 80 W/m K are reported.[109,110]

Thermal stabilities are approximately the same as the temperatures at which they are processed, approx. 300–325 °C important for subsequent high temperature processing steps such as non-lead-containing solders, solder reflowing, and hermetic package sealing. The absence of a polymer matrix in the final processed adhesive also results in essentially no decomposition or outgassing. On the negative side, the high processing temperatures cannot be used where devices are to be attached to plastic substrates or where other components of a circuit are temperature sensitive as, for example, some hybrid microcircuits.

By virtue of their high silver content, silver-glass adhesives have high electrical conductivities, typically 1.3×10^{-5} ohm-cm. Other typical properties are flexural modulus of 350 kpsi, die shear adhesion of 900 psi, and CTE of 20 ppm/°C.[110]

Applications for silver-glass adhesives are primarily for the attachment of power semiconductor die, LEDs, and devices near the engine in automobiles.

Table 3.16 Typical Properties of some film adhesives for semiconductor die attachment

Product/Supplier	Type	Shelf Life @ −40 °C (unless indicated)/Pot Life @25 °C	Cure Schedule	% Wt Loss at 300 °C	Bond Strength Al-Al lap shear (psi)	Modulus of Elasticity MPa (psi) (25 °C)	CTE, ppm/°C		T_g°C	Volume Resistivity, ohm-cm	Thermal Conductivity, W/m K	Moisture Absorption% *	Ionic Impurities (ppm)
							α_1	α_2					
Ablefilm® ECF561E/ Henkel	Silver-filled, rubberized epoxy, glass-fabric carrier	12 mos/N/A	1 hr @ 150 °C	0.55	2,000	N/A	100	380	47	N/A	1.6	N/A	N/A
Ablefilm® E5025E™/ Henkel	Silver-filled epoxy	6 mos (5 °C)/3 mos	30 min @ 150 °C	N/A	≥2,000	3,700 (530,000)	65	150	90	$\leq 5.0 \times 110^{-4}$	6.5	0.5	Cl^-, 50 Na^+, 30 and K^+ 5
Ablefilm® ECF 564A™/ Henkel	Silver-filled epoxy, glass-fabric carrier	12 mos/3 days	2 hours @ 150 °C	0.26	2,200	N/A	30	N/A	150	4.0×10^{-4}	3.8	N/A	Cl^- 5, Na^+ 20, and K^+ ND
Ablefilm® 506/ Henkel	Flexible thermally conductive epoxy glass-fabric carrier	6 mos/24 hours	1 hour @ 150 °C	N/A	1,200	N/A	45	350	3	7.0×10^{13}	0.9	N/A	N/A
Tack Film RTC8750/AI Technology	B-staged silver-filled epoxy	12 mos (0–5 °C)/ N/A	1 hour @ 150 °C, 3–5 psi pressure	N/A	1,000	N/A	N/A	110	−25	$<4.0 \times 110^{-3}$	6.4	N/A	N/A
ESP8450-WL/AI Technology	Flexible epoxy film, silver filled	12 mos (0–5 °C)/ N/A	2 hours @ 125 °C, 8–10 psi	N/A	N/A	N/A	N/A	110	−60	$<4.0 \times 10^{-4}$	6.4	N/A	N/A
Staystik® 501/ Cookson Electronics	Thermoplastic, silver-filled	12 mos, (25 °C)/NA	300–375 °C, 1–50 psi	N/A	N/A	2,483 (360,000)	N/A	N/A	180–185	$<5.0 \times 10^{-4}$	>3.0	N/A	N/A
Staystik® 581/ Cookson Electronics	Thermoplastic, silver-filled	12 mos, (25 °C)/ NA	160–175°C, 1–50 psi	N/A	N/A	2,414 (350,000)	N/A	N/A	98	$<5.0 \times 10^{-4}$	>3.0	N/A	N/A

* After 85/85 exposure.
N/A: Not available ND: None Detected.
Source: Compiled from suppliers' technical data sheets (not to be used for specification purposes).

Table 3.17 Typical properties of some film adhesives for substrate attachment

Product/ Supplier	Type	Shelf Life @ −40 °C/Pot Life @ 25 °C	Cure Schedule	% Wt. Loss at 300 °C	Bond Strength Al-Al lap shear, psi	Modulus of Elasticity, MPa (psi), 25 °C	CTE ppm/°C α_1	CTE ppm/°C α_2	T_g°C	Volume Resistivity, ohm-cm	Thermal Conductivity, W/m K (121 °C)	Moisture Absorp-tion, % at saturation 85% RH/85 °C saturation	Ionic Impurities, ppm
Ablefilm® 550K™ /Henkel.	White epoxy, thermally conductive, glass-fabric carrier	12 mos/6 mos	30 minutes @ 150 °C	0.37	3,300	N/A	50	300	102	7.2×10^{12}	0.8	N/A	N/A
Ablefilm® 550™/ Henkel.	Translucent epoxy, glass-fabric carrier	12 mos/6 mos	30 minutes @ 150 °C	1.1	5,700	3,793 (550,000)	95	480	105	1×10^{14}	0.2	2.2	Cl^-45, Na^+ 10, K^+ 1
Ablefilm® 564AK /Henkel.	Opaque white epoxy, thermally conductive, glass fabric	12 mos/72 hours	120 minutes @ 150 °C	0.16	2,100	N/A	29	N/A	137	2.8×10^{14}	1.4	N/A	Cl^-30 Na^+ 15, K^+ 5
Ablefilm® 561 K /Henkel.	Thermally conductive epoxy glass fabric	12 mos/6 mos	30 minutes @ 150 °C	0.34	3,300	4,345 (630,000)	85	300	55	9.1×10^{12}	0.9	0.9	Cl^- 110, Na^+ 35, K^+ 9
Ablefilm® 5020 K /Henkel.	Thermally conductive epoxy glass fabric	12 mos/6 mos	60 minutes @ 150 °C	0.3	3,000	N/A	45	270	109	8×10^{14}	0.7	N/A	Na^+ 20, K^+ 1 Cl^{-1} 60
Emerson Cuming™ CF3350 /Henkel.	Thermally conductive epoxy, polyester carrier	9 mos/(5 °C), 3 mos	30 minutes @ 150 °C	N/A	3,400	2,400 (348,090)	65	150	90	2.0×10^{-4}	7.0	N/A	Cl^- 50, Na^+ 30, K^+ 5

(*Continued*)

Table 3.17 Typical properties of some film adhesives for substrate attachment—*cont'd*

Product/ Supplier	Type	Shelf Life @ −40 °C/Pot Life @ 25 °C	Cure Schedule	% Wt. Loss at 300 °C	Bond Strength Al-Al lap shear, psi	Modulus of Elasticity, MPa (psi), 25 °C	CTE ppm/°C α_1	CTE ppm/°C α_2	T_g °C	Volume Resistivity, ohm-cm	Thermal Conductivity, W/m K (121 °C)	Moisture Absorp-tion, % at saturation 85% RH/85 °C saturation	Ionic Impurities, ppm
Tack Film TK7758/AI Technology	Epoxy, aluminum nitride filled	12 mos/N/A	30 minutes @ 150 °C	N/A	1,000	N/A	N/A	110	−25	$> 1 \times 10^{14}$	3.6	N/A	N/A
TP8550/AI Technology	Thermoplastic epoxy, silver filled	12 mos (25 °C)/ N/A	0.5–5 seconds @ 150 °C, 5 – 10 PSI	N/A	N/A	N/A	N/A	65	-70	$<4.0 \times 10^{-4}$	N/A	N/A	N/A
ESP 7455/AI Technology	Tack Free, alumina filled epoxy	12 mos/-	60 minutes @ 150 °C	N/A	N/A	N/A	N/A	N/A	N/A	$>1 \times 10^{14}$	6.50	N/A	N/A
Staystik® 571 /Cookson Electronics	Thermoplastic, silver filled	N/A	175–200 °C (1–50 psi)	N/A	>3,000	414 (60,000)	N/A	N/A	75	$\leq 5.0 \times 10^{-4}$	3.0–3.5	N/A	N/A
Staystik® 672 /Cookson Electronics	Thermoplastic, aluminum nitride filled	N/A	175–200 °C (1–50 psi)	N/A	>3,000	414 (60,000)	N/A	N/A	75	$>1.0 \times 10^{9}$	1.0–1.5	N/A	N/A
Staystik® 872 /Cookson Electronics	Thermoplastic, alumina filled	N/A	175–200 °C (1–50 psi)	N/A	>3,000	414 (60,000)	N/A	N/A	180–185	$>1.0 \times 10^{9}$	0.5–1.0	N/A	N/A

Source: Compiled from suppliers' technical data sheets (not to be used for specification purposes). N/A: Not Available, ND: None Detected.

References

1. Zarrow P, Kopp D. On the Forefront: Conductive Adhesives. *Circuits Assembly.* Apr. 1996.
2. Lee H, Neville K. *Handbook of Epoxy Resins.* McGraw-Hill; 1967. Reissued (1982).
3. Skeist I, ed. *Handbook of Adhesives.* 3rd ed. Van Nostrand Reinhold; 1990.
4. Ellis B, ed. *Chemistry and Technology of Epoxy Resins*, Blackie Academic and Professional. London:1993
5. Tess RW, Poehlein GW. Applied Polymer Science. *ACS Symp. Series.* 1985;285:560.
6. Dow Epoxy-Liquid Epoxy Resins, http://epoxy.dow.com/epoxy/products/prod/liquid.htm (accessed 7/7/2010)
7. Epoxy Resins Comparative Cross-reference Guide, Publication SC 432-91. *Shell Chemical Co*; Oct. 1991.
8. Lee H, Neville K. *Epoxy Resins.* McGraw-Hill; 1957.
9. Lee H, Neville K. *Handbook of Epoxy Resins.* McGraw-Hill; 1967 [Chapter 10].
10. Functional Products, www.cognis.com. May 6, 2008.
11. Lau JH, Wong CP, Prince JL, Nakayama W. *Electronic Packaging: Design, Materials, Process, and Reliability.* McGraw-Hill; 1998.
12. Tess RW, Poehlein GW. Applied Polymer Science. *ACS Symp. Series* 1985;285:934.
13. *Materials and Applications Resource Guide.* CDS 4415, G.E. Silicones; Jul. 1997.
14. *Gelest Reactive Silicones: Forging New Polymer Links.* Gelest Inc; 1999.
15. Licari JJ. *Coating Materials for Electronic Applications.* Noyes Publications; 2003: 119–125 [Chapter 2].
16. Mittal KL, ed. *Polyimides, Synthesis, Characterization, and Applications.* Vols. I and II: Plenum Press; 1984.
17. *Epo-Tek®R1011, Technical Data Sheet Rev III.* Epoxy Technologies; Aug. 2007.
18. Buchwalter LP. Adhesion of Polyimides to Metals and Metal Oxides. *J Adhesion Sci Technol.* 1987;l(4):341-347.
19. Ying L. A Novel Approach - Thermoplastic Die Attach Adhesive. *Solid State Technol.* Sep. 1987.
20. Eagle Specialty Products, www.espinc.US/Urethane.htm (accessed July 27, 2010).
21. *The Chemistry of Polyurethane Coatings: A General Reference Manual.* Publication 7178 10.0, Bayer Corp; Nov. 1998.
22. *Private communication with Pierre Villanueva.* Bacon Industries; Sep. 7, 2001.
23. Bayer Chemicals Product Information, http://www.bayerus.com. Nov. 7, 2001.
24. Chien IY, Nguyen MN. Pursuing the Perfect Paste: A Novel Cyanate Ester Die Attach Adhesive for Plastic Packaging. *Solid State Technol.* Nov. 1994.
25. Shah T, Danzinger S, Moores K, Joshi Y. Cyanate Ester Die Attach Material for Radiation Hardened Electronic Packages. *Proc. 3rd Intl. Conf. Adhesive Joining and Coating Techniques in Electronic Mfg.* Sep. 1998:49-54.
26. Hanafiah NM, Rasiah I, Jaenicke S. *Catalytic Reaction in a Cyanate Based Adhesive, Applied Chem. Honors Project, Dept. of Chemistry.* Natl. Univ. of Singapore; 2000/2001.
27. Bauer J, Hoper L, Bauer M. Cyclotrimerization Reactivities of Mono-and Difunctional Cyanates. *Macromol Chem Physics.* 1998.
28. Bauer M, Bauer J. Aspects of the Kinetics, Modeling, and Simulation of Network Buildup During Cyanate Ester Cures. In: Hamerton I, ed. *Chemistry and Technology of Cyanate Ester Resins.* Glasgow, UK: Chapman & Hall, Blackie Academic and Professional; 1994:55-86 [Chapter 3].

29. Hamerton I, ed. *Chemistry and Technology of Cyanate Ester Resins*. Glasgow, UK: Chapman & Hall, Blackie Academic and Professional; 1994.
30. Nguyen MN, Chien IY. An Ultra Low Moisture Polymer Adhesive for BGA Packages. *Intl. J. Microcircuits and Electronic Packaging*. 1998;21(2).
31. Nguyen MN, Edwards CS, Chien IY. Low Moisture Adhesive for Microelectronic Packages. *HDI Magazine*. Aug. 1998.
32. Nguyen MN, Chien IY, Koll PM. Recent Advancements in Olefin Thermoset Adhesive. *Proc 50th Electronics Components and Tech. Conf.* 2000:878-881.
33. Low Moisture Polymer for Die Bonding Fights Delamination. *Semiconductor Intl.* Johnson-Matthey Publication, Mar. 1998.
34. Nguyen MN. Low Stress Silver-Glass Die Attach Material. *IEEE Trans. Components, Hybrids, and Manufacturing Tech* Sep. 1990;13(3).
35. Bjorneklett A, Halbo L, Kristiansen H. Thermal Conductivity of Epoxy Adhesives Filled with Silver Particles. *Intl J Adhesion.* Apr. 1992;12(2).
36. Bolger JC, Morano SL. Conductive Adhesives: How and Where They Work. *Adhesives Age.* Jun. 1984.
37. Swanson DW. Qualification of a High-Purity, Low Outgassing Conductive Adhesive for Hybrid Microcircuit Assembly. *Proc 3rd Annual SAMPE Electronic Materials and Processes Conf.* Jun. 1989.
38. Haynes WM, ed. *CRC Handbook of Chemistry and Physics*. 91st ed. CRC Press; 2010.
39. Halbo L, Ohlckers P. Electronic Components, Packaging, and Production, *Dept of Physics*. Oslo, Norway: Univ of Oslo; Dec. 1995.
40. *PolarTherm® Thermally Conductive Boron Nitride Fillers For Polymeric Materials.* Advanced Ceramics Co., www.advceramics.com; 2001. accessed July 28, 2010.
41. AI Technology Catalog, AI Technology, Princeton Junction, NJ.
42. Technical Data Sheets for Adhesives, Ablestik Electronic Materials; Dec. 2003.
43. McNeilly K, Pernice RF. Silver Flake Increases Performance of Conductive Adhesive. *Advanced Packaging*. Jun./Jul. 1999.
44. Szegvari A, Yang M. *Attritor Grinding and Dispersing Equipment Seminar on Dispersion of Pigments and Resins in Fluid Media*. Kent State Univ; Apr. 29, 1999.
45. Markley PL, Tong QK, Magliacca DJ, Hahn TP. Characterization of Silver Flakes Utilized for Isotropic Conductive Adhesives. *Proc IMAPS, Intl Symp. On Advanced Packaging, Materials, Processes, Properties, and Interfaces.* 1999:16-20.
46. Ryszard K, Mosicicki S. Electrically Conductive Adhesives in SMT - The Influence of Adhesive Composition on Mechanical and Electrical Properties. *Proc. IMAPS Poland Conf.* 2000:199-206.
47. Lu D, Wong CP. Properties of Conductive Adhesives Based on Anhydride-Cured Epoxy Systems. *J. Electronics Mfg.* Sep. 1999;9(3):223-232.
48. Lyons AM. Electrically Conductive Adhesives: Effect of Particle Composition and Size Distribution. *Society of Plastics Engineers Annual Tech. Conference.* 1990:843-845.
49. Pandiri SM. The Behavior of Silver Flakes in Conductive Epoxy Adhesives. *Adhesives Age.* Oct. 1987.
50. Jost EM, McNeilly K, Sexton P. Silver Flakes for State of the Art Conductive Adhesive Applications. *Proc. 7th Intl. Microelectronics Conf.* 1992:422-428.
51. McNeilly K, Schlag S. Effect of Lubricant Solvent System on Rheological Properties of Silver-filled Polymers. *Proc. 3rd Intl. Seminar on Hybrid Microelectronics and Hybrid Packaging.* Apr. 1996:93-99. IMAPS-Korea.
52. Harper C. *Electronic Packaging with Resins*. McGraw-Hill; 1961:132.

53. Formulating with Dow Epoxy Resins: Casting, Potting, and Encapsulation, Dow Form 296-363-883.
54. Lee H, Neville K. Reissued (1982), *Handbook of Epoxy Resins*. McGraw-Hill; 1967. Chapter 14.
55. Wiles QT. Adhesive Compositions Containing Glycidyl Ethers and Alumina, U.S. Patent 2528933. Nov. 7, 1950.
56. *Cab-O-Sil Fumed Silica in Adhesives and Sealants*. Cabot Corp. Publication; Jun. 1990.
57. McNeilly K, Jost EM, Sexton P. Effect of Lubricant Solvent System on Rheological Properties of Silver-Filled Polymers. *ISHM Proc. 8th Intl. Microelectronics Conf.* 1994:308-317.
58. Korkin A, Gusev E, Labonowski JK, Luryi S, eds. *Nanotechnology for Electronic Materials and Devices*. Springer; 2007.
59. Introduction to Nanotechnology, www.nanowerk.com/nanotechnology/introduction/introduction_to_nanotechnology_30.php.
60. Das RN, Lauffer JM, Markovich VR. Polymer Nanocomposites for Electronic Packaging. *Advancing Microelectronics*. May/June 2009.
61. Sun Y, Meguid SA, Liew KM, Ong LS. Design and Development of New Nano-reinforced Bonds and Interfaces. *NSTI-Nanotech*, www.nsti.org; 2004:3.
62. Gilbert EN, Hayes BS, Seferis JC. Nano-alumina modified epoxy based film adhesives. *Polymer Engineering and Science* May 2003;43(5):1096.
63. Meguid SA, Sun Y. On the tensile and shear strength of nano-reinforced composite interfaces, *Materials and Design*. Elsevier; 2004.
64. Ghosh PK, Nukala SK. Properties of Adhesive Joint of Inorganic Nanofiller Composite Adhesive. *Indian Journal of Engineering and Materials Science* 2008;15.
65. Conde M, et al. Effect of Filler Content on Cohesive Strength of an Experimental Adhesive System. *Journal of Dentistry* May 2009;37(5).
66. Qu L, Dai L, Stone M, Xia Z, Wang ZL. Carbon Nanotube Array with Strong Shear Binding-on and Easy Normal Lifting-off. *Science*. Oct. 10 2008.
67. Han Z, Diao C, Doug L, Zhang X. Effect of surface-treated nanosilica on thermal behavior and flame retardant properties of EVA/ATH composites. *International Conference on Solid Dielectrics*. July 2007.
68. Das RN, Egitto FD, Markovich VR. Nano- and Micro-filled Conducting Adhesives for Z-axis Interconnections. *Circuit World* 2008;34(1).
69. Borders J, McGovern MR. Cyclic olefin Polymeric Nanocomposite Optical Plastic Article and Method of making same. *U.S. Patent 6552111*, www.patentstorm.us/patents/6552111/fulltext.html; April 22, 2003.
70. www.malvern.com/LabEng/industry/nanotechnology.
71. Bunshah RF, ed. *Handbook of Deposition Technologies for Films and Coatings*. 2nd ed. Noyes Publications; 1994.
72. Kloeppel JE. *Linear arrays of nanotubes offer path to high-performance electronics*. Univ of Illinois Press Release; March 26, 2007.
73. Harris TF. *Carbon nanotube science, synthesis, properties, and applications*. Cambridge Univ. Press; 2009.
74. Guo T, et al. Catalytic growth of single-walled nanotubes by laser vaporization. *Chem Phys Lett*. 1995;243.
75. Wang H, Northwood DO. Synthesis of Ni nanoparticles by hydrolysis of Mg_2Ni. *Journal of Materials Science*. Feb. 2008;43(3). Springer.
76. Trouiller B, Reliene R, Westbrook A, Solaimani P, Schiestl R. Titanium dioxide nanoparticles induce DNA damage and genetic instability in vivo in mice. *Cancer Research*. Nov. 15, 2009;69(22).

77. Nanotechnology Safety Act of 2010, U.S. Congress Bill S2942; Jan. 1, 2010.
78. Mitsubishi product data sheets, www.micchem.com.
79. Tess RW, Poehlein GW. Applied Polymer Science. *ACS Symp. Series* 1985;285:950.
80. Swanson DW, Enlow LR. Stress Effects of Epoxy Adhesives on Ceramic Substrates and Magnetics. *Proc. IMAPS Conf.* Oct. 1999.
81. Swanson DW, Enlow LR. Stress Effects of Epoxy Adhesives on Ceramic Substrates and Magnetics. *Microelectronics Reliability* 2001;41:499-510.
82. Underwriters Laboratory UL-94, Tests for flammability of plastic materials for parts in devices and appliances; June 2, 2006.
83. Rotival C, Renacco E, Arfi C, Pauli AM, Pastor J. Gases emitted during thermal decomposition of a polypropylene film and a polyurethane adhesive. *J of Thermal Analysis and Calorimetry* June 1994;41(6):1514-1527.
84. Rosler RK. Rigid epoxies, Electronic Materials Handbook. *Packaging ASM Intl.* 1989;1:810-816.
85. Pecht MG, Ardebili H. *Encapsulation Technologies for Electronic Applications.* Elsevier; 2009:102-108.
86. Levchik SV, Weil ED. Thermal decomposition, combustion and flame retardancy of epoxy resins - a review of recent literature. *Polymer International.* 2004;53(12): 1901-1929.
87. Producing flame retardant adhesives, www.specialchem4adhesives.com/resources/print.aspx?id=2836; June 3, 2009.
88. Lu S, Hamerton I. Recent developments in the chemistry of halogen-free flame retardant polymers. *Progress in Polymer Science* Oct. 2002;27(8):1661-1712.
89. Bopp RC. U.V. light stabilizer compositions comprising cyclic aliphatic epoxy, UV screener, and polyalkyldipiperidine (HALS) compounds. *U.S. Patent 4843116.* June 27, 1989.
90. Petrie EM. Antioxidants for adhesives, www.specialchem4adhesives.com; Aug. 11, 2004.
91. *BNX 1010TF, Antioxidant and thermal stabilizer, Product Data Sheet.* Mayzo Corp; March 8, 2005.
92. Rocheleau H. Frozen Adhesive Advantages in Electronic Manufacturing. *Electronic Mfg.* July 1988:18-19.
93. Test Methods and Procedures for Microelectronics; Evaluation and Acceptance Procedures for Polymeric Adhesives, MIL-STD-883, Method 5011.
94. Performance Specification, Hybrid Microcircuits, General Specification, MIL-PRF-38534.
95. Shores AA. Adhesive Bonding Hybrid Microcircuit Substrates with PEEK Thermoplastic Film. *ISHM Proc.* 1988;11(2).
96. Electronic and Electrical Component Parts, MIL-STD-202, Test Method Standard.
97. Jensen RJ. *IEEE Trans. Components, Hybrids, and Mfg Tech, CHMT-7.* 1984;384(4).
98. Brown R. Materials and Processes for Microwave Hybrids. *ISHM Monograph*; 1991.
99. Silver-Filled Cyanate Ester Die Attach Material, Product Data Sheet for JMI7000, Johnson-Matthey (undated).
100. EX-1502-1 Paste Adhesive, Bryte Technologies Product Data Sheet (May 1999).
101. *Highly Accelerated Temperature and Humidity Stress Test, JEDEC-STD-22-A110.* Electronic Industries Alliance; July 1988.
102. Technical Data Sheets for LC 1100 and 1200 Series Light Curing Adhesives, 3M Co. (undated).
103. Ultra Light-Weld® 9-20269 Wire and Component Tacking Adhesive, Dymax® Corp. (undated).

104. Surface Mount Technology (SMT) Adhesives, Dymax developmental MULTI-CURE® 997, Dymax Corp. (undated).
105. Damico DJ. Reactive acrylic adhesives. In: Pizzi A, Mittal K, eds. *Handbook of Adhesive Technology*. Marcel Dekker; 1994 [Chapter 28].
106. *Ablefilm® 561K, Technical Data Sheet*. Henkel; Mar 2009.
107. *Ablefilm® 550, Technical Data Sheet*. Henkel; Mar 2009.
108. *Henkel Electronics Assembly Solutions, Product Catalog*. Henkel Corp; May 2009.
109. *Henkel Hysol QMI 3555R, Technical Data Sheet*. Henkel Corp; Sept 2003.
110. DM 3030, Silver/glass die attach paste, *Thermally Conductive Adhesives Selection Guide*. Diemat, a Namics Co, www.diemat.com; 2003.

4 Adhesive Bonding Processes

Chapter Outline

Adhesives Technology for Electronic Applications. DOI: 10.1016/B978-1-4377-7889-2.10004-X

Process steps for bonding have undergone significant changes since the early 1970s. Most processes then were manual or semi-automatic, at best. Adhesives were used to attach small die in hybrid microcircuits or to attach large components to sparsely populated printed-wiring boards (PWBs). Adhesives were manually dispensed from a syringe or dabbed with a probe, and components were picked up and placed with tweezers or with a vacuum pick-up tool. With the recent proliferation of electronics in consumer products, high production attachment and curing processes have been developed along with precision automated equipment for cleaning, dispensing adhesives, placing of components, and curing of the bonded parts. New high-density devices and packaging formats have also been introduced such as chip-scale packaging, flip-chip devices, ball-grid array (BGA) packages, chip-on-board (COB), and chip stacks. Challenges have had to be met in mixing different technologies in the same assembly; examples are assembling wire-bonded die alongside flip-chip-attached die, large molded inductors or capacitors along with bare-chip devices, and soldered components together with components attached and connected with electrically conductive adhesives. Indeed, in the drive to reduce costs and increase circuit densities, the trend is to use a mixture of parts and assembly technologies.

Added to cost and reliability goals, a manufacturer must now comply with a myriad of ever stringent regulations for solvent emissions, toxicity, and energy conservation. To meet Occupational Safety Health Administration (OSHA), Environmental Protection Agency (EPA), and a multitude of other federal, state, and local government requirements, changes have had to be made in cleaning solvents and processes and in adhesive materials, bonding, and curing methods. This chapter addresses the following key bonding processes:

- Cleaning
- Surface treatment
- Dispensing of adhesives
- Placement of devices or parts
- Curing
- Reworking (if necessary).

A summary of process and material requirements for each of these steps is given in Table 4.1.

4.1 Cleaning

The most important requirements to assure good adhesion between surfaces are the cleanliness and chemical and physical condition of the surfaces. The adhesive chosen must be able to wet the surfaces so that a contact angle of less than 90° results, as

Table 4.1 Material and process requirements for bonding

Category	Process requirements	Material requirements
Cleaning	Batch process, high throughput	High exchange rate; low viscosity, low surface tension solvents leaving low residues; compliance with OSHA, EPA, and other regulations
Dispensing	Data-driven automated dispensing preferred. Stencil or mass pin transfer only for long production runs	High dot profile to fill component-to-board gaps; suitable viscosity for coverage; low stringing to avoid conductive pad contamination; long shelf life for production runs
Device placement	High throughput	Rheology suitable for dispensing
Curing	In-line compatible, low stress on flip-chip and large die	Rapid IR, UV, or convection oven cure, <3–4 minute (in-line application preferred); Low volatile organic compounds (VOC) emissions during cure
Rework	Temperatures, solvents, and chemicals that do not damage components or substrates; ease of rework	High T_g, thermally and chemically stable laminates; modest rework conditions (low T_g adhesives) for specialized applications

discussed in Chapter 2. Unfortunately, the choice of solvent and cleaning process is often made hastily without considering the nature of the contaminants or how clean the part needs to be. Inadequate cleaning prevents or reduces adhesion and residues can cause corrosion or electrical deterioration over time, but extra cleaning, if not necessary, becomes costly. It is a good engineering practice to analyze the types of contaminants that are likely to exist on a surface as a result of exposure to the ambient, handling, or previous processing steps such as plating, soldering, or etching. Analytical procedures such as infrared spectroscopy, energy-dispersive X-ray (EDX), scanning-electron microscopy (SEM), and ion analysis can be used to identify the residues. Once the contaminants are identified, the appropriate solvents and cleaning process for their removal can be selected.

Contaminants may be of three general types: particulates, highly polar (ionic) residues, and nonpolar (grease-like) residues. In many production environments, all three types are present and several solvents and cleaning processes may be required to remove them. Both the choice of solvent or cleaning solution and the cleaning process are critical, and several theoretical principles may be followed in their selection. Ultimately, cleaning is more an art than a science, and the procedure chosen should be experimentally verified in each case. Qualitative and quantitative tests can be performed to determine the efficiency of a selected solvent and process.

4.1.1 Solvent selection

Solvents, like contaminants, may be polar or nonpolar. As a general rule, polar solvents dissolve polar residues while nonpolar solvents dissolve nonpolar residues. Thus, ionic residues such as chlorides, salts, acids, acid fluxes, and alkalis are best dissolved and removed with polar solvents such as water, isopropyl alcohol, ethanol, or methylethyl ketone. Greases, oils, silicones, rosin flux, and low-molecular-weight monomers are best dissolved and removed with solvents such as hydrocarbons, Freons®, hydrochloro-fluorocarbons, xylene, terpenes, and naphtha. To remove both polar and nonpolar residues, a two-step process using both types of solvents may be used or, more conveniently, an azeotrope mixture of the two solvents can be used in a one-step process. Most of the chlorofluorocarbon solvents (Freons®) and their azeotropes with alcohols, methylene chloride, or ketones are being phased out due to their high ozone-depletion potentials. Solvent blends and azeotropes of hydrofluoroethers and hydrochlorofluorocarbons (HCFC) are now replacing these solvents.

4.1.1.1 Organic solvents

Purely organic solvents or mixtures are effective in removing rosin flux residues and other organic residues such as greases and oils. The highly polar alcohols such as isopropyl alcohol (IPA) and ethanol are also fairly effective in removing ionic and salt residues. Organic solvents have low surface tensions, good surface-wetting ability, high solvency for surface contaminants, and are easily vaporized, leaving the surfaces and electronic components dry. In fact, rinsing with fresh organic solvent is often used as the last step in aqueous cleaning to augment the removal of moisture and improve drying.

In the early days of electronics, an organic mixture found effective in removing both ionic and grease-like residues from printed-wiring assemblies was a 50/50 mixture of isopropyl alcohol and naphtha. For decades and up until recently, the chlorinated and chlorofluorocarbon (CFC) solvents were used alone or as mixtures with alcohols or ketones. DuPont's Freon® TF and other Freons® were among the best cleaning solvents in the electronics industry. They served the industry well because of their nonflammability, high solvency, and inertness to other circuit components. However, the phasing out of the chlorinated solvents, because of their high VOCs and ozone-depletion properties, has created a challenge for the industry to develop replacements that are equally effective as solvents, yet meet EPA and OSHA requirements. Solvents in use today for cleaning fragile, non-hermetically sealed assemblies are the hydrofluorocarbons (HFC), hydrofluoroethers (HFE), hydrochlorofluorocarbons (HCFC), and n-propylbromide. Even among these, some, like the HCFCs, are being allowed only for a few more years, and n-propylbromide, because of its high polarity and high boiling point, may be limited in use on plastic parts. As a result, aqueous and semi-aqueous solvents are finding more uses as alternates to organic solvents. At the same time, research has produced water-soluble fluxes and "no clean" fluxes, the latter leaving only minor residues that are not considered corrosive so that final cleaning, in some cases, can be eliminated.[1,2]

Organic solvents are most effective as blends of polar and nonpolar solvents that have a constant boiling point known as *azeotropes*. Azeotropes are mixtures of two or more solvents used primarily in vapor-degreasing equipment.

HCFC, HFC, and HFE solvents are available as azeotropes with alcohols such as ethanol. Besides their effectiveness as solvents, these hydrohalocarbons are desirable because of their low toxicity, nonflammability, heat stability, and inertness to components and materials on circuit-board assemblies. Examples of azeotropes include DuPont's HFC 43-10mee (2,3-dihydrodecafluoropentane) and 3M's HFE-7100 (methylnonafluorobutyl ether) both of which are comparable or better than CFC-113.

Table 4.2 lists some commercially available organic solvents, azeotropes, and mixtures.

4.1.1.2 Aqueous and semi-aqueous cleaners

Aqueous and semi-aqueous cleaners are used primarily to clean printed-circuit boards prior to assembly, during application of the solder paste and adhesive (if they have been misprinted or smudged), and after assembly and solder reflowing. Highly automated multi-stage cleaning consoles are used in production where the parts are cleaned in a series of baths ending up with one or more deionized water rinses.

Water, being a highly polar solvent, is ideal for the removal of ionic and polar residues. Alone, as a co-solvent, or as an emulsion with organic nonpolar solvents, water is widely used in cleaning printed-circuit boards both prior to and after assembly. Aqueous and semi-aqueous cleaners are nonflammable and relatively inexpensive compared with organic solvents. There is also no hazardous waste or handling concerns and little or no ozone-depletion potential. Aqueous solvents may be purely aqueous (water with detergents, saponifiers, surfactants, neutralizers, or other additives) or semi-aqueous (water with an organic polar or nonpolar solvent). Examples of organic solvents used with water in semi-aqueous or co-solvent systems include alcohols, esters, and terpenes. *Co-solvents* may consist of mixtures of two solvents that are completely miscible in each other or, if not, can be rendered miscible by the addition of an emulsifier. Emulsion cleaners combine water with a hydrocarbon-based solvent. By emulsification, the hydrocarbon becomes suspended as tiny droplets in the water. The main advantages are the small amounts of organic solvent used and the reduced odor and air emissions.

Thermal-phase cleaning, also known as *microphase cleaning* (MPC®, a trademark of Zestron Corp.), is yet another variation of aqueous cleaning.[3] The thermal-phase cleaner consists of water to which is added a material such as a modified alcohol that is fully miscible at room temperature. Heating the clear solution, typically to 40–50 °C, separates the organic alcohol as a turbid mixture of fine droplets suspended in water. The phase transition that occurs produces a mixture that exhibits the properties of both solvent and water-based cleaners, thus it is capable of dissolving both ionic and nonpolar residues. The onset of the phase change can be tailored by changing the composition of the mixture; thus, increasing the percent of alcohol increases the temperature at which phase change occurs. Thermal-phase cleaners may be used as a drop-in replacement in most equipment designed for liquid cleaning

Table 4.2 Organic solvents and azeotrope cleaning materials and properties

			Properties					
Product/ Supplier	**Description**	**Application**	**Kauri-Butanol (KB) Value**	**Boiling Point[a], °C**	**Surface Tension, dynes/cm**	**ΔHvap cal/g**	**Flash Point, °C**	**Vapor Pressure, mm[b]**
Lenium® FEC/ Petroferm	Hydrofluorocarbons and methanol mixture	Open top and vacuum vapor degreasers for circuit assembly cleaning, and flux and ionic residue removal	N/A	36	N/A	54	None	497
Lenium® FRA/ Petroferm	Binary azeotrope of fluorocarbons for use as a vapor-degreasing co-solvent	Batch or in-line vapor degreasers	N/A	42	14.8	41.4	None	398
Megasolv® NOC/ Petroferm	Solvent blend including 2-methoxy-methylethoxy propanol and 1-methyl-2-pyrrolidone	Room temperature cleaning of electronic assemblies, surface-mount adhesive cleaning (misprints and stencils). Used in pressure spray, immersion, and ultrasonic applications	N/A	180-202	Low	N/A	82	<1 (20 °C)
Bioact® EC-7R Defluxer/ Petroferm	Bio based (citrus terpene mixture) semi-aqueous spray cleaner	Ambient temperature removal of high-temperature solder flux residues from SMDs	N/A	N/A	N/A	N/A	47	<2 (20 °C)
Bioact® EC-7M/ Petroferm	Bio based (ultra high purity citrus terpene mixture) semi-aqueous spray cleaner	Hand wipe and immersion cleaning for removal of a broad range of flux residues from surface-mount devices	N/A	N/A	25	N/A	48	1.6

Bioact® EC-15/ Petroferm	Semi-aqueous mixture of esters, modified alcohols, and surfactants	Semi-aqueous and immersion cleaning at room or elevated temperatures for flux residue and ionic contamination removal	N/A	N/A	N/A	N/A	154	<1.0
1,1,1-trichloro-ethane/DuPont	$C_2Cl_3H_3$	Vapor degreasing	124	74	25.6	57.5	None	28
TCE /Du Pont	Trichloroethylene (C_2HCl_3)	Vapor degreasing	129	87	26.4	57.8	None	57.8
CFC-113/DuPont	Trichlorotrifluoroethane $C_2Cl_3F_3$	Vapor degreasing	32	48	17	35	None	334
Methylene chloride/DuPont	CH_2Cl_2	Vapor degreasing	136	40	28.2	78.7	None	349
Vertrel® SMT/ DuPont	Azeotrope of Vertrel XF (2,3 dihydrode-cafluoropentane) with trans 1,2-dichloroethylene	Vapor degreasing	N/A	37	15.5	207.1	None	0.639
Vertrel® XMS Plus/DuPont	Mixture of Vertrel® XF, 1,2 dichloroethane, cyclopentane, and methanol	Vapor degreasing	N/A	38	14.9	255.2	None	0.628
Abzol® EG and JEG /Albemarle	Alkyl bromide mixture	Degreasing electronics and printed-circuit boards, ionic contaminant removal	N/A	70	N/A	N/A	None	N/A
Zestron® FA/ Zestron	Modified alcohol	Ultrasonic, spray-under immersion, or centrifugal for solder paste and low	N/A	162–190	29.7	N/A	75	0.47

(*Continued*)

Table 4.2 Organic solvents and azeotrope cleaning materials and properties—*cont'd*

			Properties					
Product/ Supplier	**Description**	**Application**	**Kauri-Butanol (KB) Value**	**Boiling Point[a], °C**	**Surface Tension, dynes/cm**	**Δ Hvap cal/g**	**Flash Point, °C**	**Vapor Pressure, mm[b]**
		solids and rosin flux removal						
Zestron® VD/ Zestron	Modified alcohol	Solder paste removal in closed-loop vapor degreasing type systems	N/A	170–175	26.3	N/A	62	1.0
Zestron® SD 300/ Zestron	Modified alcohol	Solder paste, SMT or conductive adhesive, thick-film paste in explosion-proof, spray-in-air systems	N/A	120–150	27.0	N/A	41	6.2
Zestron® SD 301/ Zestron	Modified alcohol cleaning agent, low-odor and faster drying time version of SD 300	Solder paste and SMT adhesive removal from PWBs and stencils/screens in spray in air systems	N/A	150–170	26.0	N/A	47	2
AK 225/Asahi Glass Co.	Dichloropenta-fluoropropane ($C_3F_5HCl_2$)	Vapor degreasing, flux removal	31	54	16.2	34.6	None	0.385 (25 °C)
AsahiKlin AK-225 FPL/Asahi Glass Co.	Mixture of AK-225 and fluorocarbon cleaning solvents	Cleaning processes compatible with plastics	13	58	14.0	34.9	None	0.31 (25 °C)
			41	52	16.8	40.6	None	291 (25 °C)

AsahiKlin AK-225 AES/Asahi Glass Co.	Azeotrope of AK-225 with ethanol and a stabilizer	Substitute for CFC-113/alcohol systems						
AsahiKlin AK-225T/Asahi Glass Co.	Mixture of AK-225, ethanol, hydrocarbon, and stabilizers	Removes rosin-blend fluxes while preventing white residue formation	63	54	17.6	47.2	None	427 (37.8 °C)
HFE-71DA/3M Co.	Low viscosity and low surface tension HFE-7100 with trans 1,2-dichloroethylene and ethanol (azeotrope)	Vapor degreasing, flux removal and enhanced removal of ionic contaminants	33	40	16.4	N/A	N/A	381
HFE-71DE/3M Co.	Low viscosity and low surface tension and increased solubility mixture of Nonafluorobutyl ether azeotrope and trans 1,2-dichloroethylene	Immersion or vapor-degreasing applications	27	41	16.6	N/A	N/A	383
HFE-7200/3M Co.	Low viscosity and low surface tension ethoxy nonafluorobutane	Vapor-degreasing or cold cleaning applications	10	76	13.6	30	N/A	109
Micronox® MX2501/Kyzen	Mixture of fluorinated materials and trans 1,2-dichloroethylene	Precision cleaning of electronic parts in removing no-clean and rosin flux residues in vapor degreasing and cold cleaning	N/A	44	N/A	N/A	None to boiling	225

N/A – Not Available.
[a] Some boiling points are for concentrated solutions.
[b] 25 °C unless indicated.
Source: Compiled from suppliers' technical data sheets.

Table 4.3 Test results for thermal phase vs. conventional cleaners[a,3]

Material	Thermal Phase Cleaner with Saponifier	Neutral Thermal Phase Cleaner	Alcohol with Organic Saponifier	Inorganic Saponifier	Deionized Water
No-clean flux (Interflux 2005M)	0.28	0.28	0.35	0.44	0.40
Water-soluble paste (Heraeus F530)	0.28	0.54	0.56	0.58	0.31
No-clean rosin-based paste (Heraeus F380)	0.29	0.33	0.28	0.49	0.28

[a] In μg $NaCl/cm^2$. A value of 0.28 is baseline for an uncontaminated bare board.

including spray and ultrasonic equipments.[4] Some reported test results comparing the efficiency of thermal-phase cleaning with conventional cleaners are given in Table 4.3.[3]

Aqueous cleaners should be used with caution when cleaning assemblies that contain highly polar polymeric materials or thin-film metallization. Plastic materials such as epoxy-glass or polyimide-glass laminates, used extensively for circuit boards, absorb moderate-to-high amounts of water which, if not subsequently baked out, degrade the adhesion strength of bonded components. Moisture retention also contributes to various corrosion and electrical failure mechanisms, including leakage currents, electrical opens and shorts, silver migration, and copper migration (conductive anodic filaments or CAF).[5]

Retention of moisture in polymer substrates or adhesives, especially in the presence of chloride or other ions, corrodes thin-film metallization such as nichrome resistors or aluminum conductors (further discussed in Chapter 6). Other limitations in using aqueous or semi-aqueous solvents include:

- Solutions must be heated to accelerate removal of contaminants.
- Solutions may have to be agitated, brushed, sprayed with pressure, or exposed to ultrasonics to improve their cleaning efficiency.
- Co-solvents are required to remove contaminants having a high organic content (oils, greases, fluxes).
- Bath lives are short and require monitoring, replenishment, and filtering.
- Expended solutions must be neutralized prior to disposal.

Table 4.4 lists commercial cleaning solutions currently used in the electronics industry for cleaning printed-circuit boards, assemblies, ceramic printed circuits, thick-film interconnect substrates, electronic components, wiring harnesses, and stencils and screens (used in applying adhesives and solder pastes).

4.1.1.3 *Measuring solvency and cleaning ability*

An important consideration in selecting solvents or solvent-based cleaners is their solvent-cleaning abilities, which may be predicted or experimentally measured by

Table 4.4 Aqueous and semi-aqueous cleaning solutions

			Properties				
Product/ Supplier	**Description**	**Application**	**Boiling Point[a] (°C)**	**Viscosity (cps, 25°C)**	**Surface Tension (dynes/cm)**	**Flash Point (°C)**	**Vapor Pressure[b] (mm)**
Axarel®32/ Petroferm	pH neutral semi-aqueous defluxer with easy water rinsing; mixture of highly refined solvents and surfactants (KB value = 64)	Spray and immersion cleaning for ambient or heated defluxing and ionic contamination removal	N/A	2.8	N/A	96	<0.1
Axarel®36/ Petroferm	Semi-aqueous water rinsable defluxer; mixture of highly refined solvents (KB Value = 64)	Spray and immersion cleaning for ambient or heated defluxing and ionic contamination removal	N/A	2.8	N/A	96	<0.1
Axarel®2200/ Petroferm	Fast-drying solvent; mixture of highly refined polar and non-polar solvents (KB value > 150). Replacement for isopropyl alcohol, hydrocarbons, hydrofluoroethers, hydrofluorocarbons, and hydrofluorochlorocarbons	Hand wipe and immersion cleaning for removal of no-clean water soluble and rosin solder paste, white residue associated with isopropanol cleaning, and flux residues	N/A	1.9	N/A	44	3.54
Hydrex®LF/ Petroferm	High performance alkaline, low foaming, aqueous spray; mixture of solvents, saponifiers, and surfactants	Aqueous spray and immersion cleaning of uncured surface-mount adhesives and solder pastes from stencils and defluxing	**N/A**	~1	N/A	85	N/A

(*Continued*)

Table 4.4 Aqueous and semi-aqueous cleaning solutions—*cont'd*

Product/ Supplier	Description	Application	Properties				
			Boiling Point[a] (°C)	Viscosity (cps, 25°C)	Surface Tension (dynes/cm)	Flash Point (°C)	Vapor Pressure[b] (mm)
Hydrex® A-Plus/ Petroferm	Neutral pH cleaner compatible with many substrates; mixture of modified alcohols and surfactants	Ambient or heated stencil and surface-mount misprint cleaning and defluxing applications	N/A	~1	N/A	>105	<1
Hydrex®WS/ Petroferm	pH neutral blend of solvents and deionized water	Aqueous or solvent cleaning equipment for removing uncured surface mount adhesives, solder pastes from stencils and misprinted assemblies as well as solder residues	N/A	~1	N/A	None	>1 (estimate)
Vigon® SC/ Zestron	Fast drying and surfactant-free water-based cleaner using microphase cleaner technology	Stencil cleaning in spray in air or ultrasonic equipment	98	N/A	35.2	None	31.38
Vigon® SC 200/Zestron	pH neutral aqueous cleaner, with long bath life	Removes baked-on fluxes from solder pallets and condensation traps using spray in air, ultrasonic, or air-assisted cleaning equipment	95-212	N/A	N/A	None	18.1
Atron® SP 200/ Zestron	Aqueous-alkaline surfactant cleaner, with long bath life	Removes baked-on fluxes from solder pallets and condensation traps using spray in air, ultrasonic, or air-assisted cleaning equipment	100	N/A	N/A	None	~23

Aquanox® A 4241/Kyzen	Aqueous alkaline cleaner with inhibitors for enhanced removal without etching or darkening parts	Spray batch, spray in-line, and stencil cleaning processes	99	N/A	N/A	None to boiling	0.07
Aquanox® A 4703/Kyzen	pH neutral aqueous cleaner effective at low concentrations with inhibition technology for superior material compatibility	Spray batch and spray in-line cleaning systems for OA, no-clean, and RMA pastes and lead-free flux residues	138	N/A	N/A	<104	0.1–0.5
Aquanox® A 4625B/ Kyzen	Easy to use, long-tank life, alkaline aqueous cleaner designed for batch washers	Removes all types of flux residues including lead-free in batch washing	170	N/A	N/A	~100	0.05
Aquanox® A 4651US/ Kyzen	Easy to use, low pH aqueous cleaning solution for ultrasonic immersion cleaning systems	Flux residue removal in ultrasonic immersion cleaning systems requiring no sump side additives	~99	N/A	N/A	None to boiling	0.17
Ionox® BC/ Kyzen	Easy to use, aggressive and material compatible, applied as received, semi-aqueous solvent blend	Immersion, hand-wiping, or coarse spray	175	N/A	N/A	72	0.18
Ionox® FCR/ Kyzen	Easy to use, applied as received, water soluble solvent blend requiring water rinse and heated dry	Immersion, ultrasonic, and inverted spray for removing paste and adhesives, rosin flux, no-clean flux and assembly residues	178	N/A	22 - 32	73	0.20

(Continued)

Table 4.4 Aqueous and semi-aqueous cleaning solutions—*cont'd*

			Properties				
Product/ Supplier	**Description**	**Application**	**Boiling Point[a] (°C)**	**Viscosity (cps, 25°C)**	**Surface Tension (dynes/cm)**	**Flash Point (°C)**	**Vapor Pressure[b] (mm)**
Ionox® 13302/ Kyzen	Easy to use at full strength, semi-aqueous solvent blend	Immersion cleaning followed by hot deionized water rinse for removal of difficult paste and flux residues	150	N/A	N/A	82	0.19
Lonox® L 5005/Kyzen	East to use, mild alkaline aqueous amine blend	Removal of all types of flux and paste residues from stencils	104	N/A	18–28	None to boiling	0.12
Lonox®L5020/ Kyzen	Environmentally friendly, easy to use, highly alkaline aqueous amine blend	Batch, inline, immersion heated systems for flux removal from printed wiring assemblies	N/A	N/A	N/A	None to boiling	0.31
Micronox® MX2628/ Kyzen	Environmentally friendly, versatile aqueous solution with long tank life	Removes latest lead-free technology from advanced packaging in batch or in-line cleaning machines	170	N/A	N/A	93	0.05

N/A – Not Available.
[a] Some boiling points are for concentrated solutions.
[b] 25 °C unless indicated.
Source: Compiled from suppliers' technical data sheets.

several methods. In comparing hydrocarbon solvents for their effectiveness in dissolving residues, several tests may be used. One, the Kauri-Butanol (KB) test, is widely used to quantify solvent behavior and to rank the potencies of solvents.[6] The *Kauri-Butanol value* is the maximum amount of solvent that can be added to a stock solution of kauri (a fossil copal resin) in butanol before cloudiness occurs. Kauri resin is readily soluble in butanol, but only slightly soluble in hydrocarbon solvents. The resin-butanol solution will accept only a certain amount of the solvent before precipitation of the kauri resin occurs, as indicated by a cloudy appearance. Strong solvents such as toluene can be added in larger amounts (higher KB value) than weaker solvents such as hexane. In general, the higher the KB value, the better the solvent's cleaning ability.[7] The solvency of cleaning materials varies from neat (unmixed) solvents (KB values, approx. 10) to azeotropes (KB = 27 to 33) to co-solvents (KB > 150) (Figure 4.1). These solvents are able to remove light flux, rosin mildly activated (RMA) flux, and "no-clean" flux, respectively. Examples of azeotrope solvents used in cleaning electronic circuits include hydrochlorofluorocarbons with ethanol (for example, Asahi AK 225AES) and hydro fluoroethers with ethanol and 1,2-dichloroethylene (for example, 3M Co. HFE-71DA). Azeotropes are very efficient solvents when used in a vapor degreasing process (see Sec. 4.1.2) and widely used for cleaning flip-chip assemblies, hybridmicrocircuits, multichip modules, and other fragile assemblies.[8]

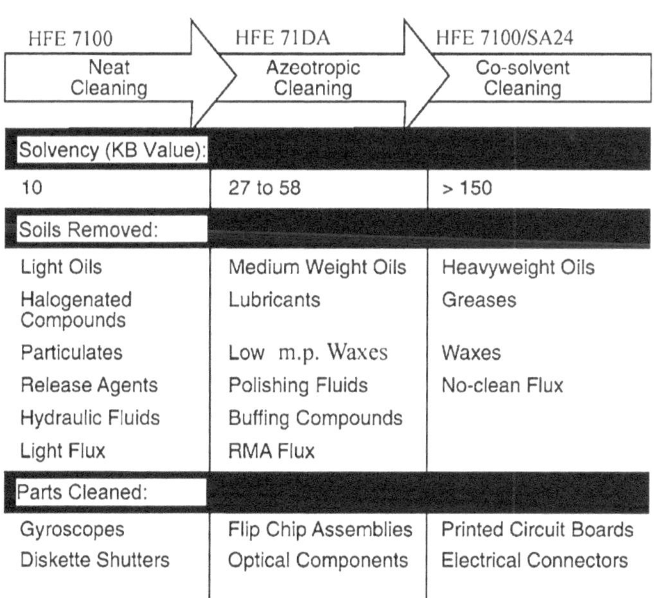

Figure 4.1 Relative solvency of 3M Co.'s Novec® hydrofluoroether cleaning solvents. (*Copyright 2000, 3M Co., reproduced with permission.*)

Table 4.5 Kauri-Butanol values for several cleaning solvents

Solvent/Supplier	Description	KB Value
Perchloroethylene/Various	C_2Cl_4	90
Trichloroethylene/Various	C_2HCl_3	129
Methylene chloride/Various	CH_2Cl_2	136
CFE-113/Various	Trichlorotrifluoroethane	32
1,1,1-TCA/Various	1,1,1-Trichloroethane	124
ABZOL® cleaner/Albemarle	Alkyl bromide mixture	125
AK-225/Asahi Glass Co.	Dichloropentafluoropropane	31
C-OS-10/Dow Corning	Hexamethyl disiloxane	17
HFE-7100/3M Co.	Perfluorobutylmethyl ether	10
Vertrel® XF/DuPont	Decafluoropentane	9
Lenium® ES/Petroferm	n-Propylbromide with IPA	>150

Sources: Supplier data sheets and *CRC Handbook of Solubility Parameters and Other Cohesion Parameters*, 2nd Ed., (A. Barton, ed.), CRC Press (1991).

Kauri-Butanol values for some solvents and cleaners are listed and compared in Table 4.5. Values for other cleaners may be found in Table 4.2.

Solvency and cleaning ability may also be evaluated as a function of density, viscosity, and surface tension. The lower the viscosity and surface tension and higher the density, the better the wettability, hence the ability to dissolve and remove residues. Surface tension is a measure of the penetrating power of a solvent; low surface tension solvents are better able to penetrate and wet tight spacings. The surface tensions of non-linear alcohols are typically low (20–22 dynes/cm) whereas that of water is high (72.8 dynes/cm).

The surface tension, viscosity, and density of a solvent are related by a figure of merit called the *wetting index* (WI). The WI is proportional to the density of the solvent and inversely proportional to the viscosity and surface tension as follows:

$$WI = (\text{density})(1,000)/(\text{viscosity})(\text{surface tension}) \quad (4.1)$$

The higher the value of WI, the better the solvent will penetrate into tight spaces and dislodge particles. Representative values of wetting indices for several solvents are given in Table 4.6.[9] The effect of solvency in removing tenacious residues such as solder fluxes has been extensively studied. The removal of flux residues using a hydrofluoroether azeotrope (3M Co. HFE-71DA) versus a co-solvent (3M Co. HFE-7100 blend with Solvating Agent 24 from Petroferm) is compared in Figure 4.2.[8] Residual ionic extracts from PWBs that were soldered using various fluxes then cleaned with the two solvents showed that the cleaning ability of the azeotrope was comparable to that of the co-solvent.

In addition to selecting an efficient solvent, manufacturers must now contend with environmental and safety regulations imposed by federal, state, and county jurisdictions. Solvents must meet volatile organic compounds (VOC) limits and, of course, there are also toxicity and flammability hazards that have to be addressed.

Table 4.6 Comparison of wetting indices of solvents[9]

Solvent	Name/Supplier	Density, gm/cm^3	Surface Tension, dynes/cm	Viscosity, Pa s	Wetting Index
CFC-113	Freon®TF/DuPont	1.48	17.3	0.70	122
Isopropanol	IPA	0.79	21.7	2.4	15
Deionized (DI) water	DI water	0.997	72.8	1.00	14
DI water + surfactant	DI water + surfactant	0.998	29.7	1.08	31
Hydrofluorocarbon (HFC)	Vertrel® XF/DuPont	1.58	14.1	0.67	167
Hydrofluorocarbon (HFC)	Vertrel® XP/DuPont	1.53	14.2	0.68	158
Hydrochlorofluorocarbon	HCFC-141B/DuPont	1.24	19.3	0.43	149
Hydrofluoroether	HFE-7100/3M Co.	1.50	13.6	0.61	181
Hydrofluoroether	HFE-7200/3M Co.	1.43	13.6	0.61	172

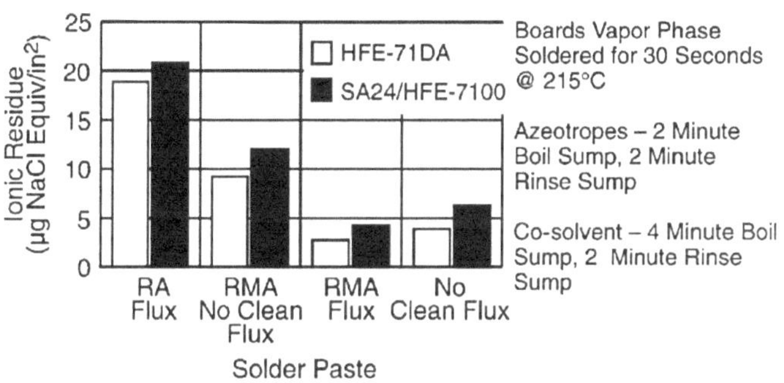

Figure 4.2 Residual ion-extract analysis of PCBs cleaned with co-solvent HFE-7100/SA24 and with azeotrope HFE-71DA. [8] *(Copyright 2000, 3M Co., reproduced with permission.)*

4.1.2 Cleaning methods

Cleaning methods may involve solvents or cleaning solutions (wet types), gaseous plasmas, or solid CO_2 (dry ice), a general comparison of which is shown in Table 4.7. Widely used wet-type cleaning processes include vapor degreasing, spraying, immersion with agitation (ultrasonic cleaning), immersion without agitation, and centrifugal cleaning. Plasma cleaning generally uses argon, oxygen, or mixtures but may also use other gases such as fluoro compounds. Carbon dioxide (CO_2) jet-spray cleaning (also referred to as snow cleaning) utilizes solid carbon dioxide (dry ice).

4.1.2.1 Vapor degreasing

Vapor degreasing involves the vaporization of a solvent or an azeotrope and its condensation as a liquid onto the part, thus flushing the part with freshly distilled solvent. The part is cleaned with freshly distilled solvent until its temperature increases to the temperature of the vapor. Dissolved residues are carried with the

Table 4.7 Comparison of wet, dry, and solid cleaning media

Cleaning Medium	Advantages	Limitations
Organic solvents	Remove adhesive and solder paste residues; have a wide-process window; can clean at room temperature; do not oxidize or corrode metals	Solvents and solvent vapors may be flammable; high cost; may be slow drying; limits on VOCs and ozone-depleting solvents (ODSs); require explosion-proof equipment
Aqueous alkaline cleaners	Relatively inexpensive; nonflammable; no ODSs or VOCs; mild odor	Do not remove adhesive residues; short bath life; narrow-process window; require an aqueous rinse and drying
Aqueous micro-phase cleaners (MPC)	Remove adhesive residues; wide-process window; nonflammable; rapid drying; long bath life	Require agitation (spray, ultrasonic, spray under immersion and overflow)
Plasma cleaning	Dry process (no solvents or chemical solutions); active oxygen plasma is an aggressive cleaner for organic residues; effective in removing adhesive bleedout	Electron-free plasma is required to avoid damage to sensitive components; cleaning surfaces with plasma prior to adhesive cure enhances bleedout
Solid CO_2 snow cleaning	Leaves no residues, removes submicron particles, removes thin films of organic materials, rapid cleaning (seconds), can be used on sensitive electronic and optical components, low-cost equipment	Condensation of ambient moisture on cooled parts during cleaning

condensed liquid back into the sump. Vapor degreasers have been widely used for over 40 years to clean fragile electronic assemblies. The process is benign and efficient since the vapors penetrate spaces that are inaccessible using other cleaning methods. The very low viscosity and low surface tension of vapor-phase solvents allow penetration into all areas of fine-pitch and low-standoff components. Vapor degreasing is very effective in cleaning hybrid microcircuits, multichip modules, and flip-chip assemblies. Azeotropes of both polar and nonpolar solvents are used to remove both ionic contaminants and greases and oils. Examples of some vapor-degreasing equipment and applications are listed in Table 4.8. Many vapor degreasers are equipped with a spray wand so that the parts may be sprayed with fresh liquid solvent after removal from the condensation portion of the equipment. Some of the many advantages of vapor degreasing are the following:

- No cleaning solutions to mix, maintain, or dispose.
- Solvent is reclaimed automatically within the unit; solvent waste is minimal.

Table 4.8 Examples of vapor degreasers and ultrasonic cleaning equipment

		Solvents/Cleaners		
Equipment/ Supplier	**Description**	**Aqueous**	**Semi-aqueous**	**Solvent**
B252R/Branson Ultrasonics Corp. B-452R/ Branson Ultrasonics Corp.	Environmentally sound, cost-effective precision degreaser	N/A	N/A	Methylenechloride, trichloroethylene, perchloroethylene, HCFCs, HFCs, hydrofluoro ethers, n-propyl bromide, and chlorinated solvents
Forward F-200 series/Crest Ultrasonics	Economical solution to precision cleaning applications using ceramically enhanced transducers	N/A	N/A	HFC, HFE, n-propyl bromide, and HCFC solvents
Forward F-300 series/Crest Ultrasonics	Open top cleaning systems (size range from 12" × 10" up to 24" × 20")			
WRD 2518-2T/ Stoelting® WRD 1812-2T/ Stoelting®	High performance wash-rinse-dry two (2T) and three tank (3T) consoles	N/A	N/A	N/A

- Process is simple with few variables to manage, and operator independent.
- Relatively low equipment, operation, and maintenance costs.

Some disadvantages include:

- Flammability and toxicity of some solvents require containment and safety features for the equipment.
- Many of the most efficient solvents are being disallowed by federal and local regulations because of their high VOC and ozone-depleting properties. Even some of the new replacement solvents may be phased out in the future.
- Vapor degreasers are becoming more expensive because of the added features required to prevent solvent evaporation into the air.

4.1.2.2 Spray cleaning

Spray-wash systems are primarily used to clean printed-wiring boards. Equipment consists of several zones including prewash, wash, drag-out, heated rinse, and final rinse followed by multiple drying zones.[10] Drag-out or isolation zones are designed to minimize contamination of the rinse water from the cleaning chemicals. A list of spray/wash equipment, manufacturers, and applications is given in Table 4.9.

Table 4.9 Aqueous and semi-aqueous cleaning equipment

		Solvents/Cleaners		
Equipment/Supplier	**Description**	**Aqueous**	**Semi-aqueous**	**Solvent**
Optimum Console™/Crest Ultrasonics	Versatile ultrasonic precision cleaning system for aqueous or semi-aqueous processes with all-in-one rinse and dry operations	Mild alkaline aqueous amine blend such as Lonox® 5005	Mixtures or highly refined solvents alone or with surfactants such as Axarel® 32 or 36.	No
Optimum Console Junior™/ Crest Ultrasonics	Four-station ultrasonic cleaning system using aqueous and semi-aqueous chemistries	Heavy duty ultrasonic biodegradable alkaline cleaner such as Chem-Crest® 275	Mixtures or highly refined solvents alone or with surfactants such as Axarel® 32 or 36.	No
Electrovert® Aquastorm 100®/ Speedline Technologies	In-line aqueous single-tank cleaning system for the removal of water-soluble OA[a] and water-based no-clean fluxes	Water only	No	No
Electrovert® Aquastorm 200®/ Speedline Technologies	In-line aqueous or semi-aqueous dual-tank cleaning system for the removal of RMA[b] and water-soluble OA fluxes	Alkaline (9–10 pH) aqueous cleaner such as Micronox MX2628	Semi-aqueous blend of highly refined polar and nonpolar solvents such as Axarel® 2200	No
Electrovert® Aquajet 200®/ Speedline Technologies	Aqueous batch cleaning and drying system	Water only	No	No
Trident High-Yield Defluxing Systems/Aqueous Technologies Corp.	Aqueous cleaner for removal of rosin, water-soluble, and all no-clean flux, both eutectic and lead-free applications, from printed-circuit assemblies	Alkaline aqueous cleaner such as Micronox® MX2628 or Lonox® L5020	No	No

Smart Sonic® Series 1500/ Smart Sonic®	Stencil (29-in maximum, <2 min) and pallet cleaning equipment for RA[c] RMA[b] no-clean synthetic and water soluble[a] solder paste (ambient), solder flux residue and SMD adhesives removal (32–43 °C)	SMT detergent 440-R (10%–12% for RA solder paste, 10% for RMA solder paste, 6%–12% for lead-free solder paste, and 10% for SMD adhesives and post-solder flux [rosin and OA] buildup)	N/A	N/A
Series 2300/Smart Sonic®	Dual tank system for stencil (29 inch maximum) and pallet cleaning RA, RMA, no-clean synthetic and water soluble (OA) flux residues, solder paste (ambient), and wet SMT adhesives (38–43 °C, ultrasonics)	SMT detergent 440-R (0%–2% for wash and rinse OA solder paste, and 10% for wash and rinse [wet and dry] solder paste, SMT adhesives, and post-solder flux residue)	N/A	N/A
Omnijet CBW-224/Stoelting®	In-line aqueous cleaning system	Yes	No	No
Advanced Packaging In-line Aqueous Cleaning System (APILACS)/Stoelting®	In-line semiconductor device cleaning (high volume)	Yes	No	No

[a] Organic acid.
[b] Rosin mildly activated.
[c] Rosin activated.

Spraying may be conducted at relatively low pressures of 50 psi or at high pressures up to 1,000 psi. Pressures must be adjusted so that fragile assemblies such as fine wires and wire bonds are not damaged. Parts cleaned in immersion systems are usually rinsed with fresh solvent and the solvent is then recycled by filtering and/or distilling.

4.1.2.3 Immersion cleaning/ultrasonic cleaning

Immersion cleaning using aqueous or semi-aqueous cleaners is usually assisted by agitating the solvent bath and heating. Agitation may be provided by ultrasonics or by pressure spraying under immersion. The removal of misprinted uncured adhesive from printed-circuit boards is more difficult than the removal of solder paste and requires the aggressive cleaning provided by ultrasonics. Ultrasonic cleaning, after electronic devices have been assembled, can be damaging to devices and, if used, the level of energy must be controlled. Ultrasonic cleaning of unsealed hybrid microcircuits and multichip modules has caused fine 1-mil-diameter wires to break and glass diodes to crack.

4.1.2.4 Centrifugal cleaning

Centrifugal-cleaning systems use centrifugal force to direct streams of solvents at parts in a rotating chamber. The centrifugal force (f) is imparted to the cleaning solvent in a cylinder of radius r that is moving at an angular velocity (ω).

$$f = m\omega^2 r \tag{4.2}$$

Forces generated are 60 G for a 16-inch-radius chamber, and 100 G for a 24-inch-radius chamber. Because of these high G forces, fast cycle times are realized, for example, 0.5–2 minutes for the wash cycle, 1–2 minutes for the rinse cycle, and 2–3 minutes for the dry cycle.[11] Solvents may be heated to a maximum of 82 °C in the wash cycle and to 60 °C for rinsing. Examples of solvents used in centrifugal cleaning are: Axarel® (Petroferm), Bioact® EC-7R (Petroferm), Bioact EC-15 (Petroferm), Ionox® FCR (Kyzen), and Zestron® FC (Zestron Corp.). Centrifugal cleaning equipment is available from Speedline Technologies and EPAK Electronics (Table 4.10).

A variation of centrifugal cleaning is EPAK Electronic's SuperFuge® system in which centrifugal shear carbon dioxide is used instead of aqueous solvents. The very low surface energy of carbon dioxide (0–5 dynes/cm) provides high solvating power and penetration beneath tight-pitch, low standoff, and densely populated flip-chip devices and packages.[12,13] Centrifugal cleaning is especially effective in cleaning low-standoff parts such as chip-scale packages (CSP), BGAs, and flip-chip devices.[14] The major benefits are:

- High efficiency
- High penetration of solvent at cleaning temperatures where viscosity and surface tension are low
- High penetration of low-standoff parts because of directed flow

Table 4.10 Centrifugal cleaning equipment

Manufacturer	Model	Application	Solvents/Cleaners		
			Aqueous	**Semi-aqueous**	**Solvent**
Speedline Technologies	Accel™ MicroCel™ Centrifugal Cleaning System; temperature ranges: wash 25–82 °C; rinse 25–60 °C; dry 25–204 °C	Electronic circuit assemblies, hybrid circuits, MCMs, bumped wafers, and advanced packages such as flip chips, BGAs, and CSPs	Aqueous (water, detergents and saponifiers), Bio T200A and, Axarel® 36	Solvent/water and emulsions (Bioact® EC-7R and EC-15), Ionox® FCR and HC (aqueous alcohols), and Zestron® FC	Chlorocarbons, chlorofluorocarbons, and alcohol-based solvents and terpenes
EPAK Electronics Limited	SuperFuge® liquid and SCFa CO_2 cleaning systems SF-1000 and SF-1200	Quartz and ceramic substrates, flip-chip bonded parts, and semiconductor devices	N/A	Micronox® MC 2301	N/A

[a] Super-critical fluid.
N/A – Not available Ref: Accel MicroCel Product Brochure, "Incomparable Cleaning Technology".
Reference: Accel MicroCel Product Brochure, "Incomparable Cleaning Technology".
Developed by Deflex, Inc.

- Compatible with fragile substrates
- Zero solvent discharge (closed-loop system)
- Fast cycle times.

4.1.2.5 Plasma cleaning

Plasma cleaning is a dry cleaning process that avoids the use of organic solvents or water solutions. Contaminant organic residues are oxidized and volatilized by subjecting them to a gaseous plasma produced by an RF energy source in oxygen or in an oxidizing gas such as sulfur hexafluoride. Oxide contaminants may also be removed by chemical reduction using a hydrogen plasma. Whether oxidizing or reducing gases are used, in practice, they are generally admixed with argon. Plasma cleaning may also be performed entirely in an inert-gas ambient, such as argon, where the interaction with contaminants is more a physical dislodging than a chemical reaction. Plasma cleaning of electronic assemblies prior to adhesive attachment or after attachment and curing has the following benefits:

- Clean, safe, and cost-effective process
- Avoidance of organic solvents and the consequences of toxicity and air pollution
- Generally fast cycle times
- Commercially available equipment
- Improved adhesion avoiding delamination in area-array packages
- Enhanced adhesion for die-attachment adhesives
- Void-free underfill (plasma is effective in penetrating tight gaps in high-density arrays, increasing surface energy, minimizing voids, and increasing underfill throughput)
- Removal of bleedout that occurs during curing of some epoxy adhesives.

Some limitations are the following:

- Removal of heavy residues requires long cycle times
- RF energy may damage some circuit components such as bare chip ICs
- Ultraclean surfaces may promote adhesive bleedout during adhesive curing; however, bleedout is easily removed by plasma cleaning after curing
- Some discoloration of silver-filled adhesives may occur due to surface oxidation of the silver. This is considered more a cosmetic issue than a reliability problem; however, the tarnish can be removed by subsequent exposure to an argon plasma.

4.1.2.5.1 Plasma-cleaning mechanisms

Unlike solvent and aqueous-based cleaning, plasma cleaning is based on reactions with surface contaminants at the molecular or atomic level.[15] Two mechanisms are possible depending on the level of RF energy and the nature of the plasma gas.

1. An ablation mechanism similar to sand blasting whereby the surface is roughened and its area increased by the formation of peaks and valleys. This mechanism also dislodges surface particles and ionic or salt-like residues. Argon gas is usually used to form a high-energy plasma consisting of positively charged argon particles and electrons.[16]

$$Ar + e^- \rightarrow Ar^+ + 2e^- \quad (4.3)$$

2. A chemical reaction whereby a gaseous plasma may either oxidize or reduce contaminants depending on the nature of the plasma used and the type of contaminant. To remove organic residues, such as oils, greases, silicones, and epoxies, a gas mixture typically of oxygen and argon is used. The oxygen is converted to an energized highly reactive form of monatomic oxygen that quickly oxidizes any organic residues converting them into volatile carbon dioxide and water as follows:

$$O_2 + e^- \rightarrow 2O^* + e^- \tag{4.4}$$

$$[-C_xH_y-]_n + O^* \rightarrow CO_2 + H_2O + e^- \tag{4.5}$$

Typical plasma-cleaning conditions for removal of organic residues prior to wire bonding consist of 100 W of RF power, 10–20 mTorr oxygen/argon pressure, and a 10-minute dwell time. During oxygen plasma cleaning, oxides may be formed on silver-filled conductive adhesives. These may subsequently be removed using a purely argon plasma.

To remove metal oxides (MeO) or other inorganic oxide surface contaminants, a reducing gas such as a mixture of hydrogen and argon is used. In the RF plasma, hydrogen molecules dissociate into highly active hydrogen atoms that reduce the oxides to metal (Me) and water. Simultaneously, the argon plasma physically dislodges oxide and reduced oxide residues from the surface.

$$H_2 + e^- \rightarrow 2H^* \tag{4.6}$$

$$MeO + 2H^* \rightarrow H_2O + Me \tag{4.7}$$

4.1.2.5.2 Plasma-cleaning equipment

Plasma equipment is extensively used for cleaning electronic parts and altering surface topographies to improve adhesion.[16] Plasma is also used to remove thick layers of photoresist, polyimides, or other organic materials. In its basic form, the equipment consists of a reaction chamber, a power supply, and a vacuum pump. Parts are placed in the chamber and evacuated. A gas or mixture of gases is then introduced. Gases generally consists of argon, oxygen, hydrogen, carbon tetrafluoride, or sulfur hexafluoride or mixtures. The plasma is generated by subjecting the gas to an RF frequency of 13.56 MHz and about 400 volts. Process and equipment parameters must be well controlled since some semiconductor devices can be damaged by the electrons produced, especially if placed within the active plasma zone. However, equipment is now available that generates a down-stream plasma and is modified so that electrons can be deflected from the part. Also, RF frequencies of 40–50 kHz are being used to avoid heating the parts that occurs at the higher 13.56 MHz frequency.[17] Some plasma cleaners are parallel-plate units that can be operated at 1,000 W or 500 W in both the active and electron-free plasma modes. The horizontal flow of plasma gas through the magazines assures uniform cleaning.

Commercially available equipment can be operated in three modes: active plasma for routine direct cleaning of surfaces, aggressive directional cleaning (reactive ion etching) for non-ESD sensitive parts, and downstream electron-free plasma for cleaning bare chip semiconductor devices and other electrically sensitive circuits.[16] The down-stream plasma is the most benign mode, capable of removing thin layers of residues 10–50 angstroms thick.

Plasma-cleaning equipment range from small laboratory models having 6-inch-diameter chambers to large-capacity automatic production units. Equipment from March Instruments (a Nordson Co.) is reported to enable a user to process 500–1,000 leadframes per hour.[18] Table 4.11 provides further detail for some commercially available plasma cleaners.

4.1.2.6 Solid carbon dioxide (CO_2)

In CO_2 jet cleaning, also known as "snow cleaning", a jet spray of solid carbon dioxide impinges on a part and dislodges and sweeps away particulate residues and some organic material such as greases and oils. The "snow" is produced from liquid pressurized carbon dioxide as it expands rapidly through specially designed nozzles. The CO_2 is constrained to pass through a triple-point where solid, liquid, and gas co-exist. At the triple-point, the aerosol turns into solid CO_2 creating "snow flakes." The velocity, density, and size of the snow flakes can be controlled to provide aggressive, moderate, or benign cleaning conditions. For example, high velocity flakes, approximately 1–2 μm diameter, provide aggressive cleaning while low-velocity flakes 0.5 μm in diameter are used for benign cleaning of sensitive optics and microelectronic components. Solid CO_2 cleaning was developed as an alternative to using chlorinated and fluorinated solvents and other organic solvents that are environmentally damaging.

4.1.3 Methods for testing and assuring cleanliness

Several quantitative and qualitative tests are used for qualification of a solvent/cleaning process and for in-line quality assurance. Some tests, like the water-drop spreading test, are quick and simple. In this test, a 2–4 μl droplet of water is applied to a substrate, and the contact angle it makes with the surface is measured using a goniometer.[19] A contact angle of less than 40° indicates improved wetting and a relatively clean surface free of oily, nonpolar residues. In one study, 20-mm-square flip-chip devices having silicon nitride passivation had contact angles of 40° before cleaning. After cleaning in an oxygen plasma, the contact angle was reduced to 20°. This drastically reduced the flow-out time for the underfill adhesive from 60 seconds to 22 seconds, an important factor in increasing throughput of a production line.[20] In another study, after plasma cleaning copper, aluminum, and epoxy surfaces, the contact angles were reduced by 50–74% (Table 4.12).[21] A silicon surface that is well cleaned and free of contaminants has a contact angle below 10° and results in considerable spreading of a water drop. When contaminated with hydrophobic residues, contact angles increase to 45°–90° and little or no spreading occurs. It should be

Table 4.11 Plasma-cleaning systems

Manufacturer	Model	Description	Power (W)/ Frequency (MHz)	Application	Capacity
March Instruments (a Nordson Company)	AP-1000 HTP	High throughput system with reduced process time and improved treatment uniformity	600/ 13.56	PC bond pad cleaning Fixture and component cleaning	20 leadframes/ magazine and 12 magazines (maximum) depending on size
	AP-1500	Effective plasma treatment in an extra-large batch system	2,000/13.56	N/A	
	FlexTRAK™	Highly configurable, high-throughput plasma system	600/13.56	PC bond pad cleaning Fixture and component cleaning	
Plasma Etch, Inc.	PE-200	Benchtop plasma cleaner	300/13.56	N/A	
	PE-1000	Automatic, in-line plasma system	1,000/13.56	N/A	
	PE-2000R	Reel-to-reel plasma etching system	3,000/N/A	N/A	
SAMCO International, Inc.	PC-300	Automatic plasma cleaner for surface cleaning and treatment RIE and plasma modes		Improvement of adhesion Surface cleaning of plastic packages and leadframes Organic contaminant removal	
	PC-1100	Parallel plate plasma cleaning system Three processing modes: RIE, plasma, and downstream		Surface cleaning of: • Plastic packages • LCD substrates • Multichip modules • Hybrid ICs	Multiple processing shelves

(*Continued*)

Table 4.11 Plasma-cleaning systems—*cont'd*

Manufacturer	Model	Description	Power (W)/ Frequency (MHz)	Application	Capacity
	PC-5000	Continuous automatic cleaning process		Surface cleaning of: • Plastic packages • Glass substrates	500 × 500 mm substrates (maximum)
	PVA-110	High throughput plasma cleaning of electronic parts		Cleaning of leadframes	Throughput: up to 50 × 200 mm substrates at About 12 seconds/ substrate
Yield Engineering Systems	YES-G1000 YES-G500	Plasma cleaning and etching systems	40–50 kHz 40–50 kHz	Surface cleaning electronic devices, circuit assemblies Wire bond surface preparation Contaminants & flux removal	4 powered shelves 2 powered shelves
	YES-CV200RF YES-CV200RFS	Plasma strip systems Automatic strip/descum	40–50 kHz Variable power settings : 150–1,250 W	Stripping of thick photoresist or polyimide layers; surface etching; removal of organic materials	Strip rate of 6,000–7,000 A/minute

Table 4.12 Effect of plasma cleaning on contact angles[21]

	Contact Angle (°)	
Surface	**Before Plasma Cleaning**	**After Plasma Cleaning**
Copper	88.4	31.7
Aluminum	64.5	32.0
Epoxy	85.0	22.0

noted, however, that the contact angle is useful only for determining the presence of grease-like or nonpolar residues. The presence of ions or hydrophilic residues will also give low contact angles and must be measured by other means. Colorimetric tests may also be used qualitatively to detect residues of organic materials.

The test most widely used in the printed-circuit industry to measure the efficiency of a cleaning process is the resistivity of solvent extracts test known as ROSE. According to this test, after the assembly has been cleaned, the part is flushed with a solution of 75% isopropanol and 25% deionized water and the electrical resistivity of the extracts measured.[22] Ionic residues dissolved in the extracts reduce the resistivity of the original solution. The resistivity is then correlated to total ionic residues removed and reported as equivalent to μg of NaCl/cm^2. Extraction methods may be manual, dynamic, or static. Stand-alone, modular equipment such as the Ionograph® or Omegameter® (Specialty Coating Systems, Cookson Electronics) are widely used in printed-board shops for periodic testing, but can also be integrated with the cleaning console for continuous monitoring. Although quantitative, the ROSE test measures only ionic or ionizable residues, not organic and nonpolar residues.

The level of surface cleanliness may also be evaluated by measuring the *surface-insulation resistance* (SIR) of a separate test pattern. The SIR test sample consists of an interdigitated comb pattern of conductor lines and spacings on a printed-circuit laminate representative of the actual circuit. The IPC B-25 or IPC B-24 test patterns (Figure 4.3) are standard in the industry to evaluate the degree of cleanliness of a surface. If the test pattern is well cleaned and no ionic residues remain, the insulation resistances between conductor lines will be 10^8 ohms or greater and will not drop appreciably under moist ambients. If residues remain, leakage currents occur between conductor lines reducing the insulation resistance. The effects of ionic residues are dramatic when exposed to moisture where the ions are easily mobilized. Inadequate cleaning has been reported to cause a drop of seven orders of magnitude in SIR values in less than one-hour exposure to humidity.[23] As with ROSE, SIR is best used to evaluate the presence of ionic residues. Organic residues may remain on the surface and still give high insulation readings. Further information on SIR may be found in IPC-9201 *Surface Insulation Resistance Handbook*[24] and in IPC-TR-468, *Factors Affecting Insulation Resistance Performance of Printed Circuits.*[25] Minimum SIR values of 10^{12} ohms may be required for some RF and high-speed circuits. [Figure 4.3]

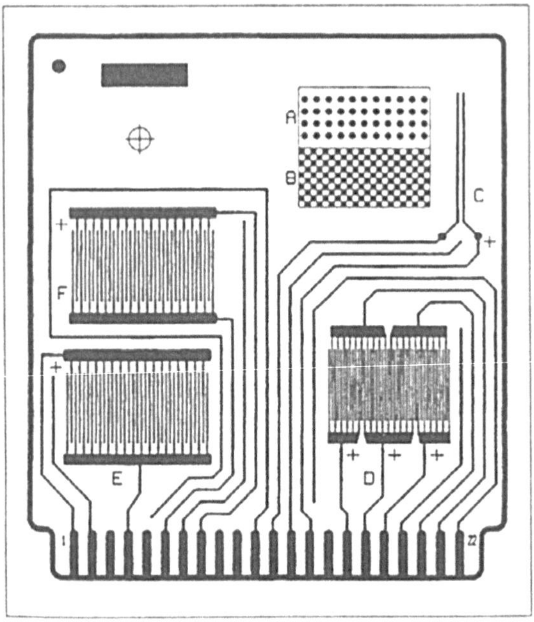

Figure 4.3 Surface insulation resistance (SIR) test pattern (IPC B-25).

4.2 Surface treatments

In some cases, even the best solvent, cleaning process, and adhesive are not sufficient to produce a reliable bond. Chemically inert surfaces such as gold, other noble metals, and nonpolar plastics such as Teflon®, polyolefins, and parylene are extremely difficult to bond and require additional surface treatment besides cleaning. Roughening a gold-plated bonding pad lightly using an abrasive or scuffing the surface with an electric eraser have been found to increase the adhesive strength of epoxy adhesives. Surfaces may also be sputter etched or ion milled to remove surface contaminants and to physically etch the surface providing greater surface roughness and area for adhesives to wet and adhere. Plastic surfaces treated with a corona discharge have also shown improved wetting and lower contact angles. Contact angle measurements of corona discharge treated plastic are defined in ASTM D5946-96.[26]

Primers recommended by the adhesives supplier are also useful. Primers are generally not necessary for epoxies because of their high polarity and affinity for most surfaces. In fact, solvent-thinned epoxies have been used as primers to improve the adhesion of other polymer types. Primers are often required for one-part and two-part RTV silicone adhesives. Primers are low-solids, solvent-based, air-drying coatings that can be applied by spraying, brushing, dipping, or spin coating. Solvent solutions of various silanes are among the best primers.

Finally, chemical alteration of surfaces can improve adhesion. Semi-crystalline materials have low polarity and low surface tension that must be increased to enable adhesives to bond. Oxidizing processes such as corona discharge, plasma, flame treatments, or UV irradiation in the presence of air or oxygen may be used to alter normally nonpolar surfaces, rendering them polar and increasing their wettability. Treating polymer surfaces in a plasma of reactive polar molecules results in chemically altering the surfaces rendering them more compatible with adhesives. For example, in an amine plasma such as allyl amine, highly polar amine groups may be attached and chemically bonded to an otherwise non-wetting surface. Hydroxyl groups from an allyl alcohol plasma may also be grafted producing a more wettable surface for adhesives.[27,28]

Teflon and other fluorocarbon surfaces may be altered by chemical treatments using solvent solutions of sodium naphthalene complex or molten sodium. Teflon etchants, for example, Tetra-Etch®, a tradename and product of W. L. Gore Associates, are effective in improving adhesion to Teflon. Oxidation treatments for metals such as copper or aluminum also produce highly polar oxide surfaces that are more easily bondable. The reader is referred to several sources for a more comprehensive discussion of surface treatments for adhesives.[29,30]

4.3 Adhesive dispensing

Adhesive dispensing methods consist of either mass-transfer or selective dispensing. Stencil printing, screen printing, pin transfer, daub transfer, and stamping are examples of mass-transfer types. They are also referred to as tooling-related processes since hard tooling such as stencils and screens are required. In comparison, selective-dispensing methods are computer data driven, such as automated syringe or needle dispensing and jetting processes. Automated dispensing, currently used in over 80% of all production processes,[31] is a continuous process where individual adhesive dots, lines, or patterns are deposited in selected areas of an interconnect substrate using pressurized syringes or pumps. The main advantage of the data-driven processes is flexibility, accommodating a large variety of printed-circuit board types, configurations, and design changes. Selective dispensing is also the only method that can be used once components have been assembled onto a board since it does not require a flat surface as does screen or stencil printing.

Screen printing of conductive and insulative adhesives has been used for decades in the assembly of hybrid microcircuit and multichip modules. Since screen-printing processes are also used to deposit thick-film conductors, resistors, and dielectrics, the processes are fully compatible and cost effective in the production of hybrid microcircuits. Screen printing and stencil printing are both batch processes that are usually less expensive and faster than selective dispensing for large production runs where few changeovers are expected. As an example, 1,000 dots can be stencil printed in 30 seconds compared to 1.5 minutes for a selective dispenser operating at 40,000 dots per hour.[31]

In surface mounting of components, both solder paste and adhesive need to be dispensed. The solder paste is usually screen printed or stenciled first, followed by selective dispensing of the adhesive between the solder pads. The prime function of surface-mount adhesives (SMAs) is to maintain the position of the components during subsequent reflow or wave soldering. The adhesives may be partially or fully cured prior to solder reflowing or may be cured in the same step as solder reflowing. With the trend toward mixed assembly technologies, such as solder-attached components on the same board as adhesive-attached devices, selective dispensing must be used. In such cases, the solder operations are conducted first since they require high temperatures and aggressive cleaning to remove flux residues. A summary of the advantages and limitations of the main dispensing methods is given in Table 4.13,[32,33] and a cross-reference of dispensing methods for general adhesive types is given in Table 4.14.

4.3.1 Screen printing and stencil printing

Of the mass-transfer dispensing methods, screen printing and stencil printing are the oldest and most widely used. Screen printing has been used for over 40 years in the electronics industry to apply thick-film conductors, resistors, and dielectrics in fabricating circuits on ceramic and plastic-laminate substrates. Screen printing is also used as a batch process for depositing electrically conductive and insulative adhesives to interconnect devices on thin-film and thick-film hybrid microcircuits.

Screen printing is a simple, low-cost process for both small and large production runs. Patterns are formed on a stainless-steel mesh through a photosensitive emulsion process in which portions of the screen mesh are coated while other areas, through which the paste is to be deposited, are left open. The photolithographically formed screen patterns, however, wear out after many runs and the screens must be reprocessed or replaced. Also, separate screens are required for different circuit configurations, which presents a logistics problem in their storage, identification, and retrieval, especially where hundreds of part types are to be fabricated.

Stencils differ from screens in that the apertures are directly etched or cut into a metal sheet or foil. In both stencils and screens, the paste is squeegeed through apertures onto a substrate. Both screen printing and stencil printing can be used to deposit adhesives or solder pastes, but stencil printing produces finer dimensions and closer spacings. The key advantage of stencils over screens, especially for the laser cut and electroformed types, is the finer dimensions and smaller pitch dots that can be printed. Stencils have become essential in depositing adhesive and solder paste for small-pitch devices of less than 0.5 mm. Conductive epoxies can be printed as small as 3-mil-diameter dots on 5-mil pitch.[34] Differing from pin transfer, stencil printing can deposit small dots of various sizes and heights from a single stencil. Dots smaller than the stencil thickness can be deposited through apertures that act like tiny dispensing tips.

Stencil printing is a fast efficient process for long production runs and is widely used in surface mounting of components.[35] In surface mounting, the solder paste is generally dispensed first by stenciling, followed by needle dispensing the adhesive; however, these processes may be reversed or both the solder and adhesive can be selectively dispensed by needle.

Table 4.13 Advantages and limitations of adhesive dispensing methods

Type	Description	Advantages	Disadvantages
Pin transfer	Surface mount, hybrid, or MCM/bed-of-nails tooling	Very fast, consistent	Difficult/expensive retooling; adhesive handling issues
Stamp transfer	Single, variable surface area tooling for adhesive pickup	Excellent in high-speed applications, Allows large areas to be bonded in a single application	Tooling intensive; adhesive handling issues
Screen, stencil printing	Surface mount, hybrid, or MCM/ stencil or screen printer	Fast, multiple dot size, Dots as small as 3-mils diameter on 5-mil pitch can be printed	Stencil printing requires a new stencil per design; complex cleaning procedures for small aperture stencils; cannot be used with other components on boards
Automated dispensing	Time/pressure dispensing	Data-driven process, rates to 50,000 dph, flexible, simple operation, and easy cleanup	Higher speeds result in less consistent dots
	Auger	Data-driven process, flexible, handles wide range of adhesives, improvement over time/pressure	Sensitive to viscosity changes; some effect of speed on consistency
	Piston pump	Data-driven process, Viscosity changes do not affect flow rate; consistency at higher speeds; capable of larger dot sizes	Complex cleaning procedures; sensitive to air in fluid
Jet dispensing	SMT adhesive using a spring loaded pin or hammer to rapidly cycle adhesive through the nozzle	10–50% increase in dph relative to auger and piston pumps, yields higher dot profile, noncontact, less sensitive to warpage or dispense height variations, dispensing rates up to 70,000 dph, no tailing	Single dot size, complex cleanup, limited material selection
Preform/Tape bonding	Chip-on-board, hybrid or MCM, single-chip (standard and lead-on-chip)	Eliminates fluid dispensing, good thickness control	Requires die-cut preforms or manual processing

Table 4.14 Cross-reference of dispensing methods and materials[33]

	Process						
Material	**Needle**	**Fluid Jet**	**Stencil Print**	**Screen Print**	**Pin Transfer**	**Stamp Transfer**	**Film Bond**
Insulative die-attach paste	XX	X	X	X	X	X	
Isotropic die-attach film							XX
Anisotropic die-attach film							XX
Isotropic paste	X		XX	X		X	
Anisotropic paste	X	X	X	XX		X	
Underfill	XX	X	X	X			

XX = Preferred. (Copyright ©1996 PennWell Corporation. Reproduced by Permission.)

As with most processes, numerous variables affect the final product. For solder paste deposition using stencils, Chouta and Fleck[36] identified 39 variables relating to materials, 32 variables to processes and equipment, and 10 variables to personnel and environment. Among variables critical for the successful stencil dispensing of adhesives are: stencil design, the adhesive material and its rheological properties, and the printing process.

4.3.1.1 Stencil fabrication and design

Most stencils are fabricated from stainless steel, nickel, or copper foil 2–10-mils thick. Apertures are formed in the foil by one of three processes: chemical etching, laser cutting, or electroforming. Chemical etching, although the least costly of the three methods, provides the coarsest features. In chemical etching the apertures, the metal foil is coated on both sides with a photosensitive coating, exposed to UV light through a mask that has the desired aperture pattern, and developed. The part is then immersed or sprayed with a chemical etchant that removes metal only in the unprotected areas, after which the photocured coating is removed. With the trend toward finer pitch devices (less than 20-mil pitch), both laser-cut and electroformed stencils must be used. Laser-cut stencils can be produced directly from Gerber data by computer numerically controlled laser cutting. Aperture dimension tolerances of $\pm 5\ \mu m$ are reported.[37] Chemically etched and laser-cut stencils are formed by subtractive processes. On the other hand, electroformed stencils are produced by an additive process in which a metal such as nickel is electroplated onto a mandrel that has a negative photoresist image of the aperture pattern. The electrodeposited nickel containing the apertures is then removed from the mandrel and used as the stencil. Because of their high precision patterns, electroformed stencils are used for ultra-fine pitch applications.

In recent years, there has been increased interest in stencils produced from plastics such as acetal and polyethylene. The prime benefit in using plastic stencils is the thicknesses that can be used (greater than 1 mm up to 8 mm). By drilling apertures of various sizes, adhesive deposits of different heights ranging from 0.2 to 2 mm were

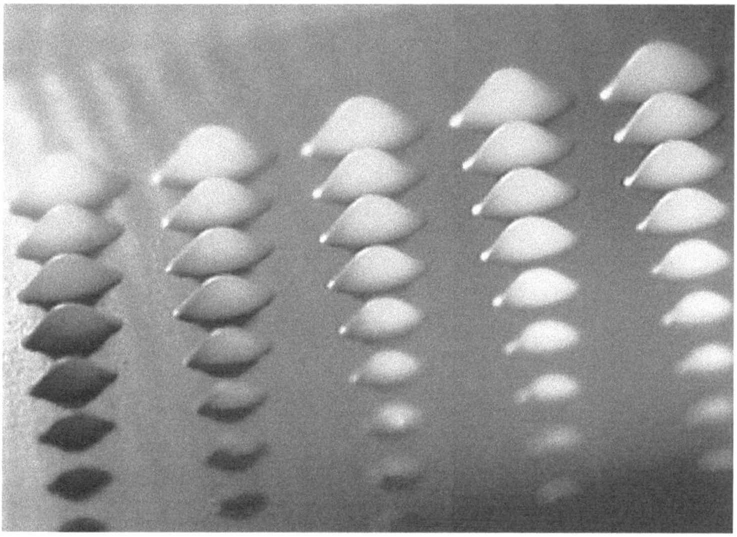

Figure 4.4 Typical range of adhesive dots produced using a 1-mm-thick plastic stencil (for apertures ranging from 0.6 to 2.0 mm).[38] *(Source: M. Whitmore, DEK Printing Machines Ltd.)*

printed from a single screen thickness in one pass.[37] Figure 4.4 shows a typical range of dot heights that were produced from a 1-mm-thick plastic stencil having aperture sizes from 0.6 to 2 mm.[38] By carefully controlling the aperture size, it is possible to produce heights for every component type with a single squeegee stroke across a printed-circuit board (Figure 4.5).

The two basic aperture shapes used to design and fabricate a stencil are dots and lines (slots). Lines or slots are used when a large volume of adhesive is required to prevent components from shifting during their placement. In determining the dot or line sizes, most design engineers use the *40% rule* according to which the size of the dot or line should be approximately 40% of the distance between the pads.[39] Examples of dot sizes for selected discrete devices and for distances between bond pads may be found in the literature.[40] Double dot or slot designs are most widely used for two-pad devices (Figure 4.6, top). Apertures smaller than the stencil thickness produce small dots. Circular patterns having areas roughly equal to the substrate area yield the largest dots (Figure 4.6). An array of larger dots is recommended for high-peaked dots.

Stencil thickness should be determined by the highest dot size needed to make contact with the part. In most cases, the dot height should be 1.5–2 times the standoff height. Most passive components require stencils that are at least 6–8 mils thick. Large, small-outline ICs (SOICs) require stencils that are 10–12 mils thick.[35] A specific example in the use of these guidelines is given for a 12-mil (0.3 mm) thick stencil. If the stencil apertures are 0.3 mm (a 1:1 ratio of aperture width to stencil thickness), the surface tension between the adhesive and stencil will be so great that small dot diameters are formed. As the opening increases to 32 mils

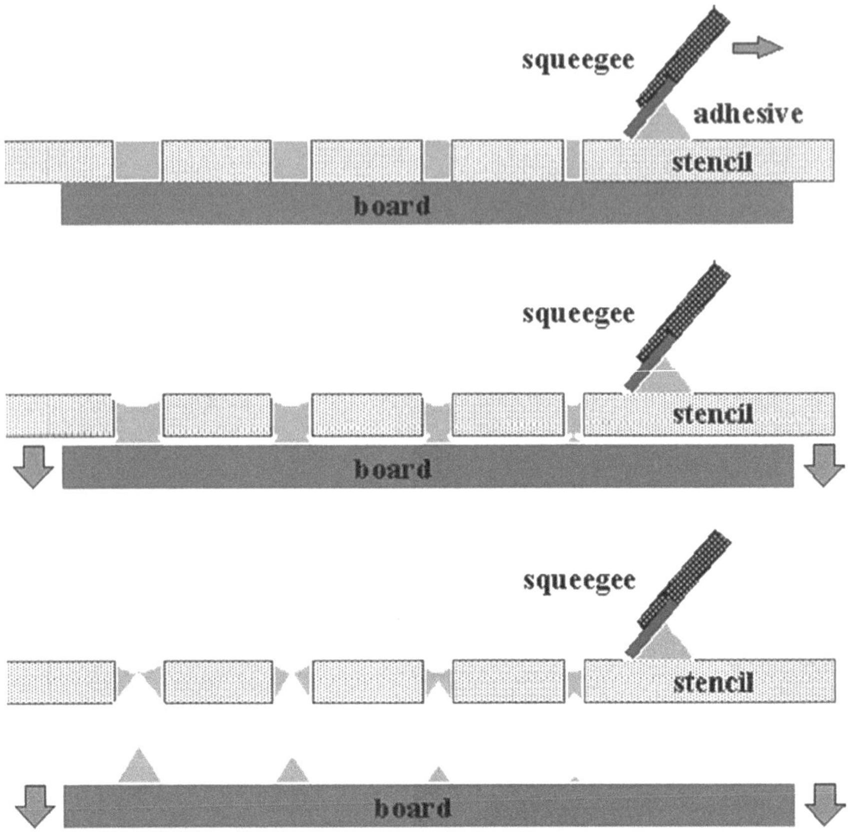

Figure 4.5 Schematic diagram of plastic-stencil screening process. (*Source: M. Whitmore, DEK Printing Machines Ltd.*)

(0.8 mm), a 2.7:1 aspect ratio, most of the adhesive will transfer from the stencil and the dot height will be high. When the ratio of aperture diameter to stencil thickness exceeds 4:1, complete transfer occurs.[41] As a general rule, acceptable paste release is obtained when the aspect ratio of aperture width to stencil thickness is greater than 1.5. Other guidelines for stencil design are summarized in Table 4.15.

4.3.1.2 Adhesive material and rheology properties

Adhesives used for screen or stencil printing in surface-mount applications are generally electrically insulative types whose functions are mechanical attachment and thermal dissipation. However, electrically conductive, silver-filled epoxies have been used for many years as ohmic contact adhesives to interconnect bare-chip devices in hybrid microcircuits and are used as solder replacements for surface mounting of components on printed-circuit boards. Regardless of their

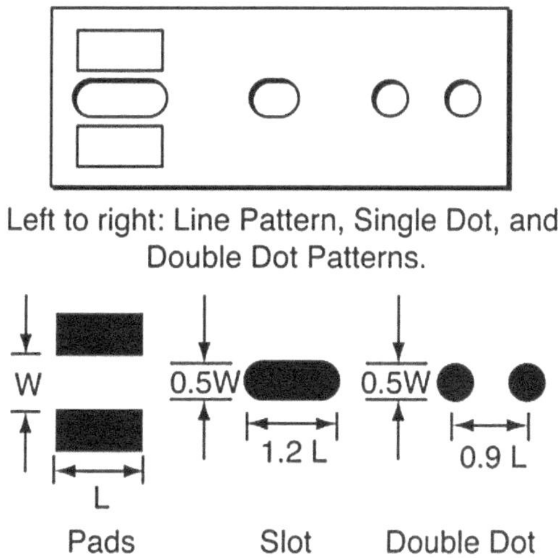

Figure 4.6 Stencil patterns used to adhesive bond two-pad devices having a double dot or slot design.

Table 4.15 Stencil design guidelines[39]

Typical Dot Patterns for Adhesive Printing	
0.76-mm round	0805 chips, SOT 23's
1.0-mm round	1206's or larger, MELFs (cylindrical components, 2 dots on 1-mm centers)
2.0-mm round	SOIC's (multiple dots for each SOIC)
40% round	40% of the distance between the pads, center
Typical Line Designs	
0.20 to 0.25 mm	Length of line determined by the width of the component
40% line	40% of the distance between the SMT pads with the length of the line determined by the width of the component
Typical Stencil Properties	
Material	Stainless steel
Thickness	0.18–0.30 mm

conductivities, adhesives applied by screen or stencil printing should have the following properties:

- Low moisture absorption and workable viscosity over a 3–5 day period of exposure on the stencil.
- Viscosity suitable for dispensing (30,000–100,000 cP at 2.5 rpm).[42]
- Thixotropic properties (ability to flow freely through the stencil/screen apertures during printing and recover quickly after printing).

- High green strength to hold components in place during component placement and reflow soldering.
- High yield value to avoid slump and produce appropriate dot geometries.

The *yield value* (yield point) is an important parameter in dispensing adhesives. It is the minimum force required to begin the flow of adhesive from rest. The yield point is determined by extrapolating the flow or viscosity curve to the intercept on the *y*-axis or shear-stress axis.[43] In general, adhesives having high viscosities also have high yield points and produce dots having high *aspect ratios* (ratio of height to diameter). Thus, adhesives having high yield values of 200–500 Pa produce tall narrow dots while those having low yield values of 150–200 Pa produce wider, shorter dots.[35] Examples of some commercially available adhesives that can be stencil printed are given in Table 4.16.

4.3.1.3 Stencil/screen printing process and equipment

The printing process involves pressing the paste with a squeegee through apertures in the stencil or screen onto a substrate. The paste is applied with the squeegee at an angle of 45° or 60° in a single stroke, or often in a double stroke to assure complete filling of the apertures. Stencil printers capable of printing solder pastes can also be used to print adhesives. An important aspect of the printing equipment is the ability to form a programmable snap-off or gap between the stencil and the substrate. The apertures of the stencil define the pattern for the deposited adhesive. Stencil and screen printers may be bench top for small to moderate runs, semi-automated, or automated floor models with magazine loading for production runs. Squeegee blades may consist of metal or hard plastics such as polyurethane or polyethylene. Printing speeds depend largely on the viscosity of the adhesive. In general, low-viscosity adhesives may be printed at higher speeds than high-viscosity types. Thus, some solder replacement adhesives having viscosities of 50,000–70,000 cP (at 10 rpm) can be printed at squeegee speeds of 6 inches/second. Surface-mount adhesives having high viscosities of 800,000 cP, as well as those of intermediate viscosities (75,000–250,000 cP), can be printed at 1 inch/second.

Although stencil printing is widely used for both adhesives and solder paste, and has many advantages over other dispensing processes, there are several limitations that must be addressed as follows:

- Tailing or distortion can occur during snap-off.
- Redeposition occurs on warped substrates, leading to contamination of adjacent pads.
- Tooling changes and added change-overtime are required for design changes and new designs. Retooling costs are high.
- Application for very-fine-pitch components is limited (sidewall influence is so great that adhesives cannot be effectively dispensed).
- Frequent cleaning of stencils is required to avoid clogging, missed dots, and misprints.
- Long exposure to air and moisture shortens the pot life and increases the viscosity of the adhesive.

Table 4.16 Examples of adhesives used for stencil printing

Product/Supplier	Applications	Viscosity, Casson Pa s @25°C	Yield Value, Pa @ 25 °C	Process Conditions
Epo-Tek® H67MP/ Epoxy Technology	Stencil/screen printing	N/A[a]	N/A	N/A
Epo-Tek® H70E/ Epoxy Technology	High-speed epoxy chip bonding	N/A[b]	N/A	N/A
Loctite™ 3612/Henkel	Metal-stencil print applications for small parts bonding	15–55	350–700	Print speed: 60 mm/s Squeegee pressure: 3–4 N/cm Separation speed: 0.1–3 mm/s Stencil and printed-circuit board gap: none (on contact)
Loctite™ 3614/Henkel	Metal-stencil print applications for small-parts bonding	320–65	350–700	Print speed: 60 mm/s Squeegee pressure: 3 to 4 N/cm Separation speed: 0.1–3 mm/s Stencil and printed-circuit board gap: none (on contact)
Loctite™ 3627/Henkel	Metal-stencil print applications for small-parts bonding	32–5	100–275	Print speed: 60 mm/s Squeegee pressure: 3–4 N/cm Separation speed: 0.1–3 mm/s Stencil and printed-circuit board gap: none (on contact)
Eccobond D125F/Henkel	Stencil application for surface-mount devices (capacitors, resistors, small outline ICs, and plastic leadless chip carriers)	N/A	N/A	N/A

[a] Brookfield viscosity: 300,000–400,000 cP (1 rpm).
[b] Brookfield viscosity: 4,000–7,000 cP (50 rpm).

4.3.2 Pin transfer

Pin- and daub-transfer processes are also mass-transfer methods, but comprise less than 10% of the adhesive-dispensing processes used in production. In pin transfer, hard tooling consisting of an array of pins corresponding to the positions on the circuit board where components are to be attached is fabricated. The pin arrays are dipped into a tray containing the uncured adhesive, then removed taking with them suspended drops of adhesive. The pins are then lowered onto the board and the adhesive transferred to the preselected areas (Figure 4.7). The diameter of the pins and design of the pin tip are critical to ensure that the adhesive dot is not distorted. For most applications, the tip is designed as in Figure 4.7.[44]

Pin transfer is the simplest mass-transfer process for high production runs of one or two part types, but limited where numerous types of different configurations and small quantities are to be produced. Because pin transfer uses hard tooling, each job requires new tooling and considerable time to set up and change to new tooling.[45] Other limitations of pin transfer are[32]:

- Risk of contaminating the adhesive contained in an open tray
- Moisture absorption due to exposure of the adhesive to the ambient
- Limited to dot printing
- Requires flat substrates
- Difficulty in applying adhesive to attach components smaller than 120 × 60 mils (1206 part types)
- Difficulty controlling the volume and shape of the dot.[46]

Furthermore, surface-mount adhesives (SMAs) used in pin transfer must have relatively low viscosities to facilitate pickup by the pins and transfer to the PCB. Because viscosity is so critical, the temperature of the adhesive reservoir must be controlled and preferably maintained between 25 °C and 30 °C.[47] Some SMAs that may be applied by pin transfer are listed in Table 4.17.

4.3.3 Stamp transfer

In stamp transfer, a stamping pad is dipped into a reservoir of liquid adhesive and the adhesive redeposited onto a substrate on contact. The amount of material transferred depends on its viscosity, the size and shape of the stamping tool, and the depth to which the stamp is immersed.[39,45] An advantage of stamp transfer is the ability to deposit adhesive over large areas in a single application. Pad areas as large as 500-mils square can be covered. However, like pin transfer, limitations of stamp transfer

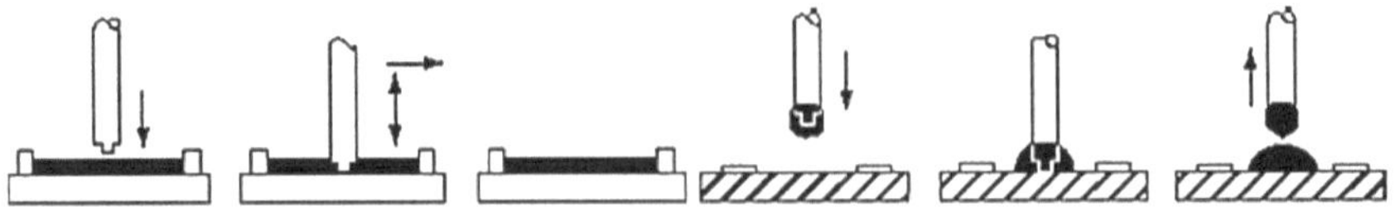

Figure 4.7 Pin-transfer dispensing steps. (The tip of the pin is modified to prevent distorting the dot.)

Table 4.17 Examples of adhesives applied by pin transfer

Adhesive/Supplier	Viscosity Brookfield (cps) 1 rpm	10 rpm	Thixotropic Index (1:10)
Eccobond D125F/Henkel	1,500,000–2,000,000	300,000–400,000	5
Loctite® 3607™ /Henkel	N/A[a]	N/A	N/A
Epo-Tek® EM127/Epoxy Technology	N/A[b]	2,500–3,500 @ 100 rpm	5.5

[a] Casson viscosity –5–20 Pa.s.
[b] Yield point 150–350 Pa.

are the inability to print small areas or dots and the high tooling cost required for each application type. Some commercial adhesives that can be stamp transferred are listed in Table 4.18.

4.3.4 Automated selective dispensing (contact type)

Automated selective dispensing of adhesives may be contact or non-contact types. In the contact type, also known as syringe, needle, or machine dispensing, the adhesive is forced through a small orifice onto the substrate so that the dispensed dot makes contact with the substrate during dispensing. The adhesive is dispensed using a *x–y–z* positioning system. Automatic data-driven equipment is used to selectively dispense reproducible amounts of adhesive with high accuracy. Large amounts of adhesive can also be dispensed onto a small area in order to assure contact with high-standoff components. Multiple small dots, 2–5 mils in diameter, can be deposited in selected

Table 4.18 Examples of adhesives applied by stamp transfer[48,49]

Adhesive	Viscosity (cP)	Thixotropic Index (1:10)	Pot Life	Cure Schedule
Epo-Tek® H20S	4,710 @ 20 rpm	2.5	4 days	Snap cure (175 °C, 45 seconds) or fast cure (15 min, 120 °C) die attach
Epo-Tek® H20S-MC	1,700 @ 100 rpm)	3.0	4 days	Snap cure (175 °C, 45 seconds) or fast cure (15 min, 120 °C) die attach
Epo-Tek® H70E	4,000–7,000 @ 50 rpm	3.0	4 days	5 min, 150 °C
Epo-Tek® P1011S	7,987 @ 20 rpm	1.8	6 mo	10–30 min. @ 70 °C plus 1 hr @ 150 °C, requires prebake for large chips

Source: Epoxy Technology technical data sheets.

Table 4.19 Effect of adhesive viscosity on dispensing and component placement time[40]

	Viscosity	
Parameter	**Low**	**High**
Number of dots required for one device	12	5
Dispense time per dot (msec)	120	150
Total dispense time (msec)	1,440	750
Placement time per device (msec)	2,000	2,000

areas at rates up to 50,000 dph (dots per hour), although typical rates are approximately 15,000–20,000 dph. These rates, seemingly high at first glance, are low compared with stencil printing where 50,000 dots or lines can be applied in 15 20 seconds.

The number of dots that can be dispensed per unit of time also depends on viscosity. The use of low-viscosity adhesives increases the number of dots that can be dispensed. However, there is a tradeoff between the speed at which the adhesive can be dispensed and the speed of device placement. Most automated component-placement equipment operate faster than adhesive-dispensing equipment. A comparison of the effect of viscosity on dispensing and placement times is shown in Table 4.19 for an 86-lead plastic-encapsulated leadless chip carrier (PLCC).[40]

Automated dispensing is now used for over 80% of all dispensed adhesives and is particularly suited to printed-circuit assemblies where some components have already been attached and, therefore, stencil or screen printing cannot be used. Another significant advantage is in bonding components smaller than 120 × 60 mils. Pin transfer and stencil/screen print methods are generally not used below this size. Adhesives used for pin transfer are also difficult to control, and screens and stencils require more maintenance due to clogging and the need for frequent cleaning. Other advantages of automated dispensing include the following:

- Ideal for both short runs and production runs
- Adhesive can be dispensed on partially assembled boards
- Dispensers can handle a wide range of viscosities
- No hard tooling required
- Minimal setup and change-over times
- Less sensitive to board warpage than stencil printing
- High dispensing speeds up to 40,000 dph
- Consistent dot profiles for a range of component sizes as small as the 0402 and 0603* passive components.

Automated dispensing is used for the application of die-attachment, surface-mount, and underfill adhesives.

* 0402 and 0603 are passive component part sizes. Throughout the book, part sizes are given in mils; thus 0402 is 40 × 20 mils and 0603 is 60 × 30 mils.

4.3.4.1 *Automated dispensing of die-attachment adhesives*

In assembling small quantities of hybrid microcircuits of many part types, manual application of the adhesive is still used, especially by smaller companies. Skilled operators are able to dispense adhesive from syringes in fairly controlled amounts that cover the bonding pad and provide enough fillet after die placement. However, for moderate-to-large runs, and even for small or prototype runs, automated dispensing provides the best method for repeatability and quality control. Automated dispensing is widely used for the assembly of integrated-circuit chips onto leadframes, for chip-scale and BGA packaging, and for bare-chip attachments to thin- or thick-film interconnect substrates. Regardless of the specific application, the dispensing patterns must be sufficient to cover the bonding pads.

The shape and size of the devices to be bonded define dispensing patterns. Patterns for square die are simpler than for rectangular die, and patterns may vary within each category as die size increases. Simple dot dispense patterns are used for small square die (less than 5 mm per side), but the volume must be estimated so that the circular area of the adhesive will spread to the edge of the die and produce a fillet.[50] Square die, larger than 20 mm per side, still require a center dot, but must also be supplemented with a cross-pattern and other dots in order for the adhesive to flow uniformly (Figure 4.8). The lines should extend far enough so that, once the die is placed, its bonded surface will be completely covered. Typically, this involves depositing the adhesive approximately two-thirds the distance to the edge of the bonding pad.

The pattern for rectangular devices is somewhat more complex as shown in Figure 4.9. The major line should be centered and its length should be the difference in length between the sides of the chip. The four remaining lines should extend approximately two-thirds the distance to the corners and make a 45° angle with the centerline. Another general rule is that the height of the centerline should be about 50% of the line's width.

Other machine parameters may require adjustments for die size and pattern. One parameter is the length of time spent before the dispensing pump begins to move, known as the *on-delay*. On-delay times that are too short result in thin deposits as dispensing begins. Further, the *shut-off* distance, the distance from the edge where dispensing stops, requires adjustment of machine settings. Stopping too close to the edge results in excessive filleting while too far from the edge produces incomplete

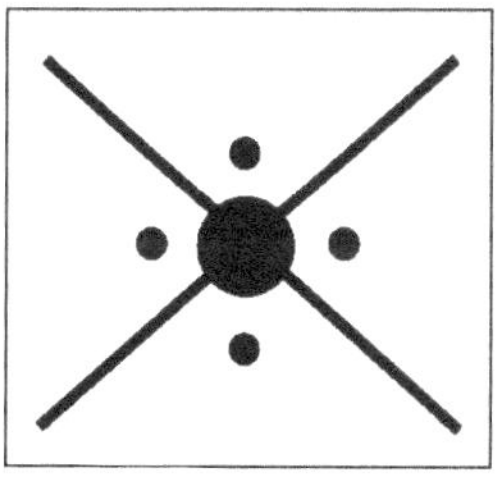

Figure 4.8 Dispense pattern for a square die (> 20mm/side). (*Copyright 1999, Reed Exhibition Corp., reproduced with permission.*)

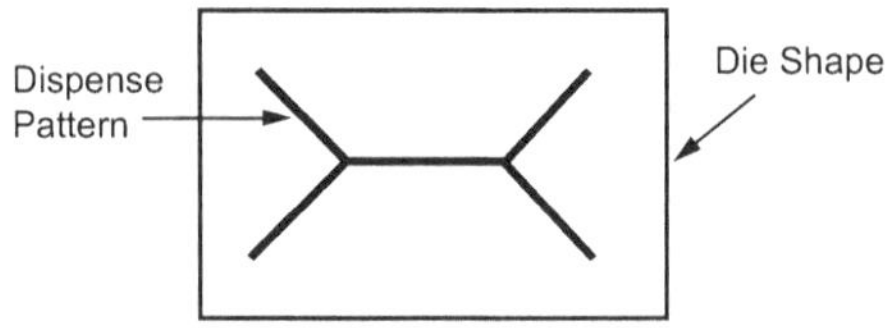

Figure 4.9 Dispense pattern for a rectangular die. (*Copyright 1999, Reed Exhibition Corp., reproduced with permission.*)

coverage. The exact shut-off value depends on the filleting requirements for the bonded device.

4.3.4.2 Automated dispensing for surface mount adhesives

Automated dispensing of adhesives for surface-mount technology (SMT) applications follows guidelines similar to those for die attachment. Patterns of dots and lines are deposited covering areas similar to those of the component sizes to be bonded. Surface-mount adhesives are applied primarily by automated dispensing, although stencil printing and pin or stamp transfer are also used. Regardless of the application method, surface-mount adhesives should meet the following requirements:

- Compatible rheology with the dispensing method (clog-free, no stringing) and with assembly requirements (dot profiles, green strengths)
- Short cure cycles and ability to be cured in-line or simultaneously with solder reflow
- Preferably one-part systems
- Long shelf life and extended working life
- Application at speeds of 40,000 dph or greater
- Consistent dot profiles for a range of components including very small components such as the 0402 and 0603 passive devices.

The behavior of adhesives under high-speed dispensing conditions is also critical.[51,52] The ability of a thixotropic adhesive to flow with an applied shear stress and to recover quickly after the stress is removed is critical for high-speed automatic dispensing as well as for screen or stencil printing. The rate at which the adhesive recovers its original high viscosity is as important as the shear-induced drop in viscosity. Figure 4.10 shows examples of the thixotropic loops for two adhesives. Both display excellent recovery since the return curves trace well with the original. Thus both adhesives are suitable for dispensing based on shear thinning. However, the first material (A) demonstrates quicker recovery and, therefore, is better suited for high-speed dispensing. It can be dispensed at a rate exceeding 50,000 dph, without stringing or tailing. Other material factors affecting dispensing are as follows:

- Low viscosity is required for fast dispensing
- High green strength is necessary for dot stability
- Cooling the adhesive may be necessary to prevent premature aging in the syringe (heating up to 70 °C can occur from piston movement).

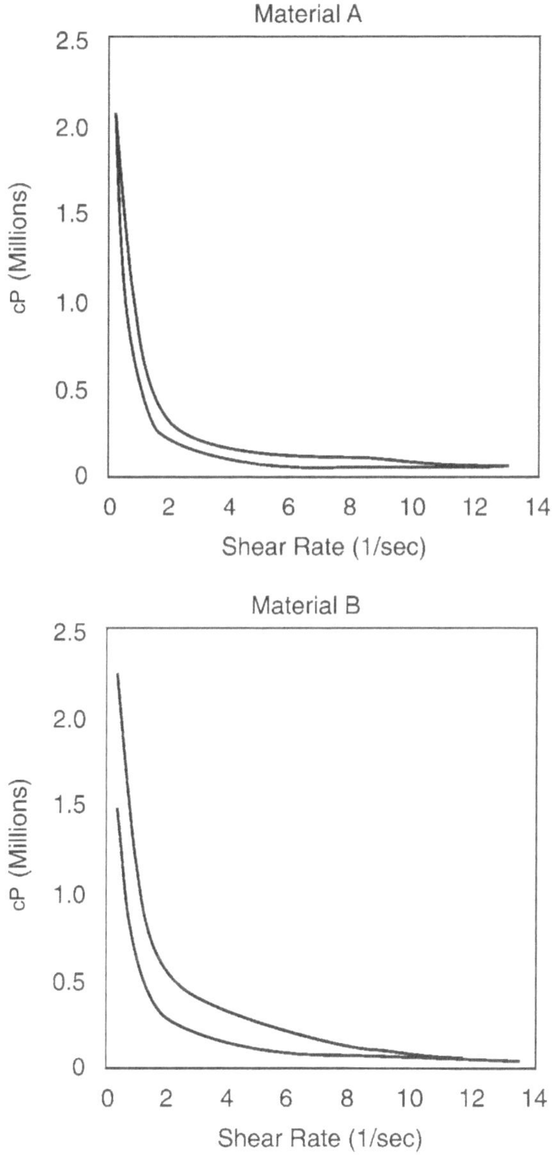

Figure 4.10 Thixotropic hysteresis loops for SMT adhesives.[40,53]

Besides premature aging, an increase in temperature during dispensing is also critical; for example, a 5 °C rise above room temperature alters the quantity of material dispensed as well as the shape of dots and patterns. Increased temperatures during automated dispensing lower viscosity and increase the mass of dispensed adhesive. The increased amount of material may interfere with adjacent circuitry or with devices already assembled. A 5 °C rise above room temperature also results in wider, shorter dots, creating insufficient dot height necessary for component contact. Other factors critical to dispensing are needle selection, equipment used, and pumping method.

4.3.4.2.1 Needle selection

Needle sizes must be selected to dispense the desired amount of adhesive. Dispensing too large an amount of adhesive can bridge components and lead to mechanical failures such as lifting during thermal cycling while insufficient adhesive results in low bond strengths. As a rule, the needle's inside diameter (ID) should be at least 50% of the size of the desired dot.[51] Thus, to dispense dots 0.7–0.9 mm in diameter the needle's ID should be approximately 0.4 mm, which corresponds to a 22-gauge needle. Needle sizes are listed in Table 4.20, and a guide for selecting needle sizes for various components is provided in Table 4.21.

Composite "needle maps" have been generated for various adhesives and may be available from either the adhesives supplier or the manufacturer of the dispensing

Table 4.20 Needle gauge sizes[50]

Needle Gauge	ID (mm)	OD (mm)
18	0.84	1.24
19	0.68	1.06
20	0.58	0.88
21	0.51	0.81
22	0.41	0.71
23	0.33	0.63
25	0.25	0.55
27	0.20	0.40
30	0.25	0.30
32	0.1	0.225

Table 4.21 Needle selection guide based on part type

	Gauge				
Part Type	**27**	**25**	**23**	**22**	**21**
0402	X	X			
0603		X	X		
0805		X	X		
1206		X	X		
SOIC		X	X	X	
PLCC		X	X	X	
QFP			X	X	X

Source: GPD Global, Application Notes, Version b1.1.

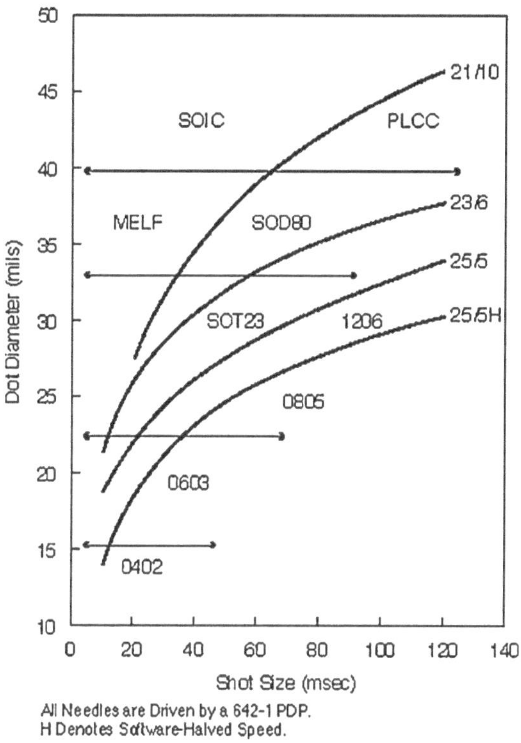

Figure 4.11 Example of a "needle map" for Loctite® 3609 adhesive dispensed from a Camalot® dispenser. (*Reproduced with permission from Speedline Technologies.*)

equipment. The maps are a ready reference guide in selecting the needle size and shot size for a specific adhesive and dot size. Needle maps correlate the shot size in milliseconds with the dot diameter for various needle gauges and standoffs and for various component types. An example of a needle map for a high-speed surface-mount adhesive is given in Figure 4.11. In the figure, the needle gauges and standoffs are given at the end of each curve; thus, 21/10 indicates a 21-gauge needle with a standoff of 10 mils. For example, a 0402 part type requires a dot diameter of 0.25–0.38 mm and shot sizes of 5–50 msec. As one goes up the scale, the dot diameters become larger as do the shot sizes and thus larger diameter needles are required. Therefore, a plastic leadless chip carrier (PLCC) would require dot diameters of 1–1.3 mm and shot sizes above 60 msec using a 21-gauge needle.

4.3.4.2.2 Equipment

A wide variety and numerous suppliers of automated dispensing equipment are available as listed in Table 4.22. An important consideration in selecting an automated, contact-type dispensing system is the pump used in the system, of which there are three types: time-pressure, auger, and linear piston or positive displacement.[54]

Table 4.22 Commercially available adhesive dispensing systems

Dispensing System/ Supplier	Description	Fluid delivery method(s)	Dispensing speed or throughput	Applications	Examples of Compatible Adhesives
Century® C-718/ Asymtek®	Mid-range, general-purpose dispensing system	Dispense jet and rotary positive displacement pump	20,000 dots per hour (0.5 mm [22 mil] or larger) on 3 mm grid	Mid-range SMT production lines	Surface-mount adhesive, solder paste, solder mask, and RTV dispensing
Axiom™ X-1010/ Asymtek®	Mid-range throughput for in-line, high-volume, high-accuracy environments	Auger, positive displacement, and time pressure	N/A	SMT assembly	Thermal grease, solder paste, and surface-mount adhesives
Axiom™ X-1022/ Asymtek®	High volume, dual-line dispensing system	N/A	N/A	Advanced electronics packaging and semiconductor assembly	Flip chip and CSP underfills, die-attach adhesives for advanced electronics and semiconductor packaging
Spectrum® S-910/ Asymtek®	Scalable dispenser for high-volume applications	Jet dispensing, auger, and pressure time	Flow rates up to 500 mg/second and 200 dots/second	High speed assembly	Surface-mount adhesives, silver epoxy, solder paste, and corner and edge bonding
XyFlexPRO®DLM™/ Speedline Technologies	Increased throughput applications, especially for multiple pass underfill applications	Multi-piston and auger pump	35K dots/hour (3 mm pitch), 38K/hour (2 mm pitch), and 25K/hour (IPC9850 Test PCB)	Surface mount, semiconductor, COB, die attach	Die encapsulants, dam and fill materials, epoxies, and other hybrid materials
XyFlexPRO®HVB™/ Speedline Technologies	High volume batch dispensing system	Auger and multi-piston pumps	N/A	Underfill on large die (multiple flow pass)	SMT adhesive, solder paste, or underfill materials

XyFlexPRO®+/ Speedline Technologies	High volume batch dispensing system	Positive displacement	45K/hour (3mm pitch), 50K/hour (1.27 mm pitch), and 35K/hour (IPC9850 Test PCB)	Ideally suited for lower viscosity materials	Adhesives, including epoxies, and solder paste (dot DU version), and underfill and encapsulants (Line DU version)
Champion 3700/ Creative Automation	Stand-alone automated fluid dispensing system for a wide range of manufacturing processes	Positive displacement piston pump	30,000 dots per hour	Dispensing of materials with a wide range of viscosity	Adhesives, sealants, solder pastes, and lubricants
Champion 6809/ Creative Automation	Automated dispensing system designed for precision processes	Positive displacement piston pump	90,000 deposits per hour to form high definition dots, lines, fills, and builds	Dispensing of materials with a wide range of viscosity	Adhesives, sealants, solder pastes, and lubricants
Champion 8300/ Creative Automation	Automated dispensing system designed for in-line conveyorized processes	Regular and low-viscosity positive displacement piston pumps	45,000 deposits per hour to form high definition dots, lines, fills, and builds	Dispensing of materials with a wide range of viscosity	Underfill and encapsulants
MicroMax/GPD Global	High-throughput dispensing system for non-heated applications	N/A	38,000 dots/hour (adhesive)	SMT, hybrid microcircuits, or semiconductor assemblies	SMT adhesives, solder pastes, conductive epoxies, solder masks, and no-flow underfill materials
MicroMax II/GPD Global	High-throughput dispensing system for heated applications	N/A	38,000 dots/hour (adhesive)	SMT, hybrid microcircuits or semiconductor assemblies	SMT adhesives, solder pastes, conductive epoxies, solder masks, and no-flow underfill materials

4.3.4.2.3 Time-pressure pump method

Time-pressure, the simplest and oldest of the methods, is still in use today. According to this method, pulsating high-pressure air actuates a plunger that forces the adhesive through a syringe.[55] A needle-valve controls the amount of adhesive dispensed. Advantages of this method include lower cost and simplicity of operation compared with the auger and the linear piston types. Some disadvantages are:

- Pressure changes occur as the syringe empties due to compression of air in the syringe.
- High-speed dispensing causes heating of the adhesive which changes its viscosity, limits its pot life, and changes the volume of material dispensed. (Viscosity variations in adhesives require process adjustments during dispensing.)
- Stringing of adhesive can occur under some conditions.

4.3.4.2.4 Auger pump

A second type of pump is the auger pump. In this system, air pressure is used to feed adhesive into a reservoir that contains a feed screw or auger. As the auger turns, it shears the adhesive and forces it down the reservoir and out through the needle. Auger pumps are designed to minimize variability when dispensing adhesives at high production rates. Compared to the time-pressure pumps, the effects of changes in viscosity are reduced because the auger is mechanically feeding the needle. Some limitations of the auger pump approach are the following:

- Self heating of highly-filled, thermally-conductive adhesives at high shear rates affects the amount of material dispensed.
- Higher cost than time-pressure approach.
- Complicated cleaning process for the auger and reservoir.

4.3.4.2.4 Linear-piston pump

The third type of pump used in dispensers, linear piston, is a true positive-displacement pump. This pump uses a piston to change the volume of a reservoir that is fed from the main syringe. The displacement of the piston produces an equivalent positive displacement of fluid through the pump. Changes in viscosity have little or no effect on the repeatability of dispensed mass. Of the three methods, the repeatability of the linear-piston type is the best ($\pm 1\%$, except for very small dispense volumes of less than 20 μl).[54] Repeatability of the time-pressure method is $\pm 10\%$ and for the auger method is $\pm 4\%$ (short term less than 30 min).

4.3.4.3 Automated dispensing of underfill adhesives

In addition to die attachment and surface mount, automated selective dispensing is also used to apply adhesives as underfills for flip-chip devices, chip-scale packages, and ball-grid array packages. The adhesive is dispensed along one or more sides of the device and flows by capillary action beneath the devices filling in the volume around the solder balls. Some underfill materials have high flow rates and are "self filleting." In other cases, an additional "seal" pass is required on the dispense side to create

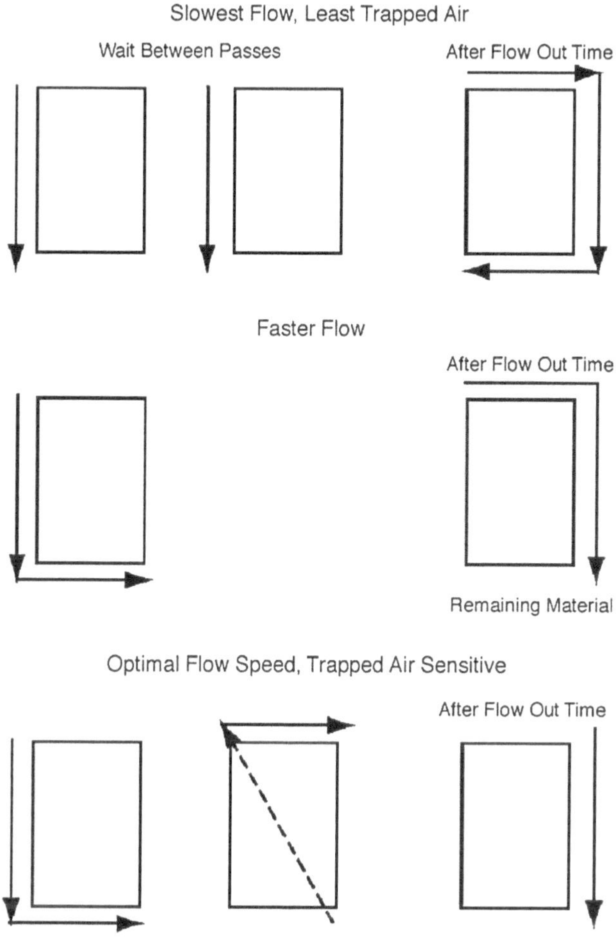

Figure 4.12 Dispense patterns for capillary-flow underfill adhesives[57]: Top: "I" shape; Middle: "L" shape; Bottom: modified "U" shape.

a uniform fillet around the device's perimeter.[56] In general, the needle tip should be 0.025–0.075 mm off the substrate and offset by 0.025–0.075 mm from the edge of the device.[54]

A range of fill patterns that can be used is shown in Figure 4.12.[57] The specific dispensing pattern used depends on the shape of the part (square or rectangular) and the size of the device.

The simple "I" dispense pattern is widely used for square devices. Multiple applications, however, may be needed followed by a fillet or seal at the other three sides using a fine gauge needle. A limitation of this approach is the time required to underfill large die. For die smaller than 3-mm square, a single pass usually suffices. Yet another option for small die is to apply a single dot at one edge.[56]

An "L" dispense pattern is also used (Figure 4.12). After the adhesive flows out, a fillet is dispensed at the other two sides. The L pattern may be started from one corner using a single pass or from a common corner using a two-step process.[58] Die greater than 3-mm square, but less than 6-mm square, may require L-shape dispensing and 2–3 more passes after allowing sufficient time for the adhesive to flow-out beneath the device. L-shaped dispensing is also used for rectangular devices such as flip-chip ICs in which case the dispensing is started along the longer side of the device.[57] Both L-and I-shaped dispense patterns work well with small flip-chip and chip-scale package devices. Large devices, regardless of type, require the L-shape pattern.

The modified U pattern is fastest to apply but, of the three methods, has the highest risk of air entrapment. The adhesive is dispensed in two strokes forming a U pattern, then is followed by a seal along the last side (Figure 4.12).

4.3.4.4 Material factors affecting automated dispensing

In dispensing adhesives for underfill, as well as for die attachment and surface mount, flow properties and viscosity are paramount. Low viscosities are essential for rapid dispensing and efficient capillary flow beneath the devices. Dispense temperatures of 70–85 °C may be used to reduce viscosities of the fluid to less than 500 cP from otherwise 20,000 cP or greater at room temperature.[59] Most adhesives experience an increase in viscosity of 25–100% when their pot lives have been exceeded. An example of the increases in viscosity that occurred for a snap-cured epoxy used in surface mounting is given in Table 4.23. Over time, this increase in viscosity results in a decrease in the amount of material that can be dispensed. To compensate for this, one or more machine adjustments must be made such as increasing the air pressure, increasing the dwell time for dot patterns, or reducing the line speed of the dispensing head.

Several defects associated with the automated dispensing of adhesives are possible, among which are stringing, tailing, missing dots, satellite dots, and inconsistent dot heights. A common defect, known as *stringing*, occurs if, after dispensing, the adhesive does not break quickly and cleanly from the needle (Figure 4.13). Stringing may contaminate adjacent surfaces or components. The incidence of adhesive stringing is high if the adhesive volume is too low for the needle's diameter and for the standoff height. Using a smaller-diameter needle and reducing the standoff

Table 4.23 Viscosity changes vs pot life for a surface-mount adhesive[60]

	at 0.5 rpm	**at 5 rpm**
Initial viscosity (cP)	919,143	403,286
After aging 3 hours at r.t. (cP)	1,065,714	533,438
% change	16	32
Thixotropic index	2.3	2.0

Values are averages of seven batches of Aptek Laboratories' surface-mount epoxy adhesive: DIS-A-PASTE 2311 PMF. (*Source: Aptek Labs. Inc.*)

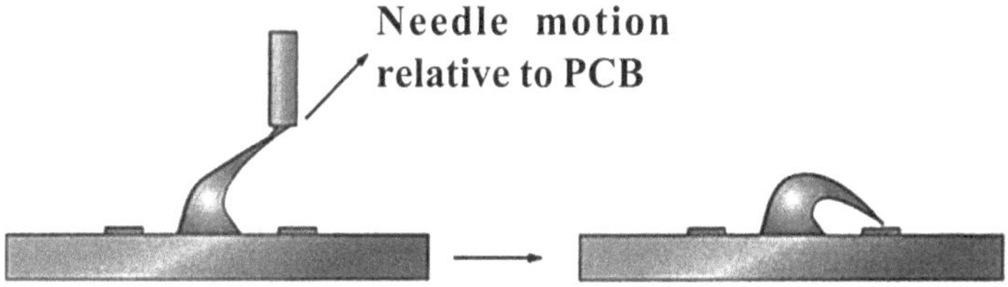

Figure 4.13 Adhesive stringing during dispensing.[51] *(Courtesy PennWell Corp.)*

height usually solves the problem. A maximum standoff height of one-half the needle's ID is used; above this, stringing and inconsistent dispensing occurs. Other causes of stringing include electrostatic charge on the board, incorrect z-stroke height, board warpage, and insufficient support for the board.[51]

Tailing occurs when the dot size becomes less than twice the inside diameter of the needle. With a 0.4 mm ID needle, dot size should be at least 0.75 mm; with a 0.6 mm ID needle, the dot should be greater than 1 mm.[52] Tailing is also reduced by increasing the retraction speed.[50]

Missing dots result from insufficient line pressure, clogged needles, or air bubbles in the adhesive causing the adhesive to be dispensed intermittently.[61] Missing dots result in improper parts placement, incorrect orientation, and degraded bond strengths.[51]

Satellite dots are extra, mislocated dots caused if the standoff height is too high or if the volume dispensed is too great for the needle. Satellite dots result in bond pad contamination or low-strength bonds. They can be eliminated by reducing the pressure or by switching to a needle having a larger inside diameter.[51]

Inconsistent dot heights and sizes may result from a needle standoff post that is landing on solder pads or by increasing in the viscosity of the adhesive during its working life. A replacement needle with a different post usually solves the first problem. Also, increasing the dispensing pressure as viscosity increases and reducing the dispensing speed generate more reproducible dot sizes.

4.3.5 Jet dispensing (non-contact type)

Jet dispensing, also known as jetting, is a direct-dispensing process whereby the adhesive paste is forced through a fine nozzle and programmed to flow onto selected areas controlled from a computer database. It is an automated method that differs from syringe/needle dispensing in that contact with the surface is not required to produce the dots, lines, or patterns.

Jet-dispensing equipment uses a closed-loop, positive shut-off piston to dispense the adhesive. The adhesive is pressurized at the syringe to assure constant flow. Using a spring-driven ball and seat mechanism, the adhesive fills the void left when the ball retracts from the seat. When the ball returns, the force due to acceleration breaks the stream of adhesive, which is then jetted through the nozzle in precisely controlled amounts onto the printed-circuit board (Figure 4.14). Unlike contact automated

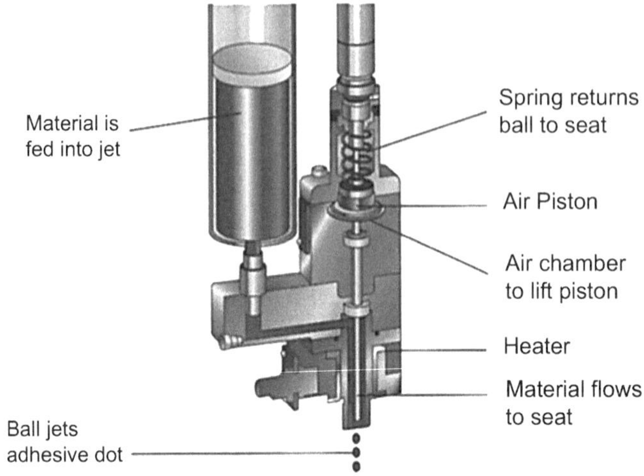

Figure 4.14 Jet-dispensing concept, DispenseJet® DJ-2100 valve.[64] *(Courtesy Asymtek®, a Nordson Co.)*

syringe dispensers, jetting requires no *z*-axis motion. The time between shots is approximately 15 msec versus 90 msec, and the uniformity and shape of the dots are not affected by substrate warpage or by the surface tensions of the needle and board.

Jet dispensing was developed for SMT primarily to increase throughput by increasing the speed at which dots can be dispensed up to 70,000 dph. Jet dispensing also produces better quality dots and much smaller dots than are possible with syringe dispensing, thus meeting requirements for attaching very fine-pitch small devices. Droplets of adhesive in picoliter volumes can be dispensed but, if larger dots are required, they can be produced by dispensing additional shots over the first dot in rapid succession.[62] Jet dispensing has also simplified and improved the dispensing of epoxy underfill adhesives. With jet dispensing, there is no risk of a needle clipping and damaging the die or of the needle itself being damaged since there is no *z*-direction motion of the dispense head. Because of the absence of *z*-direction motion, a 20% improvement in throughput has been reported.[63]

Adhesives for jet dispensing must have relatively low viscosities of approximately 120,000 cP at 5 rpm, or 20,000–40,000 cP at 10 rpm, and fast recovery of thixotropic properties after dispensing (narrow hysteresis in viscosity loop).[65] Examples of SMT adhesives that are jet dispensible are Henkel's® Loctite® 3621 and Epoxy Technology's Epo-Tek® H70E.[66]

In summary, the key advantages of jet dispensing are:

- High throughputs of 70,000 dph compared with less than 40,000 dph for single-head syringe dispensing. These high deposition rates are better able to match the high component placement rates.[67,68]
- Reduced cleaning operations.
- Elimination of stringing.

- Improved dot uniformity.
- Adjustable dot height and profile by stacking dots.
- Tolerance for board warpage and previously assembled components. Avoids need to compensate for height.

Some limitations include the following:

- Highly viscous materials are not suitable. Only low viscosity, highly thixotropic adhesives can be used.
- Droplets are limited to one size, however, dots can be stacked to increase their size.

4.3.6 Preform/Tape attachment

Preforms, tape, or film adhesives, unlike the paste or liquid adhesives, are already in the solid state. They may be thermoplastic or thermosetting but, in both cases, may be purchased as sheets of various thicknesses and cut to size to accommodate specific device or part sizes. In using preforms, the electronics assembler avoids problems associated with the storage, handling, and dispensing of liquid adhesives. Parts such as leadframes, lids, and substrates may also be purchased with pre-attached preforms, a convenience to the user, who then avoids having to cut and handle individual preforms. Most parts can also be stored for up to one year in ambient conditions without affecting their properties.

Thermoplastic preforms provide instant bonding since they melt, bond the surfaces, and resolidify quickly on cooling. They also have the added advantage of being easily reworked. Attachment conditions vary according to the preform used, but generally consist of melting the preform at around 200 °C with an applied pressure of about 10 psi. Thermosetting preforms are partially cured adhesives, primarily epoxies. By heating and applying pressure, they will also flow and wet the surfaces to be bonded, but then must be fully cured by further polymerization through continued application of heat. Flow may be induced at 60–100 °C followed by curing at 80–150 °C with applied pressure of 10–15 psi. Thermosetting preforms generally result in higher shear strengths than thermoplastic types and are capable of resisting higher temperatures. Die-shear strengths of 2,000–3,000 psi for thermosetting preforms are typical. Silver-filled preforms are also available and, because of their high thermal conductivities, are used to attach heat spreaders, heat sinks, and circuit substrates on PWBs or inside ceramic or metal packages. Thermal conductivities for the silver-filled preforms are typically 6 W/m K compared with 1.5–1.6 W/m·K for the electrically insulative types. Electrically conductive preforms having conductivities of 5×10^{-3} to 5×10^{-4} ohm-cm are also used for EMI shielding.[69]

4.4 Placement of devices and components

After the adhesive has been dispensed onto the PCBs or other interconnect substrate, bare die and other electronic components must be precisely placed. Of course, pick and place can be done manually using vacuum pick-up tools or tweezers, but this approach is useful only for small quantities, prototypes, or for rework. The risk of human error is high and accuracy is exacerbated as more and more very fine pitch devices are used. Even placing the device in the correct location, but in the wrong orientation can result in electrical failure. Manual placement is impractical for production quantities.

Fortunately, advanced robotic equipment has been developed and is commercially available for accurate, high-speed placement of components. Entry-level automated placement machines can place 3,000–5,000 parts per hour while mid-range equipment, that can hold more feeders, can place 8,000–10,000 parts per hour. On the high-end, placement speeds of 30,000 parts per hour and higher are being achieved.[70] In fact, speeds of the new equipment are so great (chipshooters can place up to 80,000 components per hour) that they often must be slowed in order for adhesive dispensing to catch up; alternatively extra adhesive dispensers must be added to the production line.

A wide variety of fully automatic machines exist. Generally, feeders, which may consist of tape-and-reel, stick, or tray, deliver the parts individually so they can be picked up and automatically inspected. Machines are equipped with vacuum nozzles to pick up the components and carry them through inspection, alignment, and placement. A list and description of commercially available component placement equipment is given in Table 4.24.

4.5 Curing

After dispensing the adhesive and placing the devices, the adhesive must be cured, that is, it must be advanced from a liquid or semi-solid state to a solid state for full development of its adhesive strength and dielectric properties. In the broad sense, *curing* is the processing that is involved in altering the state of the adhesive material so that it can adequately bond two surfaces. In most cases, this involves polymerization of low-molecular-weight monomers or polymers through the application of heat or other form of energy. With prepolymerized materials, such as many thermoplastics, curing simply involves evaporation of the solvent (if there is a solvent) and heating to bring the adhesive up to its melt temperature. In other cases, as with polyimides, curing involves the formation of ring structures along a prepolymerized molecular chain by heating. However, in most cases, curing involves polymerization of monomer resins by reaction with a hardener or by initiation with catalysts and, generally, requires heat or another form of energy.

Epoxy resins and other thermoset resins cure by forming a highly cross-linked structure either by condensation or addition polymerization (see Chapter 3). Some hardeners and catalysts for epoxy resins are so reactive that, upon mixing, curing takes place in a matter of minutes, even at room temperature. Among these fast curing agents are the primary and secondary aliphatic amines and the boron trichloride or trifluoride amine complexes. Other hardeners such as aromatic anhydrides require elevated cure temperatures for a period of time to achieve full polymerization. Heating is the most widely used method for curing adhesives and may be applied in a convection oven or by infrared lamps in a conveyor furnace. Localized heat may also be generated by microwave heating equipment. Other curing methods involve exposure to ultraviolet

Table 4.24 Component placement systems

Model/Supplier	Description	Placement speed (components/hour)	Components placed	Component size range
MY100LX™/Mydata	Pick and place system for high-speed, high mix production (board size from 70 mm × 50 mm [2.7" × 2.0") to (575 mm × 508 mm [22.6" × 20")	15,000	Chip (from 01005[a]), SOIC, PLCC, TSOP, QFP, BGA, flip-chip, odd-shape, surface-mount connectors, through-hole components, CSP, CCGA, DPAK, Alcap and Tantalum capacitors	Minimum: 0.4 mm× 0.2 mm (16 mil × 8 mil) for 01005 devices Maximum: 56 mm^2 (2.2 inch. square) and 15 mm (0.9 inch) thick Maximum component weight: 140 g
MY100DX™/Mydata	Pick and place system for high-speed, high mix production (board size from 70 mm × 50 mm [2.7" × 2.0") to (575 mm × 508 mm [22.6" × 20")	34,000		
MY100SX™/Mydata		21,500		
SIPLACE MultiStar/ Siemens Electronic Assembly Solutions	Pick and place system for specialized high-speed placement of very small, very large, and odd-shaped components	80,000	N/A	Minimum: 0.4 mm × 0.2 mm (16 mil × 8 mil) for 01005 devices Maximum: 50 mm × 40 mm (~2 inch × 1.6 inch) and 11.5 mm (0.45 inch) thick

(*Continued*)

Table 4.24 Component placement systems—*cont'd*

Model/Supplier	Description	Placement speed (components/hour)	Components placed	Component size range
SIPLACE SX/ Siemens Electronic Assembly Solutions	Pick and place system for minimal production run reconfigurations and maximum production line balancing (board size from 50-mm [2-in] square to ~610 mm × 560 mm [~24 inch × 22 inch])	68,000	01005 devices up to 200 mm × 100 mm (7 inch × 4 inch)	Minimum: 0.4 mm × 0.2 mm (16 mil × 8 mil) for 01005 devices Maximum: 200 mm × 100 mm (7 inch × 4 inch)
MC Series/ Manncorp™	Pick and place systems for precision and flexible production lines (board size from 20-mm [0.8-in] square to 500 mm × 380 mm [~20 inch × 15 inch])	N/A	0201, 0402, 0603, 1608 discrete devices LED, SOT, SOP, BGA, and QFP	Minimum: 0.6 mm × 0.3 mm (24 mil × 12 mil) for 0201 devices Maximum: N/A
FLX2011/Essemtec AG	Highly flexible pick and place system for fine pitch capability and high-mix production (board sizes of 400 W mm × 300 W mm [~16 inch × ~12 inch] or 700 mm by 600 W mm [~27 inch × ~24inch])	6,000	01005 devices up to 50-mm (2-in) square components	Minimum: 0.4 mm × 0.2 mm (16 mil × 8 mil) for 01005 devices Maximum: 50-mm (2-in) square components

[a] 01005 devices are 1 mil × 0.5 mil.
Source: Compiled from suppliers' technical data sheets.

Table 4.25 Comparison of curing methods

Curing Method	Advantages	Limitations
Thermal	Low cost. Uses existing equipment	Long cycle times for most formulations. Incomplete cure may occur with fast cure, snap cure adhesives, depending on thermal mass
Infrared	Uses existing SMT equipment in place. High throughput	May require post curing
Microwave	Fast curing rates (factor of 10) over forced convection heating. Reduced stress from CTE mismatched materials	Cure schedules require optimization of many parameters. Specialized equipment
UV/visible	Rapid curing increases production throughput. Good for heat-sensitive devices. Low production costs. Reduced energy requirements	Line of site process; shadow area may require a thermal post-cure. Requires photosensitive formulations
Moisture	No heat or energy required. Room-temperature cure. Good for heat-sensitive devices	Limited to only a few polyurethane and silicone adhesives. May require several days to achieve full cure. May require post heat-curing

light or to moisture. Regardless of the curing method, establishing the degree of cure and the optimum cure schedule are vital to the reliability and performance of the bonded parts. A summary of the various curing methods with their advantages and limitations is given in Table 4.25 and a discussion of them follows.

4.5.1 Thermal curing

Adhesives that are heat cured fall into three categories: conventional, fast cure, and snap-cure types. The conventional types cure in 0.5–2 hours at temperatures of 150–175 °C. Heat-cured adhesives have generally been used in high-reliability applications, especially in hermetically sealed packages for multichip modules or hybrid microcircuits where subsequent outgassing can cause failures. Curing is conducted in box ovens with either circulating air or nitrogen. In the high production of consumer and commercial products, fast-curing adhesives have been developed that are compatible with automated in-line equipment. Thus, there are fast-cure systems that cure in 15 minutes or less, and the snap-cure types that cure in less than one minute at temperatures greater than 150 °C. Curing is conducted following in-line component placement by moving the parts on a conveyer belt through an infrared furnace. Furthermore, surface-mount adhesives have been developed that can be partially or completely cured during the solder-reflowing step. Those that are partially cured must then receive a short post-cure.

In addition to heat, some adhesives require the simultaneous application of pressure. Examples include the bonding of substrates with thermoplastic film adhesives and the bonding of flip-chip and TAB devices with anisotropic adhesives. In bonding ceramic substrates to the inside of metal or ceramic packages with an epoxy film, sufficient pressure must be applied during cure so that the adhesive flows evenly, moving air and other entrapped volatiles out toward the edges and forming a fillet around the part. Often, small voids or bubbles can be seen in the fillet after curing, an indication that solvent or absorbed moisture were pushed out. This may not be serious if the number of voids within the bond line is small or nonexistent. Techniques for reducing voids include vacuum baking the surfaces prior to applying the adhesive, curing the adhesive while under vacuum and pressure, or step curing the adhesive. The degree of voiding may be quantitatively determined by scanning acoustic microscopy (SAM) or, more simply, by using glass as one of the adherends so that voids may be detected through the glass from the bottom side.

With anisotropic adhesives, pressure is necessary to squeeze the film so that electrical contacts are made between the aligned electrodes of the parts being bonded. While under pressure and while the electrodes make contact with the metal filler particles, the adhesive cures in the area surrounding the connections and fixes the positions of the *z*-direction connections.

4.5.1.1 Outgassing during cure

Solvents, low-molecular-weight additives, reactive diluents, absorbed moisture, and other constituents outgas during thermal curing. The nature of the outgassed products and their amounts vary with the adhesive formulation used and with the cure temperature. It is, of course, important that these volatiles be removed to produce sound, void-free bond lines; however, the condensation and redeposition of these volatiles (a condition called *raining*) on the interior of an oven and on other circuits being cured in the same oven presents a contamination problem. Thus, it is desirable to use adhesives that produce the least amount of outgassing during cure and that the outgassed products not be corrosive. Selecting adhesives that have the lowest weight losses during cure have been shown to produce the highest yield of good wire bonds.[71] Compounds that have been evolved from curing epoxy adhesives include methyl alcohol, carbon dioxide, acetone, methyl ethyl ketone, toluene, nitrogen heterocyclic compounds, unsaturated hydrocarbons, 2-propenol-l (allylalcohol), and 2-propenyl ester of formic acid.[72] Most of these are solvents used as part of a solvent-based formulation or are residual solvents that were used in the manufacture of the adhesive. Weight losses during cure can be as high as 25% (as for solvent-based adhesives such as polyimides) but, in most cases, range from 0.4% to 5% when cured according to the manufacturers' instructions.[73]

4.5.2 Ultraviolet/visible light curing

Photosensitive adhesive formulations based on acrylates, epoxy acrylates, and polyurethane acrylics may be cured by exposure to ultraviolet or visible light. The key benefits of UV curing are the speed of curing and the avoidance of high temperatures for heat-sensitive components. UV curing of adhesives has major advantages over heat curing in the assembly of precision microelectronic parts. It allows the operator sufficient time to position the parts after which they can be exposed to UV light and set in place. Normally, UV curing is followed by a short heat cure to assure that portions that were shielded from UV are also cured. UV-curing adhesives are finding increased use in high production assembly where only 5–20 seconds are required to cure an entire populated circuit board. The cure time depends on the formulation and on the frequency and power of the UV source. Cure times of 10–20 seconds result from using long wavelength (365 nm) from a moderate intensity source of 50 mW/cm^2. By using a higher intensity radiation source of 150 mW/cm^2, the cure time can be reduced to 5 seconds.[74] In general, optimum cures are obtained in the wavelength range of 250–400 nm and a radiant output of at least 150 mW/cm^2.[75]

Medium pressure mercury-vapor lamps are quite popular as a source of UV radiation. UV-curing equipment can emit light at specific frequencies matching the absorption bands of the photosensitive groups of the adhesive. The shape and intensity of the emitted radiation can be varied to flood a large area or focused for high-intensity, high-speed curing. Many companies that supply UV-curing adhesives also sell compatible curing equipment ranging from portable hand-held wands to conveyorized-belt systems.[76]

4.5.3 Microwave curing

Variable-frequency microwave (VFM) curing is a fairly recent development for rapid curing of adhesives, coatings, and encapsulants.[77] VFM curing differs from conventional microwave heating in using swept frequencies instead of a fixed frequency. Variable-frequency microwave makes use of three parameters: central frequency, frequency bandwidth, and sweep rate. Through the frequency-sweeping process, problems of non-uniform temperatures and arcing associated with conventional microwave ovens are avoided. VFM curing is reported to be 2–10 times faster than conventional heat curing. A further advantage is the uniform distribution and precise control of microwave energy that can heat large sections uniformly. Compared with conventional heat curing, VFM curing produced similar die-attachment stresses and reduced warpage of FR-4 printed-circuit boards onto which the components were attached.[78] Since 1994, Lambda Technologies has pioneered the VFM process and has commercialized it through the development and sale of microwave equipment.

4.5.4 Moisture curing

Moisture-curing adhesives are formulated with resins that react with atmospheric moisture to form a cured polymer. Most of these resins are one-part cyanoacrylates, silicones, or polyurethanes. The key advantage, of course, is their ability to cure without the use of heat or other external form of energy. However, in many cases, several days of ambient exposure are required to achieve full properties since the rate of curing is dependent on the relative humidity and the thickness of the adhesive. If there is too much moisture on the surface, curing of polyurethanes occurs so quickly that carbon dioxide (a significant by-product) becomes entrapped resulting in voids. Conversely, in cold temperatures, below −9 °C, moisture levels are so low that the reaction curing rate is extremely slow. Moisture-cured silicones and polyurethanes are used more as sealants, filleting, and vibration-damping compounds than as true adhesives. Because of their low moduli of elasticity and low, often negative T_g both silicones and polyurethanes are used as adhesives and sealants where stress relief is critical, especially at low temperatures. RTV moisture-cured silicones also provide both very low and very high-temperature stability.

4.5.5 Cure schedule and degree of cure

Regardless of the curing method, establishing a cure schedule that results in optimum reproducible properties of the adhesive-bonded parts is essential for reliability and quality control. Incomplete or undercure, where some of the resin or hardener remains unreacted or where the monomer resin does not fully polymerize, adversely affects the electrical, physical, or thermal properties of the adhesive. Incomplete cures result in reduced bond strengths, lower T_g, increased moisture absorption, lower electrical values, and greater outgassing. Incomplete curing can result from sub-optimal time-temperature curing schedules, non-stoichiometric amounts of a two-part adhesive or inhibition of the cure due to contaminants from the ambient or curing oven. Amines and moisture, for example, are reported to inhibit the cure of acrylic resins while organotin, other organometallic compounds, sulfur-containing compounds, and amines are reported to inhibit the cure of silicones.

Overcuring, although not as serious as undercuring, can result in embrittlement, reduced vibration damping, and reduced shock resistance. Of course, excess overcuring can also result in decomposition of the adhesive. Some semi-empirical methods used to establish an optimum cure schedule are based on measuring specific physical or electrical properties as a function of various times and temperatures. More quantitative methods establish the degree of cure using FTIR spectral analysis or differential scanning calorimetry (DSC).

Degree of cure by FTIR. Fourier Transform Infra-Red (FTIR) spectroscopy is an excellent qualitative and quantitative analytical method to measure changes in the molecular structure of adhesives before, during, and after curing. In FTIR analysis, a small amount (10–20 mg) of the sample is scanned with infrared radiation from a wavelength frequency of approximately 4,000 cm^{-1} to 500 cm^{-1} and the percent absorption or, conversely, transmission is recorded. Specific absorption bands in the near infrared correlate with groups within the polymer structure as, for example, the alkyl, hydroxy, amino, or oxirane groups of an epoxy resin whereas absorption bands in the far infrared, often referred to as the fingerprint region, are indicative of the molecule as a whole.

FTIR spectra have been useful in following the progression of polymerization as a function of temperature and time and to establish optimum cure conditions. For a two-part epoxy system one can track the reduction in intensity of the epoxy peak at

911–912 cm^{-1} or the simultaneous increase in the OH peak at 3,350–3,500 cm^{-1} as the epoxy ring reacts and forms a OH group. Omar et al.[79], for example, studied the curing of a DGEBA epoxy resin with various ratios of tetrahydrofurfuryl methacrylate (THFMA) hardener and found that the FTIR spectrum had the required resolution to distinguish between high and low percentages of epoxy in the mixture.[79] In practice, for a two-part system, small measured amounts of Part A and Part B are mixed on a KBr crystal over a hot stage and IR scans are taken at time intervals at constant temperature. Figure 4.15 shows the FTIR spectrum for commercial-grade epoxy resin DGEBA (diglycidylether of bisphenol A). In the epoxy resin cured with diethylenetriamine (DETA) (Figure 4.16) the strong epoxy absorption band at 912 cm^{-1} has essentially disappeared indicating almost complete reaction of the epoxy groups.[80] A magnified portion of the IR spectrum for the epoxy absorption band shows the progress of cure as a function of time at temperature (Figure 4.17).

FTIR spectra are also beneficial in establishing the optimum conditions for curing acrylic adhesives by exposure to UV radiation. The degree of UV polymerization of acrylic adhesives can be followed by the decay of the ethylenic ($-CH=CH_2$) absorption band at 810 cm^{-1}. Scherzer et al. studied the UV-induced crosslinking of a pressure-sensitive acrylic resin with copolymerized benzophenone groups using variable UV doses and intensities.[81] Real time FTIR spectroscopy was also used to study the effect of various photoinitiators in accelerating the photopolymerization of acrylate resins.[82]

Degree of Cure by DSC. In addition to and complementing FTIR, differential scanning calorimetry (DSC) is used to determine the degree of cure, the optimum cure schedule, and the rate of polymerization as a function of time and temperature. DSC is a very versatile test that also provides glass-transition temperatures (T_g) for various cure schedules and the kinetics of polymerization by measuring changes in the heat flow as material changes occur over a temperature range. In practice, a small sample (approximately 20 mg) of the resin or a mixture of the resin with a catalyst or hardener is slowly heated and the heat absorbed or emitted due to the chemical reactions occurring in the sample is measured and graphed.

Specific DSC test methods are described in ISO 11357[83] and IPC TM-650 2.4.25C.[83] By measuring the isothermal heat of reaction, whether exothermic or endothermic, the degree of cure is represented by the ratio of the energy at time t to the total energy.[84,85]

Thus, the percent conversion of the monomer resin to cured polymer is given by:

$$\%\ \text{Conversion} = A_t/A_{\text{total}} \times 100 \qquad (4.8)$$

where A_t is the area under the peak at any time t and A_{total} is the total area under the peak in the DSC heat flow curve.

In one study, the DSC graph for an epoxy powder heated from room temperature to 250°C at 10°C/minute was obtained. A glass-transition temperature of 74.7 °C was observed which represents an irreversible relaxation peak due to the time-dependent nature of the glass transition event. As the temperature increases and the epoxy begins to cure, an exotherm is observed. By integrating the area under the exothermic peak, the heat of reaction (J/g) is obtained. Completion of the cure is

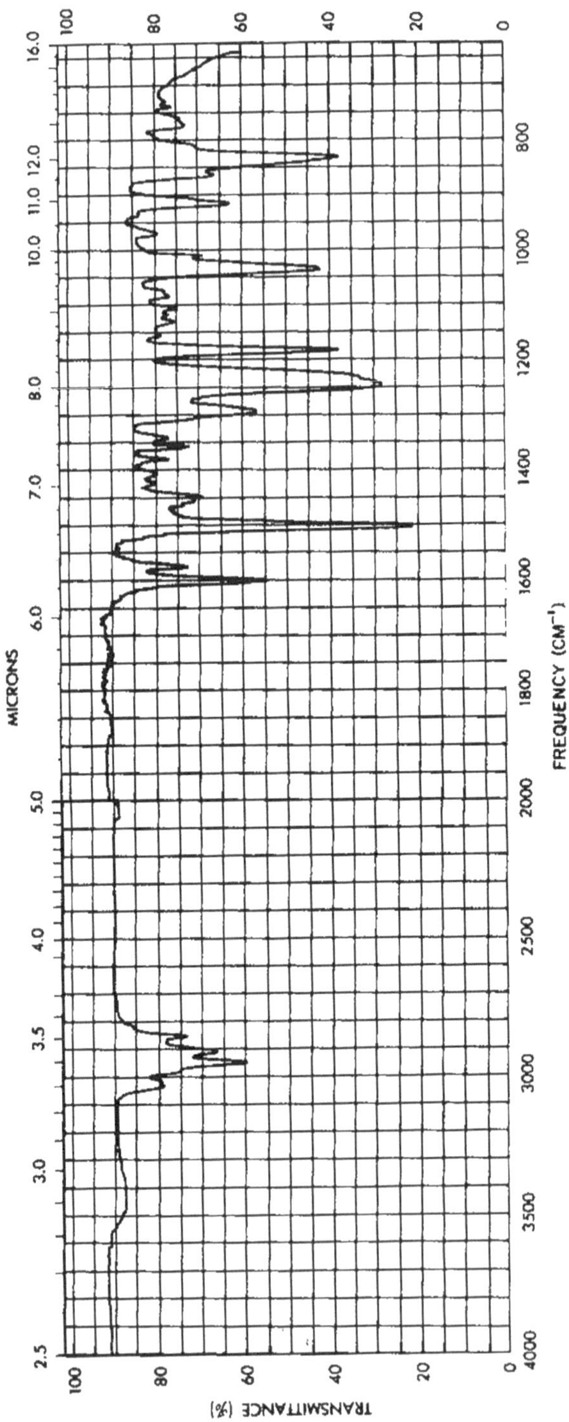

Figure 4.15 Infrared spectrum of diglycidylether of bisphenol A epoxy resin.[80]

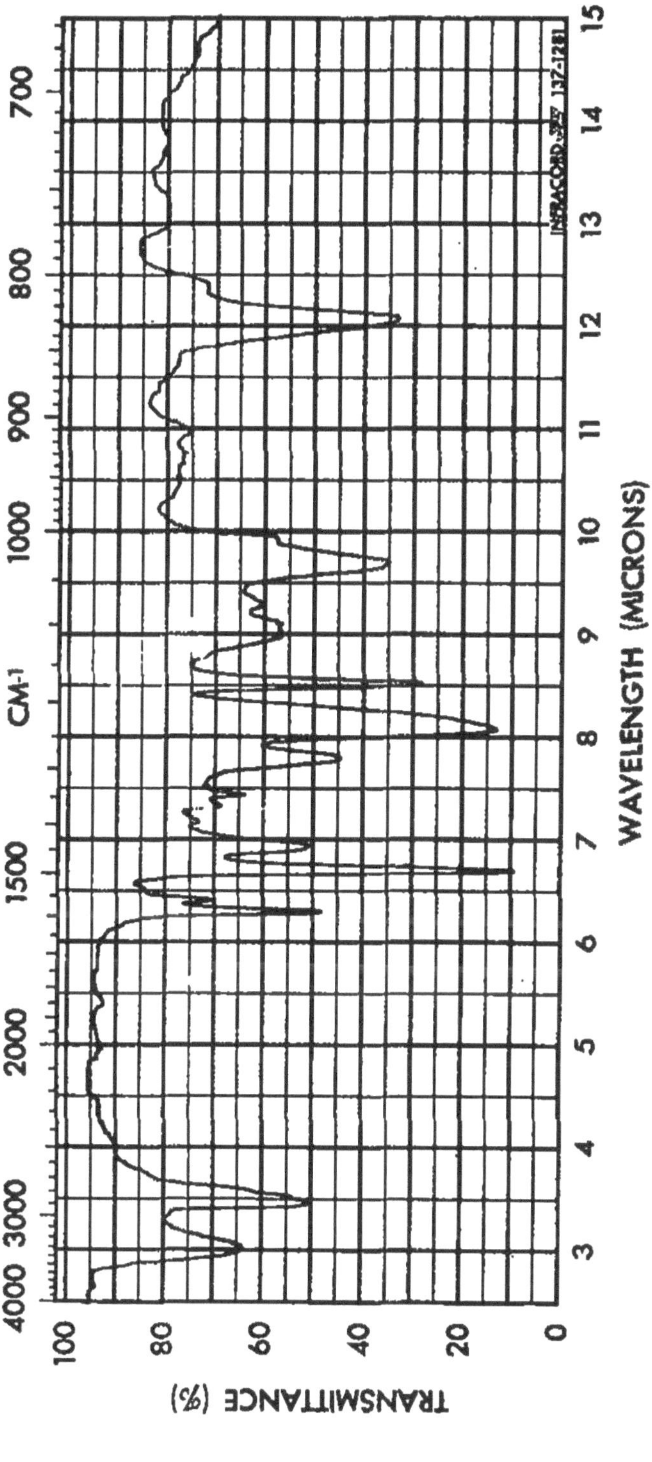

Figure 4.16 Infrared spectrum of diglycidylether of bisphenol A epoxy cured with DETA.[80]

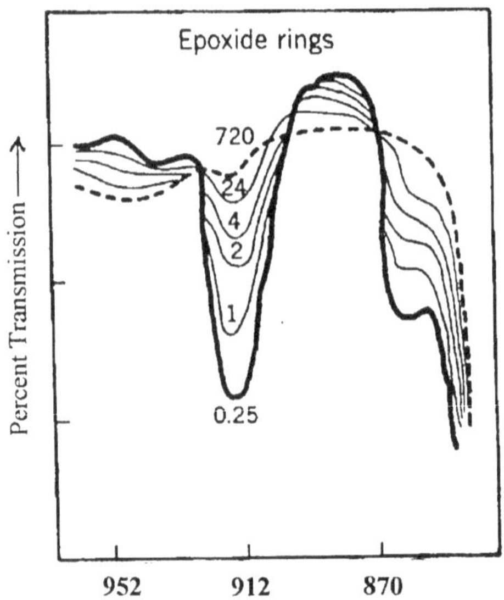

Figure 4.17 Infrared spectra showing degree of cure of DGEBA epoxy resin by changes in the 912 cm^{-1} absorption band.[80] (Number indicates the cure time in hours.)

indicated when no further residual exotherm occurs and the DSC graph returns to a linear form.[86] A typical DSC plot for a two-part epoxy adhesive is shown in Figure 4.18.

DSC may also be used to determine the degree of cure by measuring the T_g as a function of cure time at constant temperature. As an epoxy increases in crosslink density, its T_g increases to a maximum value. Figure 4.19 shows the increase in T_g as the resin continues to cure.[88] DSC has also been used to measure changes in T_g with increasing moisture content of both uncured and cured resins.

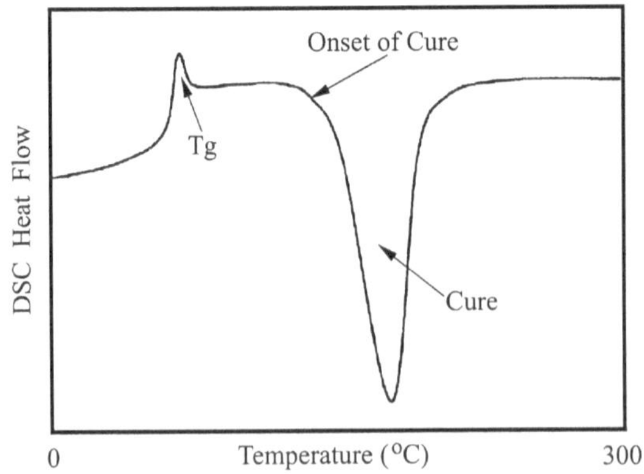

Figure 4.18 Differential scanning calorimetry plot for a typical two-part epoxy adhesive.[87]

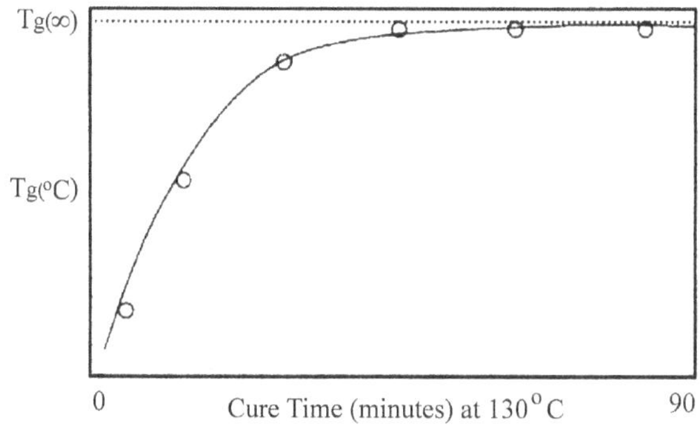

Figure 4.19 Increase in T_g of epoxy vs. cure time at 130 °C.

Using commercially available software, finite element modeling and the kinetics of curing adhesives can be performed as, for example, in the curing of silicone adhesives.[89]

DSC has been used to establish optimum cure schedules for rapid curing (snap cure) adhesives and adhesives requiring longer cures. Results showed that the snap-cured adhesive was 98% cured in 3 min, compared with over 15 min for the conventional cured adhesive.[59] The cure for the latter adhesive was verified by thermomechanical analysis (TMA) by measuring the T_g as a function of cure schedules where it was shown that up to 150 °C, the T_g had not yet reached a plateau after 30 min. At 165 °C, a plateau was reached in 30 min, and at 170 °C, the T_g peaked in 10 to 15 min.

Cure cycles for adhesives should not be interrupted. Step curing should be performed in the same oven without exposing the adhesive to ambient air between steps. Moisture absorption in partially cured adhesives affects the curing process and alters properties. The effects of interrupting the cure cycle and exposing an epoxy encapsulant to air moisture were shown by Naito and Todd.[90]

4.6 Rework

The decision to rework electronic devices or assemblies depends on the cost of reworking versus the value of the part. Consumer electronics, due to their low cost, are considered throwaways and seldom reworked. For some, such as plastic-encapsulated microcircuits (PEMs) that are produced in high volumes, rework is impractical because of the difficulty in removing the plastic encapsulant without destroying the components. However, for high-value assemblies, such as densely populated PWBs and MCMs, where the final value may be as high as $10,000 or more, the ability to rework is essential. In such cases, rework may be necessary to meet schedules when small numbers of assemblies are produced.

Rework generally consists of removal and replacement of a failed electronic component that has been detected after initial electrical testing or testing after burn-in. The actual rework procedure is more of an art than a science. Since each component or assembly is different, considerable experience and care are required to selectively remove one device without damaging adjacent ones.

4.6.1 Rework of components and chip devices

For adhesive-attached components or chip devices that are electrically connected by wire, TAB, or ribbon, the leads must first be excised. In the case of solder-connected components, the solder is heated to its melt temperature and wicked off the part.

Most electronic components are attached with epoxies. Epoxies, being thermosetting, do not melt and resolidify as do thermoplastic adhesives and solder. At sufficiently high temperatures, epoxies decompose and char and their residues are then difficult to remove. Rework procedures for thermosetting adhesives are based on selectively heating the part to a temperature slightly above the T_g of the adhesive, at which point the adhesive softens. A shear force is then applied using a wedge tool to detach the component. Where circuit components and materials are prone to thermal oxidation or decomposition, a jet of hot nitrogen gas is focused only on the part to be removed. The entire assembly or substrate is usually heated on a hot stage at moderate temperatures of 80–150 °C to facilitate the removal with the hot gas. Residual epoxy may be removed using a heated blade maintained at approximately 200 °C or by abrading the surface. After cleaning, fresh adhesive is applied, a new device attached, and the adhesive is cured.

4.6.2 Rework of high-density modules

With advancements in MCMs, where large chips (up to 500-mils square) are closely packed and where devices and interconnect substrates are fragile (for example, polyimide thin-film multilayer substrates used in MCM-D), new rework challenges have arisen. In these high-density circuits, spacings between the chips may be as small as 40 mils—too small to insert a wedge tool without damaging adjacent die. Attempts to shear off the failed device also run the risk of scraping or damaging the thin-film interconnect circuitry. Furthermore, because of the large bond line areas of the chips, the force required to dislodge the chip is high. Among new rework techniques developed for high-density thin-film circuits is one where a tool is bonded to the top surface of the die with an adhesive that has a higher pull strength than the adhesive originally used to attach the die to the substrate. The tool, called a *thermode,* is subsequently heated. Chips are removed by heating the thermode to 245 °C for 3–5 seconds while the assembly is on a hot stage at 150 °C and while a rocking motion is applied that imparts a combination of shear, tensile, and peel forces (Figure 4.20).[91] In detaching devices bonded with Ablestik® 84-1 LMIT (electrically conductive epoxy), the thermode was attached with Ablestik® 41-1. This process is a variation of the well-known Sebastian Pull Test, which uses the Sebastian Pull Tester, a nail-head tensile pull tester from Quad Group, Inc., Spokane, Washington.[92]

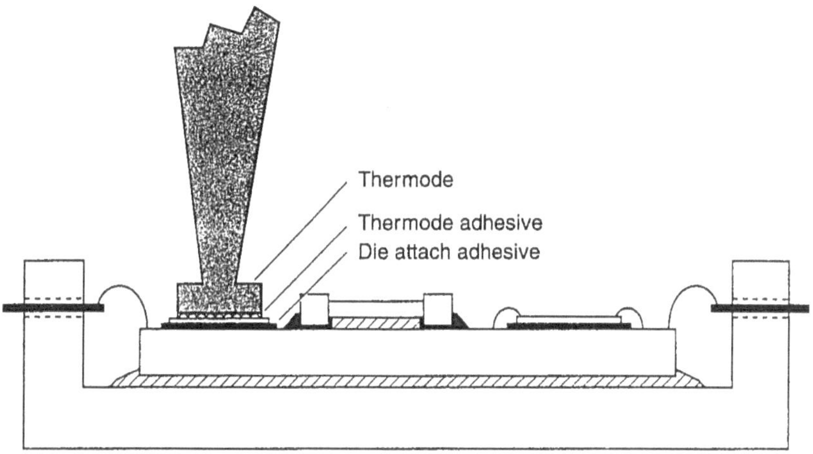

Figure 4.20 Thermode method for die removal.

4.6.3 Rework of thermoplastic adhesive-attached components

Components that are attached with thermoplastic adhesives are easiest to remove.[93] Parts to be removed are simply heated with a hot solder iron or by a directed stream of hot air or nitrogen to the melt temperature of the adhesive, then physically removed with tweezers or sheared off. A new part can be attached without having to completely remove adhesive residues since the resolidified adhesive residue will remelt on attaching the new component. Unlike thermosetting adhesives that soften only gradually after reaching their glass transition temperature and that retain much of their strength, most thermoplastic adhesives melt at a specific temperature or over a narrow temperature range and lose all their strength (Figure 4.21). Thus, they are best suited for fast-turnaround production avoiding hours of curing in controlled-atmosphere ovens.[94]

4.6.4 Rework of attached substrates or assemblies

There are occasions when it may be desirable to salvage a populated substrate that is already attached in a sealed package; for example, if the sealed package was found to be a fine or gross leaker and the costs of the substrate and components were high. In such cases, the substrate may be detached from the package by first severing the wires that connect it to the package pins, then heating the entire module while prying the substrate loose at one edge. Some more brittle adhesives may detach from the package by carefully flexing the package. However, this procedure is risky, particularly for large-area packages, since the substrate may crack and the entire circuit be damaged. Alumina substrates, being mechanically stronger than silicon substrates, are better able to survive this procedure. Thermoplastic adhesives used to bond substrates are easier to remove and desirable, provided they have the thermal and mechanical properties to withstand vibration, thermal shock, and other application-specific

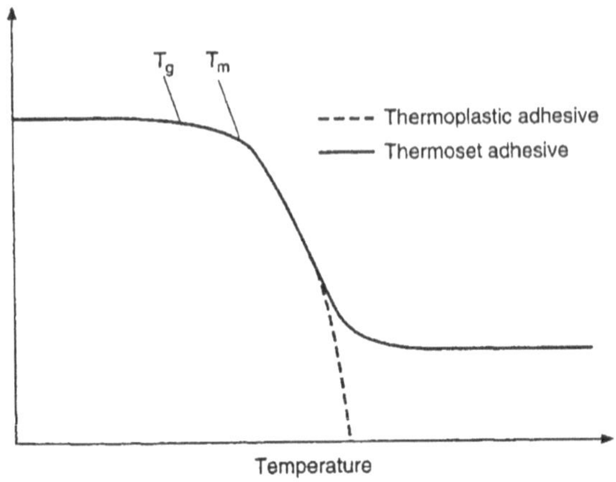

Figure 4.21 Relative strengths of thermoplastic and thermosetting adhesives (T_m is the melt temperature of the thermoplastic adhesive; T_g is the glass-transition temperature of the thermosetting adhesive).

requirements. Thermoplastic films such as PEEK® (polyaryletheretherketone) and ULTEM® (polyetherimide) have been investigated and reported to pass stringent MIL-STD-883, Method 5011 tests.[95]

Alternative removal methods have been reported, but none is widely used. Of these, one involves sliding a hot wire, through which current is passed, between the substrate and package. This process simultaneously softens and slices through the adhesive bond line. Another method consists of inserting a resistance heater on the underside of the substrate to generate additional localized heating at the adhesive interface.

4.6.5 Rework of underfill flip-chip devices and ball-grid-array packages

Reworking underfilled flip-chip devices or BGA parts also presents new challenges since both solder and adhesive are present and both must be removed. New underfill formulations have had to be developed that soften, but do not char at the temperatures required to liquefy the solder. Some equipment manufacturers have modified their flip-chip bonders to include a rework station.[96] In general, the chip undergoing rework is spot heated to melt the solder connections and simultaneously soften the underfill. Machines are equipped with bottom-side heating of the substrate to temperatures of 125–175 °C while the chip is heated from the top side at temperatures of 200 °C and higher to melt the solder. The chip is then gripped and twisted or sheared from the substrate.

References

1. Crawford T. Step-by-step SMT cleaning. *SMT*. Sep. 1997.
2. Woodgate R. No-clean the simple way. *Circuits Assembly*. Apr. 1995.

3. Koschmieder S. Thermal phase: a new technique for stencil/PCB Cleaning. *SMT*. Feb. 1998.
4. Lowe N. Removing paste and adhesive. *Electronic Packaging and Production*. Aug. 1999.
5. Bent WR, Turbini LJ. Evaluating the effect of conformal coatings in reducing the rate of conducive anodic filament formation, *Proc. APEX*. San Diego, CA: Jan. 2002.
6. Standard test method for Kauri-Butanol value of hydrocarbon solvents, ASTM D1133-02, *Book of Standards*, Vol. 6.04.
7. *Encyclopedia for the User of Petroleum Products*. Exxon Co. 1997.
8. Novec® Engineering fluids for electronics cleaning, 3M Brochure, 3M Co. Dec. 2000.
9. Lowe N. Flux removal and aqueous process. *Electronic Packaging and Production*. Oct. 1999 and Vertrel Data Sheets, DuPont.
10. *Private communication with Roger Hesse*. Stoelting. Mar. 4, 2002.
11. *Accel Microcel Product Bulletin 36276*. Speedline Technologies, Cookson Electronics; Apr. 2000.
12. Jackson D, Bixenman M. Critical cleaning of flip-chip packages using dense phase CO_2 Technology, *Proc. NEPCON West Conf.* 2000:640–649.
13. SuperFuge® Liquid and SCF CO_2 cleaning system, Bulletin SF-1000, EPAX Electronics. (Undated).
14. Rae A, Dalton S, Bixenman M. Cleaning with a centrifugal semi-aqueous process. *Electronic Packaging and Production*. Aug. 2001.
15. Mattox DM. Surface preparation for film and coating deposition processes, Handbook of Deposition Technologies for Films and Coatings. In: Bunshah RF, ed. (2nd ed). Noyes Publications; 1994 [Chapter 3].
16. The use of plasma for surface modification, Glen Technologies Application Note GT-110.
17. Plasma Cleaning Systems YES-G500 & YES-G1000, Yield Engineering Systems, www.plasmaclean.com and www.yieldengineering.com; Accessed 08.08.10.
18. Plasma for cleaning/etching. http.www.plasmod.com/ma02001.htm; March Instruments Dec. 19, 2001.
19. Ganesan GS, Lewis R, Berg HM. Characterizing organic contamination in IC package assembly. *Intl. J. Microcircuits and Electronic Packages,* 2nd Quarter 1994;17(2):152-1160.
20. Zhao J, Getty JD. Plasma processing for enhanced underfill. *Chip Scale Review*. Jul. 2004.
21. Uner J, Theriault M, Carsac C, Sindzingre T. The Application of atmosphere pressure plasma to various steps in integrated circuit packaging, Proc. 3rd Annual Semiconductor Packaging Symp SEMICON West; 2000.
22. Detection and measurement of ionizable surface contaminants by resistivity of solvent extract (ROSE), Method 2.3.25. *IPC-TM-650 Test Methods Manual*. Feb. 2001.
23. *High Precision Cleaning News*. Zestron Corp; Dec. 2002.
24. Qualification and performance of electrical insulating compound for printed board assemblies, IPC-9201. *Surface Insulation Resistance Handbook*. IPC; Jul. 1990. IPC-CC-830A(Oct. 1998).
25. Factors affecting insulation resistance performance of printed circuits, IPC-TR-468 Mar. 1979.
26. Standard test method for corona-treated polymer films using water contact angle measurements, ASTM D5946–96.
27. Gombotz WR, Hoffman AS. Functionalization of polymer films by plasma polymerization of allyl alcohol and allylamine. *Proc. ACS Div. Polymeric Materials, Science and Engineering*. 1987;56.

28. Hudis M. *Techniques and applications of plasma chemistry.* Wiley; 1974.
29. Ebnesajjad S. *Surface treatment of materials for adhesion.* Elsevier; 2006.
30. Packham D. *Handbook of adhesion.* 2nd ed. Wiley; 2005.
31. Fisher E, Hinerman JB, Dixon D. SMT Adhesive deposition: The line to success. *SMT.* Dec. 2000.
32. Piracci AF. Practical production applications for jetting technology, *APEX Conf.* Long Beach, CA: Mar. 2000.
33. Gilleo K. Adhesives/epoxies and dispensing. *SMT.* May 1996.
34. McCabe JD. Selecting the right die attach adhesives. *HDI.* Jul. 2001.
35. Breed, SR, Pernice, RF. Stencil printing SMT adhesives: a technology review, www.alphametals.com/products/techarticle 199914002.pdf. Aug. 26, 2002.
36. Chouta P, Fleck I. A new dimension in stencil print optimization. *Cookson Electronics.* 2003.
37. Whitmore M. The State of stencil technology. *SMT.* Mar. 2004.
38. Whitmore M, MacKay C, Hobby A. Plastic stencils for bottom-side chip attach, *Proc. NEPCON West.* Anaheim, CA: 1997.
39. Adhesives Printing in Surface Mount, Dorn SPE Inc. (Undated).
40. Van den Bosch A. Behavior of adhesives before curing. *SMT.* Feb. 1997.
41. Printing of SMT Adhesives, Heraeus Technical Service Bulletin. (Undated).
42. Estes R. A practical approach to die attach adhesive selection. *Hybrid Circuit Technol.* Jun. 1991.
43. *Technical Guide for Chipbonder® Products.* Henkel-Loctite, www.loctite.com/electronics/smt_viscosity.html; Mar. 2002.
44. Zarrow P. Adhesive deposition: pin, syringe, stencil. *Circuit Assembly.* Dec. 1995.
45. Piracci AF. SMA Dispensing trends. *SMT.* Jan. 2001.
46. Kastner M. Adhesive application and curing. *Printed Circuit Assembly.* Dec. 1988.
47. www.loctite.com/electronics/smt_dispensing.html (Jan. 23, 2002).
48. Hannafin JJ, Pernice RF, and Estes RH, Die Attach adhesives: a guide to material properties and applications, Technical Paper GB-48. Epoxy Technology Corp.
49. Private communication with Michael Hodgin. *Epoxy Technology Corp.* Nov. 8, 2001.
50. Lewis A, Babiarz A. Conductive adhesive dispensing considerations, *Proc. NEPCON West.* Anaheim, CA: Feb. 1999.
51. Marinelli C, Burns B. Adhesives/epoxies and dispensing: the building blocks. *SMT Special Supplement.* Jul. 2001.
52. Gossling M. The properties of SMD Adhesives,. *Circuits Assembly.* Feb. 1994.
53. Williams RE, Tirpak RE. Improved performance properties for adhesives. *SMT.* Jul. 1999.
54. Babiarz A, Huysmans F. *Adhesives in dispensing processes for flip chip underfill,* Proc. ISHM Nordic 34th Annual Conf. Oslo: Norway; Sep. 21–24, 1997.
55. Peek R. The Dawn of micro-dispensing. *SMT.* Jul. 2001.
56. Newbold J, Lewis A. The effects of new trends in flip chip packages on automating the underfill process, *Proc. NEPCON West.* Anaheim, CA: Mar. 1998.
57. Adamson SJ. CSP and flip-chip underfill. *Advanced Packaging.* Jun. 2001.
58. Lim R, Duck A. Controlling process parameters for flip-chip underfill. *Electronics Engineer.* Jul. 2000.
59. Shi DM, McKeen K, Jenson B. Advances in flip-chip material cure rates and reliability, IMAPS, Indiana Chapter Vendors' Day (Apr. 2000); *Proc. 4th Intl. Symp. and Exhibition, Advanced Packaging Matls., Processes, Properties, and Interfaces.* Braselton, GA: IMAPS; Mar. 1998.

60. Private correspondence with M. Nolan, Aptek Labs. Jan. 9, 2002.
61. Kondo M. Dispensible adhesive benchmarking, *Proc. NEPCON West Conf.* Anaheim, CA: 2000.
62. Piracci AF. *Advantages of non-contact dispensing in SMT assembly processes.* Chicago, IL: Surface Mount Technology Assoc./ATE; Sep. 2000.
63. Adamson SJ. Jet dispensing - a paradigm shift in technology. *MEPTEC Report.* Jul. 2003.
64. *ASL-90-2028, Rev. C.* Asymtek publication; Sep. 2002.
65. Private communication with Alan Lewis. *Asymtek.* Jun. 2, 2002.
66. Loctite® 3621®, Henkel technical data sheet (Nov. 2001); Epo-Tek® H70E, Revision 2, Epoxy Technology Data Sheet. Mar. 1998.
67. Zarrow P. Adhesive deposition: pin, syringe, stencil. *Circuits Assembly.* Dec. 1995.
68. Piracci AF. Ultra high-speed dispensing of surface mount materials to enhance production throughput, *Proc. APEX Conf.* Long Beach, CA: Mar. 2000.
69. Product Data Sheets for: ESP 7685-HF, TP 7150, TP 7090, ESP 8680-HF, TP 8153, and TP 8093 AI Technologies.
70. Wischoffer S. Step 6 - component placement. *SMT.* Jun. 2001.
71. Hunadi R, Bilow N, Noack L. Determining corrosion tendencies of die-attach adhesives using the sealed glass tube test. *Hybrid Circuit Technol.* Apr. 1988.
72. Benson RC, Romanesko BM, Hall BN, DeHaas N, Charles HK. Current leakage failures in hybrid microcircuits. *EEE Trans. On Components, Hybrids, and Mfg. Tech* Dec. 1986;4: 403–409. CHMT-9.
73. Ablebond® 8509 (Feb. 2001); 84–3 MVB (Nov. 2000); 2025D, 2025, 2000 (Oct. 2000); 8355F (Nov. 2000); 8360 (Nov. 2000); 8510AA(Oct. 2000); 8515 (Nov. 2000); 2100A, 826-IDS (Oct. 2000); 8290 (Jul. 2001); 8322A (Jul. 2001); 8361 (Jul. 2001), Technical Data Sheets, Ablestik Laboratories.
74. Dymax® 9-20365,9-20366, and 9–20367, Dymax Technical Data Sheets.
75. Schick S. Ultraviolet curing: a new path to productivity, *Proc. Intl. Soc. Hybrid Microelectronics.* 1986.
76. Schick S. Portable ultraviolet system provides high intensity cures. *Adhesives Age.* Apr. 1986.
77. Fathi Z, et al. Innovative curing of high reliability advanced polymeric encapsulants, *Proc. NEPCON West.* 1998.
78. Zou Y, Harris J, Ahmad I, Tucker D, Fathi Z. Comparison of die level stresses in chip-on-board packages processed with conventional and VFM encapsulant curing, *Proc. High Density Packaging and MCM Conf.* Denver, CO: Apr. 1999.
79. Omar MA, Nara KR, Obieglo A, Belk N, Schalausky R. Analysis of non-load bearing two-component epoxy adhesive to hemming process variations; thermogravimetric, differential scanning calorimetry, and FT-IR analyses. *Progress in Organic Coatings* April 2009;Vol. 65(Issue 1):104-108.
80. Lee H, Neville K. *Handbook of epoxy resins.* McGraw-Hill; 1967.
81. Scherzer T, Tauber A, Mehnert R. U.V. curing of pressure-sensitive adhesives studied by real-time FTIR-ATR spectroscopy. *Vibrational Spectroscopy.* July 2002;Vol. 29(Issues 1-2):125-131.
82. Scherzer T, Decker U. Real time FTIR-ATR spectroscopy to study the kinetics of ultrafast photopolymerization reactions induced by monochromatic UV light. *Vibrational Spectroscopy.* April 1999;Vol. 19(Issue 2):385-398.
83. ISO 11357, Plastics-Differential Scanning Calorimetry (DSC) (1997); IPC TM-650 2.4.25C, Glass Transition Temperature and Cure Factor by DSC. Dec. 1994.

84. Hale A, Bair HE, Macosko CW. "The variation of glass transition as a function of the degree of cure in an epoxy-novolac system," *Proc. SPE ANTEC*, 1987; 1116.
85. Chen Y, Chew F, Goodwin C. Determination of curing schedule for advanced packaging materials. *Intl. J. Microcircuits and Electronic Packaging.* 4th Quarter 1997;20(4):540-544.
86. Sichina, WJ, "Prediction of epoxy cure properties using Pyris DSC Scanning Kinetics Software," Perkin Elmer Instruments, Application Note, Thermal Analysis, PETech-48; 2000.
87. (www.adhesivestoolkit.com).
88. Sichina WJ. *Characterization of electronic materials using thermal analysis, Perkin Elmer Instruments.* Thermal Analysis Application Note PETech-24; 2000.
89. Zang Q, Song B, Wang S, Xu L, Liu S. Cure kinetics analysis and simulation of silicone adhesives. *Electronic Packaging Technology and High Density Packaging Intl. Conf.* Aug. 2009: 692–694.
90. Naito CM, Todd M. *Effects of processing parameters on the performance of epoxy encapsulant materials.* Dexter Technical Paper; Dec. 2000.
91. Apparatus for Removing Semiconductor devices from high-density multichip modules, U.S. Patent 5164037. Nov. 17, 1992.
92. Iwami C, Licari JJ, Nakagawa C. New Adhesive requirements for multichip modules, *Proc. 5th Intl. SAMPE Electronic Materials and Processes Conf.* Los Angeles Jun: 1991.
93. Harper PR. Thermoplastic die attach for hermetic packaging, Intl. *J. Microelectronics and Electronic Packaging.* 4th Quarter 1994;17(4).
94. Shores AA. Adhesive bonding hybrid microcircuit substrates with a thermoplastic film. *SAMPE Quarterly.* Apr. 1988;19.
95. Shores AA. Adhesive bonding hybrid microcircuit substrates with PEEK® Thermoplastic Film. *Intl. J. of Hybrid Microelectronics.* 2nd Quarter 1988;11(2).
96. Moore D, Studley G. Reworking underfilled flip-chips, Semiconductor Equipment Corp. Moorpark, CA, www.flipchips.com/tutorial08.html. May 2001.

5 Applications

Chapter Outline

Adhesives Technology for Electronic Applications. DOI: 10.1016/B978-1-4377-7889-2.10005-1

Polymer adhesives have found their place in numerous electronics' applications. Major uses include commercial/consumer products; computers; and military, space, automotive, medical, and wireless communications. Some adhesives may be used across several applications while others have been formulated to meet application-specific requirements. For example, reworkability is not a consideration for high-production, low-cost consumer products such as cell phones or calculators, but is important for high-value, high-density printed-wiring boards (PWBs) and multichip modules (MCM). Further, thermal stability at high temperatures is required for near-engine electronics in automobiles, aircraft, and for deep-well sensors, but not for office computers. The major applications for polymer adhesives are to attach and electrically insulate or to electrically connect components, devices, connectors, cables, and heat sinks to printed-circuit boards or to thin- or thick-film hybrid microcircuits. In addition, over the last several decades, new uses for adhesives have emerged for optoelectronic (OE) assemblies, microelectromechanical systems (MEMS), light-emitting diodes (LEDs), and flat-panel displays.

This chapter is structured into two segments: general applications, such as die attachment, surface mounting, and underfills that apply to many products, and specific applications, such as medical, automotive, military, space, optoelectronics, and flat-panel displays.

5.1 General applications

Some adhesive materials and processes are used across many applications. For example, adhesives are used to attach bare die, components, and substrates in assembling commercial, consumer and aerospace electronic products. Adhesives are also widely used for surface mounting components onto interconnect substrates that serve numerous functions for both low-end consumer products and for high reliability applications. Underfill adhesives are used to provide stress relief and ruggedize the solder interconnects for almost all flip-chip and area-array devices, regardless of their function as integrated circuits.

5.1.1 Surface-mount technology

Surface-mount technology (SMT) has emerged as the major assembly process for high-speed high-rate production of printed-circuit boards permitting higher-density circuits than plated-through-hole single-sided or double-sided types. Components are interconnected using solder, but are held in place with electrically insulative adhesives called *surface-mount adhesives* (SMAs) until the solder is reflowed. SMAs are also used to hold components in place for circuits that are to be wave soldered. Thus, electrically insulative SMAs play an auxiliary role in SMT solder attachment. However, electrically conductive adhesives have been investigated and used as SMAs for the primary function of replacing solder.

In the general process flow for SMT, printed-circuit boards are cleaned and dried, solder paste is stenciled or screen printed onto pad locations to which components are to be connected, adhesive is dispensed between the solder pads, and components are placed onto the wet adhesive paste. At this point, the adhesive may be fully cured, snap cured, or partially cured. In all cases, the cure should be sufficient to hold the components in place during the final solder reflow or wave soldering steps (Figure 5.1). Adhesives may also be cured during the solder-reflow step so that solder reflowing and adhesive curing occur simultaneously, thus reducing the number of process steps (Figure 5.2). In high production, these steps are performed using in-line modular equipment that is computer controlled (Figure 5.3).

After the components have been placed onto the wet adhesive, the adhesive must have sufficient "green strength" (wet strength) to hold the components in place until they arrive at the solder-reflow stage. *Green strength* is the resistance to shifting of the components that can occur during the rapid acceleration and deceleration that results from moving the boards during automated assembly. Accelerations and decelerations ranging from 5 m/sec^2 to 20 m/sec^2 result from starts and stops in automated handling equipment.[3] Thus, high green strength is necessary to maintain component orientation prior to soldering.

Green strength may be determined by measuring the lateral force that bonded parts can withstand. In one test method (Siemens Standard SN 59651),[4] a populated board is allowed to slide down an incline to a stop at the base, and the displacement of components on the board is measured (Figure 5.4). An acceleration run of 400 mm from a height of 35 mm is made. A maximum shift of 0.15 mm (50% of the track pitch for a standard 0.3 mm pitch part) is considered acceptable.[5]

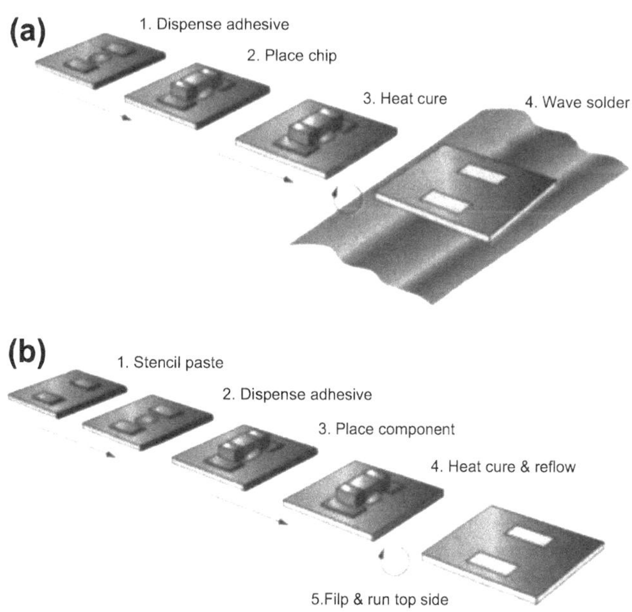

Figure 5.1 Representation of SMT for (*a*) wave soldering and (*b*) reflow soldering.[1]

Step #1 → Step #2 → Step #3 → Step #4 → Step #5

Step #1: Print or dispense solder paste.
Step #2: Apply adhesive.
Step #3: Place components.
Step #4: Cure adhesive* (Thermal and/or UV).
Step #5: Manually assemble/insert odd-form components.

Step #6 → Step #7 → Step #8 → Step #9 → Step #10

Step #6: Solder reflow.
Step #7: Inspect/test.
Step #8: Rework.
Step #9: Clean.
Step #10: Second pass assembly. (Repeat steps 1-8 as required.)

Step #11 → Step #12

Step #11: Clean.
Step #12: Final inspection/test and closeout.

* Optional and not required if adhesive cures in preheat zones of a reflow oven.

Figure 5.2 SMT process steps.

In another test, a 40-mil diameter, 10-mil thick dot is dispensed onto a test plate. A rod is lowered into the uncured adhesive and withdrawn while a load cell is used to measure the green strength in gm/cm[2,6] Green shear strengths may also be determined by mounting passive components, such as 0603*(0603, 0805, and 1206 are designations for passive components such as capacitors and resistors based on size, thus 1206 is a component that is approximately 120 × 60 mils.), 0805, and 1206 onto printed-circuit boards with uncured adhesives, subjecting the boards to the full force of a high-speed, chip-shooter table motion,[6] and measuring component displacement.

Green strength may be affected by many variables. In general, higher viscosity adhesives display greater green strength. Other factors such as dot size and profile, number of dots, and the nature of the adherend surfaces (passivation, solder maskant) also influence the results.[7,8] Moisture absorption of the adhesive prior to cure is also critical to adhesive performance. High moisture absorption shortens pot life, reduces wettability, and affects cured properties such as the T_g. The ideal characteristics for a surface-mount adhesive may be summarized as follows:

- Excellent green (wet) strength
- High viscosity for dispensing dots having high aspect ratios
- High shear thinning and flow under pressure with quick recovery and limited flow after component placement (good thixotropic properties)

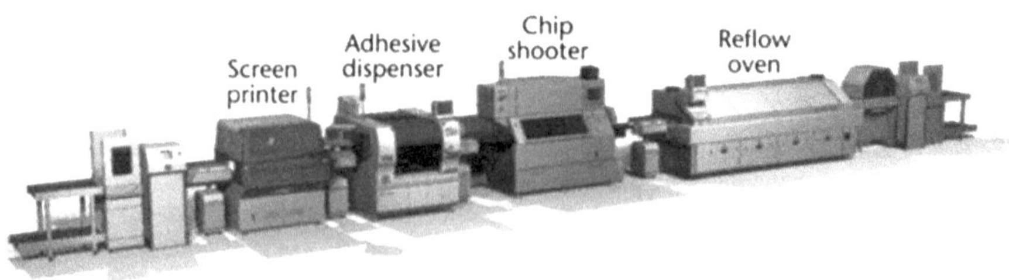

Figure 5.3 In-line modular SMT equipment.[2]

* 0603, 0805, and 1206 are designations for passive components such as capacitors and resistors based on size, thus 1206 is a component that is approximately 120 × 60 mils.

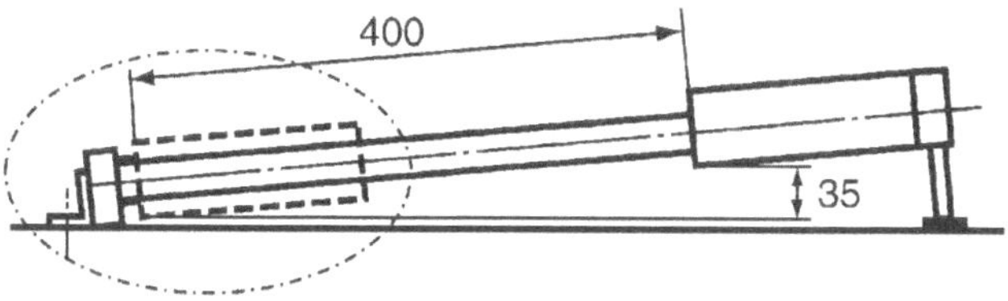

Figure 5.4 Test fixture for measuring green strength.
Source: Siemens Standard Specification SN 59651.

- Easy release from the dispensing tip with no tailing or stringing
- Ability to retain print or dot height
- Low moisture absorption
- Noncorrosive
- High strength after cure
- Low T_g (80°C–100 °C).

In surface mounting of components to plastic-laminate PWBs, curing temperatures should be chosen that do not degrade the board material, electronic components, or other parts already assembled. For epoxy-laminate PWBs, it is best to keep curing temperatures at or below 150 °C for as short a time as possible, commensurate with achieving optimum properties. Curing temperatures higher than 150 °C may be used on some high-grade epoxy boards, on polyimide boards, and, of course, on ceramic substrates. Lower temperatures are necessary for polyester and other low-cost consumer-grade laminates. Thus, in curing epoxy surface-mount adhesives, schedules of 120°C–150 °C for approximately 1–3 minutes are used, depending on the adhesive. Table 5.1 lists some specific SMT adhesives, their recommended curing schedules, and other properties. In this table, the yield value refers to the minimum force required to start an adhesive moving. Yield values are helpful in setting the equipment parameters for screen printing or stencil printing.

After the solder has been reflowed and the adhesive has been cured, the adhesive has served its primary function of holding the components in position during solder reflow. However, since the adhesive remains permanently on the board, its cured properties must be such that it will not reduce the reliability of the assembly or cause subsequent failures.

The properties of cured surface-mount adhesives that are critical to their performance include moduli of elasticity (MOE), coefficients of thermal expansion, glass-transition temperatures, and thermal conductivities. Modulus is especially critical for large components and for adherends having dissimilar expansion coefficients. High-modulus adhesives do not readily absorb stresses during thermal cycling. Expansion coefficients for surface-mount adhesives range from 45–100 ppm/°C below the T_g and 150–190 ppm/°C above the T_g. Values of T_g range from 70°C–150 °C, but some are very low and even negative, as in the case of some elastomeric polyurethanes and silicones. Low T_g values are better for stress relief and for rework. Generally, surface-mount adhesives should have CTEs below 50 ppm/°C, to avoid cracking of solder joints during thermal excursions.

Table 5.1 Properties of surface-mount paste adhesives

Product Name/ Supplier	Type	Viscosity[a] cP (rpm)	Thixotropic Index[b]	Yield value (N/m^2, Casson)	Shelf life −40 °C/Pot Life[c], (25 °C)	Cure Schedule minute @ °C	Lap-Shear Strength (psi)	CTE (ppm/°C) α_1	CTE (ppm/°C) α_2	T_g (°C)	Volume Resistivity (ohm-cm)	Thermal Conductivity (W/m·K)
EpiBond® 7275/ Cookson Electronics	Calcium carbonate filled epoxy resin	420,000 (10)	>6	N/A	9 mos (0 °C)/6 mos (20 °C)	12 @ 80 °C	>2,000	48	170	66	2.1×10^{15}	0.027
Eccobond™ D124F / Henkel	Low-temperature curing epoxy	N/A	N/A	N/A	3 mos (0–8 °C)/2 mos (25 °C)	3.5 @ 120 °C	870	55–60	N/A	85	1.0×10^{14}	0.30
Eccobond™ D125F / Henkel	Low-temperature curing epoxy	300,000–400,000 (10 rpm)	5	N/A	6 mos (0–8 °C)/2 mos (18–25 °C)	2.5 @ 120 °C (IR or convection conveyor oven)	1,160	55–60	N/A	85	1.0×10^{14}	0.30
Prima-Bond EG7655/AI Technology	Flexible, alumina-filled epoxy	300,000 (0.5 rpm)	N/A	N/A	1 year (25 °C)/ 8 hours	30 minutes @150	1,000	N/A	140	−20	$>1.0 \times 10^{14}$	1.7
Prima-Bond EG7159/AI Technology	Flexible, diamond-filled epoxy	310,000 (0.5 rpm)	N/A	N/A	1 year (25 °C)/ 8 hours	30 minutes @150	1,000	N/A	120	−25	$>1.0 \times 10^{14}$	11.4
Prima-Bond EG7658/AI Technology	Flexible, aluminum-nitride filled epoxy	305,000 (10 rpm)	N/A	N/A	N/A	30 minutes @150	1,000	N/A	120	−25	$>1.0 \times 10^{14}$	3.6
Epo-Tek® H70E/Epoxy Technology	Thermally conductive epoxy	4,000–7,000 (50 rpm)	1.17	N/A	N/A	5 @ 150, 1 @ 175	>2,000	15	64	≥80	$>1.0 \times 10^{13}$	0.9
Epo-Tek® H74/ Epoxy Technology	Thermally conductive epoxy with color cure indicator	45,000–65,000 (5 rpm)	2.14	N/A	N/A	5 @ 150, 20 @ 100	1,656	21	95	≥100	$>4.0 \times 10^{12}$	1.25

PD955 M/ Heraeus	Thermoset, solvent-free adhesive	10,000–40,000[e] (none listed)	N/A	N/A	6 mos (5–12 °C)/ >8 hours	5 @ 125	N/A	N/A	N/A	N/A	3.4×10^{11}	N/A
PD955PY/ Heraeus	Thermoset, solvent-free, yellow adhesive	50,000–150,000[e]	N/A	N/A	N/A	1.5 @ 150	N/A	N/A	N/A	N/A	1.5×10^{12}	N/A
Loctite®3609/ Henkel	High-speed dispensing epoxy	160–2,000	N/A	250–700	N/A	1.5 @ 150	1,450	N/Q	145	73	2.0×10^{15}	0.4
Loctite® 3616/ Henkel	Fast-curing epoxy	15,000–55,000[d] (None specified)	N/A	250–650	N/A	30 minutes @ 150	≥ 2,175 (steel)	86	N/A	140	4.6×10^{15}	0.2
Loctite®3619/ Henkel	Fast-curing epoxy	N/A	N/A	200–450	N/A	30 minutes @ 150	N/A	60	120	50	1.2×10^{15}	N/A
Loctite®3627/ Henkel	One-part, red, fast-curing epoxy	N/A	N/A	100–275	10 mos/1mo	2 @ 150, 3–5 @ 125	N/A	75	111	105	6.2×10^{16}	0.3
Arathane7760/ Specialty Polymers and Services, Inc.	Polyurethane with thermally conductive filler	240,000–700,000 (None specified)	N/A	N/A	3 mos/>2 hours	30 @ 100	200	29	88	−63	2.9×10^{14}	0.69

N/A – Not Available.
[a] Brookfield 10 rpm unless indicated.
[b] 1:10 rpm unless indicated.
[c] Time required to double viscosity.
[d] Casson, 0.4–30 s^{-1} unless indicated.
[e] Haak Viscometer (30 s^{-1}).
Source: Compiled from suppliers' technical data sheets. Contact suppliers for specification values.

The electrical properties of SMAs are quite good. Volume resistivities are typically above 1×10^{14} ohm-cm at room temperature in the dry state. Although volume resistivities for epoxies usually drop to 10^{6} ohm-cm under moist ambients, this is not considered serious for most applications, especially if the resistivities recover on drying.

5.1.2 Die and substrate attachment

Besides their use in SMT, other major applications for adhesives are attachment of bare die in single-chip packages, in hybrid microcircuits and in multichip modules, and attachment of substrates in packages. The requirements and properties for die- and substrate-attach adhesives have already been treated in previous chapters. Thousands of part types and circuit functions have been produced by the convenient and low-cost assembly processes of attaching chip semiconductor devices and passive chip devices and interconnecting them to high-density circuit substrates. Some applications in which die- and substrate-attach adhesives have been used are shown in Figures 5.5, 5.6, and 5.7. Examples and properties for some commercially available die- and substrate-attach adhesives, paste and film, are given in Table 5.2.

5.1.3 Underfilling

Underfilling of flip-chip devices has become a widely used process to assure that solder connections do not crack or open during thermal excursions. The use of underfill adhesives is especially necessary for attaching flip-chip devices to plastic-laminate circuit boards because of the greater CTE mismatch between solder and plastic compared with solder and ceramic boards.[9,10] As a rule, to assure reliability, almost all flip-chip attached devices are now underfilled. There are three main underfilling approaches, depending on the type of adhesive used. Adhesive types include those based on capillary flow, no-flow fluxing, and preforms. Other approaches are under development to reduce the number of steps, increase throughput, reduce cost and even obviate the use of underfill entirely.

5.1.3.1 Capillary-flow underfill adhesives

It is common practice today to underfill all flip-chip devices after they have been attached and the solder has been reflowed. Capillary-flow underfilling is the method most widely used (Figure 5.8) (see also Chapter 4). Besides flip-chip devices, capillary-flow adhesives are also used for chip-scale and BGA packages, but the solder balls of these packages are often larger than for flip-chip devices, hence solder fatigue is generally less of an issue. Although advances have been made in automated dispensing (needle and jetting), and in adhesive formulations, the additional process steps and cost of underfilling are issues in high-volume production. Thus, alternate methods are continuously being investigated to reduce process steps and cost.

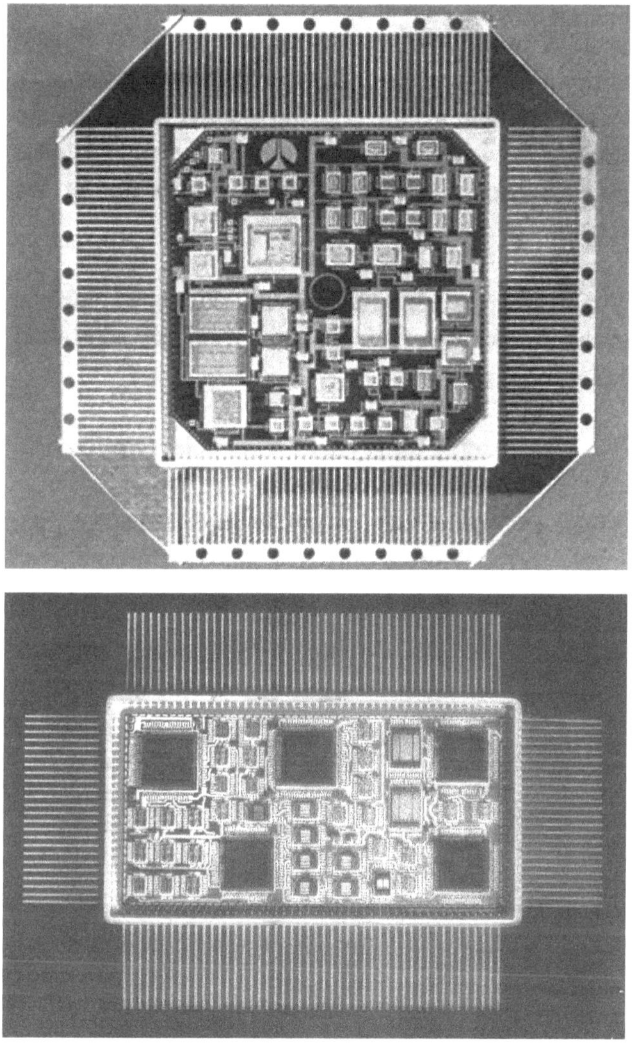

Figure 5.5 Examples of multichip modules in which die-attachment adhesives were used. Top: Aluminum/polyimide interconnect (MCM-D) on silicon substrate in Kovar package. Bottom: Multilayer thick-film (MCM-C) digital filter.

In the application of capillary-flow underfills, there are so many material, process, and equipment variables that empirical methods are widely used to establish the optimum processing conditions. The viscosity and flow properties of underfill adhesives are among the most important variables in rapidly filling different gap sizes and devices. A list of some commercially available underfill adhesives and their flow properties, as related to filler size and gap height, is given in Table 5.3.

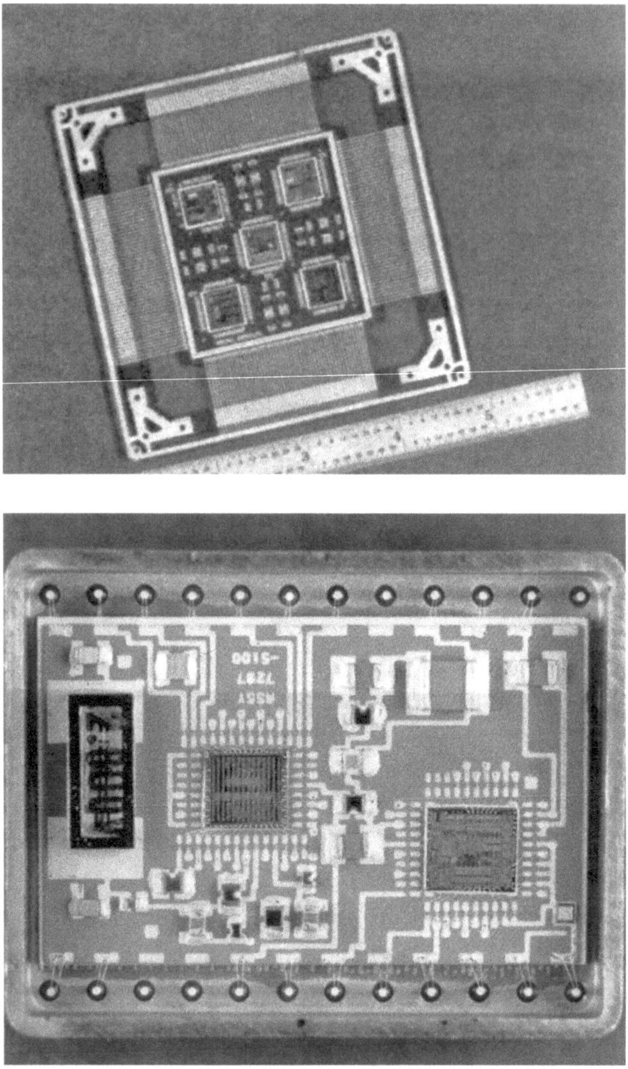

Figure 5.6 Example of a ceramic multichip module (MCM-C) and a hybrid microcircuit. Top: MCM-C fabricated from high-temperature cofired ceramic. *Source:* Honeywell Solid State Electronics Center. Bottom: Multilayer thick-film hybrid microcircuit.
Source: R. Wayne Johnson, Auburn Univ.

5.1.3.2 No-flow underfill adhesives

In addition to capillary-flow underfilling, a new approach to underfilling has been developed whereby the underfill material is applied before attaching the flip-chip or BGA device. Specially formulated adhesives, known as *no-flow underfills* (NFU) or

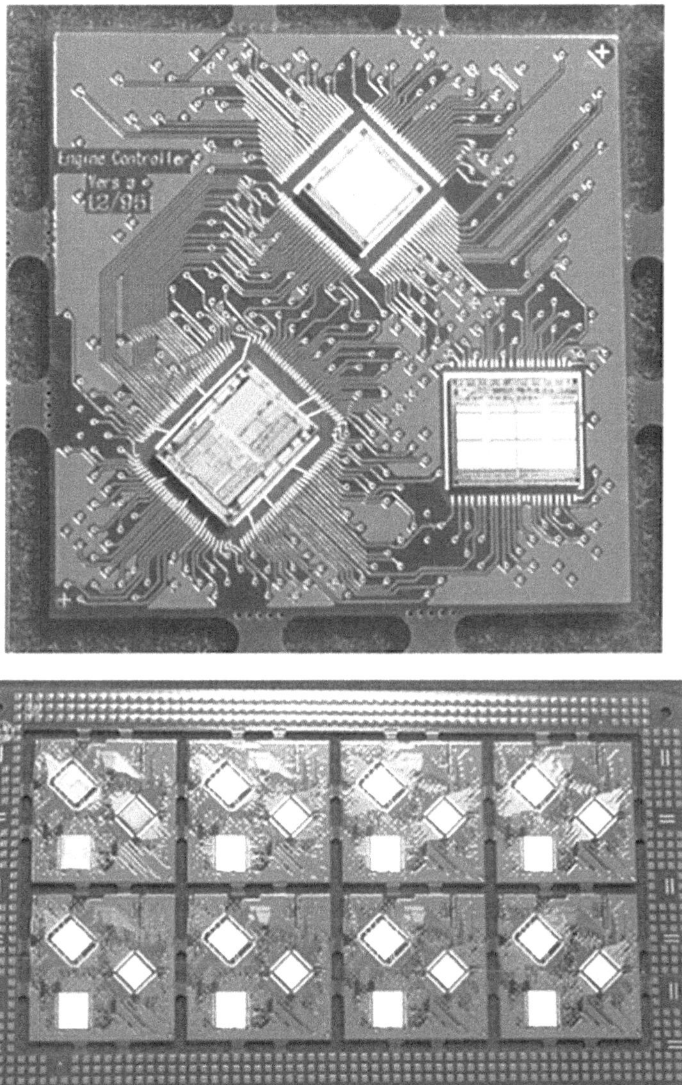

Figure 5.7 Examples of laminate multichip modules (MCM-L). Top: Engine controller module. *Source:* R. Wayne Johnson, Auburn Univ. Bottom: Engine controller, array configuration. *Source:* R. Wayne Johnson, Auburn Univ.

Table 5.2 Properties of some die- and substrate-attach adhesives

Product Name/Supplier	Type	Viscosity (cP)[a]	Thixotropic Index[b]	Shelf Life (−40 °C)/Pot Life (25 °C)	Cure Schedule	% Wt Loss @ 300 °C (TGA)	Bond Strength (psi)	MOE (MPa/psi)[c]	CTE (ppm/°C)		T_g (°C)	Volume Resistivity (ohm-cm)	Thermal Conductivity (W/m·K)	Ionic Impurities (ppm)
Ablebond® 84-1 LMI SR4/ Henkel	Silver-filled epoxy paste	8,000 (5 rpm)	5.6 (0.5/5)	1 yr/14 days	1 hr @ 175 °C	N/A	1,500	3,931/570,000	40	150	120	1×10^{-4}	2.5	Cl^- ≤20, Na^+ ≤10, K^+ ≤10
Ablebond® 84-1 LMI T1/ Henkel	Silver-filled, high thermal conductivity epoxy paste	22,000 (5 rpm)	N/A	1 yr/14 days	1 hr @ 150 °C	0.16	1,885	7.6/1,100	50	200	103	5×10^{-4}	3.6	Cl^- ≤200, Na^+ ≤50, K^+ ≤50
Ablebond® 85-1/Henkel	Gold-filled epoxy paste	N/A	N/A	1 yr/2 days	1 hour @ 150 °C	0.22	≥1,800	N/A	50	200	107	1.0×10^{-3}	2.07	Na^+ 10, K^+ 5, Cl^- 115
Ablebond® 8175/Henkel	Silver-filled, paste adhesive with high thermal conductivity	12,000 (5 rpm)	5.0 (0.5/5)	1 yr/24 hours	4 min @ 130 °C	N/A	N/A	N/A	N/A	N/A	N/A	1.0×10^{-4}	3.0	Cl^- 10, Na^+ 5, K^+ 5
Ablebond® 84-3 Henkel	Insulative epoxy paste	50,000 (5 rpm)	-	1 yr/2 weeks	1 hr @ 150 °C	0.17	2,700	N/A	40	100	85	3.5×10^{13}	0.8	Cl^- 6, Na^+ 2, K^+ 11
Ablefilm ® 5020 K/ Henkel	White epoxy film, highly thermally conductive	N/A	N/A	1 yr/6 mos	1 hr @ 150 °C	0.3	3,000	N/A	45	270	109	8×10^{14}	0.7	Cl^- 60, Na^+ 20, K^+1
Ablefilm® 566 Kapton™/ Henkel	Flexible epoxy film, polyimide carrier	N/A	N/A	1 yr/2 days	3 hours @ 90 °C	N/A	2,300	N/A	N/A	N/A	100	N/A	0.2	N/A
Ablefilm® ECF 568™/Henkel	Silver filled epoxy film (glass fabric carrier)	N/A	N/A	1 yr/24 hours	2 hours @ 95 °C	N/A	5,100	N/A	65	400	113	2.0×10^{-3}	N/A	N/A
Emerson Cuming™ CA 3150/ Henkel	Silver filled, low viscosity paste	13,000	6.0	2 mos (−20 °C)/2 days	10 s @ 130 °C (at bond line)	N/A	N/A	45.5 /6,600	N/A	N/A	N/A	N/A	N/A	N/A

QMI 516™/ Henkel	Hydrophobic, silver filled epoxy	8,000 (5 rpm)	3.0	N/A/24 hours	15 min @ 150 °C	N/A	N/A	1,800/261,000	51	112	−33	$\geq 1.7 \times 10^{-3}$	N/A	Na^+, $K^+ < 20$ CL^-, $F^- < 20$
ME 8155/AI Technology	Silver-filled epoxy paste	354,000 (5 rpm)	N/A	1 yr/N/A	60 minutes @ 150 °C	N/A	1,000	N/A	N/A	120	−20	$<4.0 \times 10^{-4}$	7.9	N/A
ME 8863/AI Technology	Silver-filled cyanate ester paste	30,000 (5.0 rpm)	N/A	1 yr/N/A	30 minutes @ 300 °C	N/A	1,000	N/A	30	N/A	240	$<1.0 \times 10^{-3}$	1.0	N/A
EG 8020/AI Technology	Silver-filled epoxy paste	150,000 (0.5 pm)	N/A	1 year (25 °C)/ N/A	2 hours @ 125 °C	N/A	>900	N/A	40	N/A	50	$<2.0 \times 10^{-4}$	5.7	N/A
Staystik® 211/ Cookson	Aluminum nitride filled thermoplastic epoxy	N/A	N/A	N/A	300°C–375 °C, 0–10 psi	N/A	>3,500	2,561/371,442	N/A	N/A	180–185	$>1.0 \times 10^{9}$	3.0–3.5	N/A
Staystik® 181/ Cookson	Silver-filled thermoplastic epoxy	N/A	N/A	N/A	160°C–275 °C, (0–10 psi)	N/A	>2,500	2,561/371,442	N/A	N/A	180–85	$<5.0 \times 10^{-4}$	3.0–3.5	N/A
Staystik® 611/ Cookson	Insulative film adhesive	N/A	N/A	N/A	3–5 minutes @ 300°C–375 °C (0–10 psi)	N/A	>3,000	2,484/360,274	N/A	N/A	180–185	>1 × 109	3.0–3.5	N/A
Staystik® 171/ Cookson	Silver-filled, low-stress paste	N/A	N/A	N/A	125°C–200 °C, (0–10 psi)	N/A	>1,800	427/61,931	N/A	N/A	25	$\leq 5.0 \times 10^{-4}$	3.0–3.5	N/A
Epo-Tek® H20E/Epoxy Technology	Silver-filled epoxy paste	2,200–3,200 (100 rpm)	4.63	1 yr/36 hours	5 minutes @ 150°	1.67	1,475	5,577/808,700	31	158	≥80	4.0×10^{-4}	2.5	$Na^+ < 2$ $K^+ < 3$ $Cl^- < 73$ $NH_4^+ \leq 98$
Epo-Tek® H35-175MP/ Epoxy Technology	Silver-filled epoxy paste	22,000–28,000 (100 rpm)	4.0	1 yr/28 days	1 hr @ 180 °C	0.28	>2,000	7,631/1,106,623	31	97	≥100	$<5.0 \times 10^{-4}$	1.46	$Na^+ < 50$ $K^+ < 50$, $Cl^- < 200$ $NH_4^+ \leq 39$
EN-4900F-1/ Hitachi	Silver-filled acrylate paste	100,000 (0.5)	5.1 (0.5/5)	N/A	60 minutes @ 150 °C	N/A	2,124	350/50,763	N/A	N/A	19	4.0×10^{-3}	2.0	N/A

(*Continued*)

Table 5.2 Properties of some die- and substrate-attach adhesives—*cont'd*

Product Name/Supplier	Type	Viscosity (cP)[a]	Thixotropic Index[b]	Shelf Life (−40 °C)/Pot Life (25 °C)	Cure Schedule	% Wt Loss @ 300 °C (TGA)	Bond Strength (psi)	MOE (MPa/psi)[c]	CTE (ppm/°C)		T_g (°C)	Volume Resistivity (ohm-cm)	Thermal Conductivity (W/m·K)	Ionic Impurities (ppm)
MD-140/ Thermoset, Lord Chem. Products	Silver-filled epoxy paste	30,000	N/A	1 yr/72 hours	150 °C and 180 °C	N/A	7,000	3,300/478,500	60	190	82	9.0×10^{-5}	12	Na^+ <5, K^+ < 1, Cl^- <50
MD-150/ Thermoset, Lord Chem. Products	Silver-filled epoxy paste	10,000	N/A	6 mos/18 hours	60 min @175 °C	N/A	N/A	5,000/725,000	54	160	93	8.0×10^{-5}	N/A	Na^+ <50, K^+ < 10, Cl^- <50
MD-161/ Thermoset, Lord Chem. Products	Silver-filled epoxy paste	40,000	N/A	6 mos (−30 °C)/ 5 days	Fast cure 150 °C	N/A	N/A	5,000/725,000	50	180	55	1.0×10^{-3}	N/A	N/A
DM4030LD F890/Diemat	Silver-filled, low-temperature processing	30,000 (10 rpm)	3.3	1 year/6 mos	150–175 °C	<0.15 (200 °C)	1,000	N/A	28	N/A	N/A	4.2×10^{-5}	15	<10 ppm total Cl^-, F^-K^+, and Na^+
DM6030Hk/ Diemat	Silver-filled, low-temperature processing	31,000 (10 rpm)	2.4	1 year/6 mos	45 minutes at 175 °C	N/A	N/A	N/A	26	N/A	N/A	8.0×10^{-5}	60	<30 ppm total Cl^-, F^-, K^+, and Na^+
DM6030SF/ Diemat	Silver-loaded (85%) epoxy for high electrical and thermal conductivity	173,000 (10 rpm)	2.8	1 year/6 mos	45 minutes at 175 °C	N/A	N/A	N/A	35	N/A	N/A	2.8×10^{-5}	12	<30 ppm total Cl^-, F^-, K^+, and Na^+

N/A – not available; ND – none detected.
[a] 10 rpm and 25 °C unless indicated.
[b] 10:1 unless indicated.
[c] Generally at 25 °C
Source: Suppliers' technical data sheets. Contact suppliers for specification values.

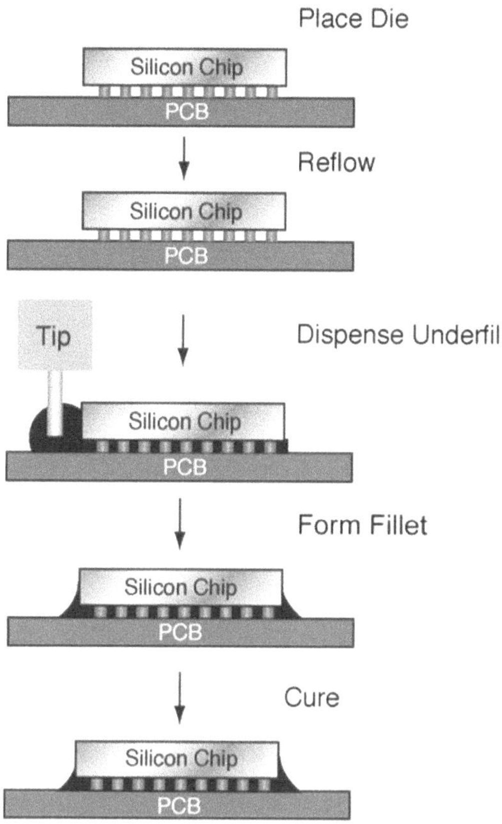

Figure 5.8 Capillary-flow underfill process steps.[10]

no-flow fluxing underfills (NFFUF) do not require the tedious and time-consuming step of having to be dispensed precisely around the sides of a device and having to flow under an already solder-attached device. NFUs are basically polymer paste adhesives such as epoxies or modified epoxies that contain a fluxing ingredient that assists in forming the solder connections during reflow. Whereas capillary-flow underfills are based on conventional bisphenol-A epoxies cured with amines or cyanate esters, NFUs consist of epoxy resins cured with anhydrides and may contain carboxylic esters. Also, unlike capillary-flow underfills, some of which are highly filled (70% by weight to reduce CTEs), NFUs are unfilled since filler particles interfere or prevent electrical contact and have an approximate CTE of 70 ppm/°C.[11,12]

In processing, the NFU adhesive is squeezed out around the solder balls during placement, allowing the solder balls to directly contact the PWB bonding pads and make good electrical connections during solder reflowing (Figure 5.9). A key advantage of NFUs is the simultaneous reflowing of the solder and curing of the

Table 5.3 Flow Properties of some underfill adhesives

Product/Supplier	Description	Filler content (%)	Max. Filler Size (μm)	Min. Gap, (μm)	Viscosity[a] (cP)	Capillary Flow: Distance, seconds (100 °C, glass to glass, 25 μm gap)
Hysol® 3500™/Henkel	Low viscosity, high flow	N/A	N/A	N/A	203	N/A
Loctite® 3513™/Henkel	Reworkable	N/A	N/A	N/A	3,000–6,000	N/A
Loctite® 3536™/Henkel	Reworkable	N/A	N/A	N/A	1,800	N/A
Loctite® 3563™/Henkel	Non-reworkable	N/A	N/A	25	5,000–12,000	35 mm, ≤25 7 mm, ≤45 4 mm, ≤150
Loctite® 3593™/Henkel	Non-reworkable	N/A	N/A	25	3,500–6,000	6.35 mm, ≤10 12.7 mm, ≤30 4 mm, ≤70
Ablefill™ UF8807™/Henkel[a]	Cyanate ester	N/A	≤50	N/A	13,000	N/A
Emerson & Cuming™ E1172A/Henkel	Epoxy	65–68	20	25	3,500–6,000	1 cm, 25 (25 and 50 μm gaps), 16 s (100 μm gap)
Emerson & Cuming™ E1216™/Henkel	Epoxy snap curable	N/A	N/A	N/A	6,000	1 cm, 9 (180 μm gap)
Staychip™ 3082/Cookson Electronics	Highly filled	52	10	<25	4,000	19 mm, 40
Staychip™ 3090/Cookson Electronics	Highly filled	55	11	<25	4,000	19 mm, 40 (50 μm gap)

N/A – Not Available.
[a] 10 rpm @ 25 °C unless indicated.
[b] Former Honeywell and Johnson Matthey products.
Source: Compiled from suppliers' technical data sheets.

adhesive, thus reducing the number of process steps. Typical reflow conditions are 220°C–230 °C for 4–6 minutes. However, some NFUs cure only partially during solder reflowing and must be followed by an additional post-cure step, usually one hour at 150°C–160 °C.[12]

Once cured, the properties of NFUs are similar to those of capillary-flow underfills. One major difference is that the no-flow materials are generally unfilled and, as a result, their expansion coefficients are higher than those of their filled counterparts. However, their lower moduli more than compensate for the mismatches in expansion coefficients. NFU adhesives have shorter shelf lives than capillary-flow types because of the incorporation of the fluxing agent into the adhesive formulation. Table 5.4 is a compilation of underfills and their properties while Table 5.5 lists examples of capillary flow and no-flow underfills and their applications.

Katze[13] evaluated four dispense patterns as shown in Figure 5.10. The X and the five-dot patterns were most susceptible to air entrapment. The single-dot and single-line patterns gave the best coverage. Initial coverage beneath the die was

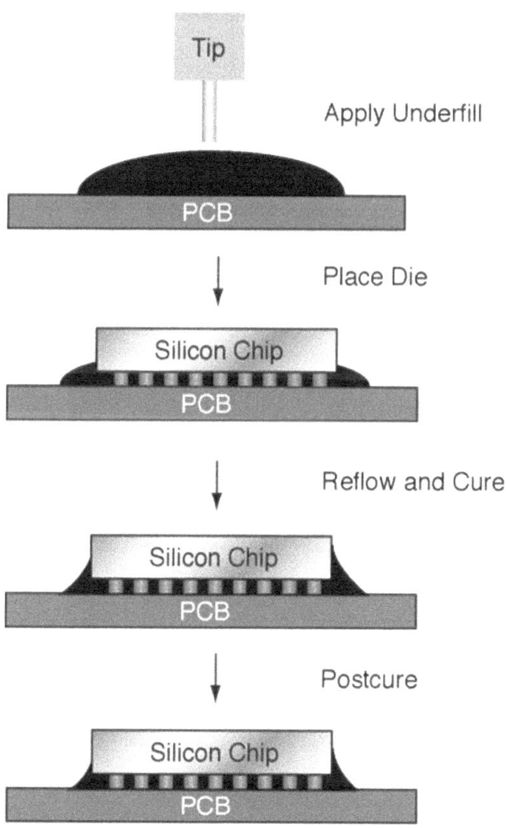

Figure 5.9 No-flow underfill process sequence.[10]

Table 5.4 Examples of underfill adhesives and properties

Product/ Supplier	Description	Viscosity cP (rpm)	Shelf Life (−40 °C)/ Pot Life (25 °C)	Cure Schedule	Bond Strength (psi) (Surfaces)	MOE (MPa/psi[b])	CTE (ppm/°C) α_1	CTE (ppm/°C) α_2	T_g (°C)	Volume Resistivity (ohm-cm)	Thermal Conductivity (W/m·K)	Moisture Absorption (%)	Ionic Impurities (ppm)
Staychip® 3082/ Cookson Electronics	Silica-filled, epoxy-anhydride underfill	4,000	6 mos/32 hours	15 min @ 165 °C	11,165 (polyimide passivated die)	4,800/696,182	38	141	125–136	N/A	N/A	<1	$Na^+ < 2$, K^+2, $Cl^- < 10$
Staychip® 3090/ Cookson	High reliability capillary underfill	5,100	6 mos/12 hours	15 min @ 165 °C	N/A	3,480/504,732	38	130	99	N/A	N/A	1	N/A
Staychip™ 2078E/ Cookson	Epoxy-anhydride no-flow fluxing underfill (rubber modified)	500–1,500 (N/A)	6 mos/3–4 hours	PCB solder reflow profiles	N/A	800/116,000	69	211	119—121	N/A	N/A	1.2	$Na^+ < 5$, $K^+ < 5$, $Cl^- < 20$
Staychip™ 811/ Cookson	Phenol-anhydride no-flow fluxing underfill (rubber modified)	19,000 (10 rpm) (N/A)	6 mos/24 hours	PCB solder reflow profiles	N/A	1,800/261,068	65	222	130	N/A	N/A	<1	$Na^+ < 1$, $K^+ < 1$, $Cl^- < 1$
Hysol® 3500™/ Henkel	Reworkable, low-temperature curing	203 (20 rpm)	N/A/14 days	2 min @ 130 °C	N/A	1,900/275,500	77	215	16	N/A	N/A	N/A	N/A
Loctite® 3513™/ Henkel	Reworkable, fast curing/Henkel	3,000–6,000	N/A/48 hours	10 min @ 150 °C	N/A	N/A	63	210	69	4.4×10^{16}	0.2	0.16	N/A
Loctite® 3536™/ Henkel	Reworkable, low-temperature curing	1,800	6 mos/14 days	5 min @ 120 °C	N/A	N/A	63	178	53	N/A	N/A	N/A	N/A
Loctite® 3563™/ Henkel	Rapid cure, fast flow underfill	5,000–12,000	N/A/8–12 hours	>5 min @ 150 °C	N/A	N/A	35	110	130	1×0^{15}	0.38	<0.7	$Na^+ < 1$, $Cl^- < 30$, $K^+ < 1$

Loctite® 3593™/ Henkel	Rapid cure, fast flow underfill	3,500–6,500	N/A/7 days	3 min @ 165 °C	N/A	N/A	50	160	110	$>9 \times 10^{16}$	0.21	<2	$Na^+ < 2$, $Cl^- < 45$, $K^+ < 1$
Ablefill™ UF8807™/ Henkel	Cyanate ester, high flow speed underfill	13,000 (5 rpm)	12 mos/8 hours	15 min ramp to 165 °C, 30 min @ 165 °C	N/A	$10{,}700/1.6 \times 10^6$	28	83	≥130	N/A	0.6	N/A	Na^+, K^+, Cl^-, <15
Emerson & Cuming™ E1172A/ Henkel	Epoxy, non-anhydride cure	6,000 (5 rpm)	6 mos/2 days	6 min @ 135 °C	N/A	N/A	27	N/A	135	N/A	N/A	1.5	$Cl^- < 35$ $Na^+ < 15$ $K^+ < 10$
Emerson & Cuming™ E1216/Henkel	Epoxy, in-line fast cure	17,000 (5 rpm)	6 mos (−20 °C)/ 5 days	3 min @ 165 °C	N/A	N/A	34	N/A	115	N/A	N/A	1.5	N/A
MC7883-UF/AI Technology	Cyanate ester, Quartz filler	N/A	N/A	10 s @ 175 °C	N/A	7,500/1,087,500	19	N/A	236	N/A	1.0	N/A	<10 (total)
MC7885-UF/AI Technology	Alumina filled cyanate ester underfill	N/A	N/A	N/A		7,500/1,087,500	20	N/A	236				<10 (total)
NF260./Indium Corporation®	No-flow epoxy underfill designed for lead-free processes	28,000 (10 rpm)	6 mos/24 hours	Lead-free solder profile	2,800	N/A	65	275	96	N/A	0.17	N/A	$Cl^- < 5$, $Na^+ < 5$ $K^+ < 5$

N/A – Not Available.
[a] Brookfield @ 10 rpm and 25 °C unless indicated.
[b] Generally at 25 °C.
[c] Now Henkel Loctite.
Source: Compiled from suppliers' technical data sheets. Contact suppliers for specification values.

Table 5.5 Examples of capillary-flow and no-flow underfill adhesives and applications

Adhesive/Supplier	Type	Gap Size, μm	Substrate or Surface Type	Dispensing	Application
Hysol® 3500™/Henkel	Epoxy, fast flow, snap cure	N/A	Components and printed-wiring boards	N/A	Reworkable CSP and BGA (Handheld devices)
Loctite® 3513™/ Henkel	Epoxy, capillary flow	N/A	Surface-mount devices to printed wiring boards	Hand, time-pressure, auger, linear piston, and jet dispensing	Reworkable CSP, flip chip, and BGA
Loctite® 3536™/ Henkel	Epoxy, rapid cure, low temperatures	N/A	Components and printed-wiring boards	Preheat substrate to 70 °C maximum to optimize flow	CSP and BGA
Loctite® 3563™/ Henkel	Rapid cure, fast flow epoxy	≥25	Surface-mount devices to printed-wiring boards	Hand, time-pressure, auger, linear piston, and jet dispensing	CSP and BGA
Loctite® 3593™/ Henkel	Epoxy, fast flow, high process speed	≥25	Electronic components to printed-wiring boards	Hand, time-pressure, auger, linear piston, and jet dispensing	Chip-size packages
Ablefill™ UF8807™/ Henkel	Silica filled cyanate ester	≥250	Organic substrates	N/A	CSPs and high-density interconnect substrates
Emerson & Cuming™ E1172A/Henkel	Non-anhydride and highly-filled epoxy	25 μm geometry	N/A	Preheat substrate to 90–100 °C	Flip-chip devices
Emerson & Cuming™ E1216™/Henkel	Snap-curable epoxy	N/A	N/A	Preheat substrate to 70–100 °C	CSP, BGA, and flip-chip devices

Staychip™ 3082/ Cookson Electronics	Fast-flow, anhydride-cured epoxy	N/A	N/A	Preheat substrate to 70–00 °C	Flip-chip in package or flip-chip on board
Staychip™ 3090/ Cookson Electronics	Epoxy, fast flow, fast cure	N/A	Ceramic, laminates	Dispense, capillary underfill	FCIP
Staychip® 2078E/ Cookson	Epoxy anhydride, fluxing underfill	N/A	Sn/Pb bumped pads and Sn/Pb, Cu/Ni/Ag, and OSP coated copper finishes	Automatic dispense or stencil	Area array on PWBs using solder-reflow cycle for underfill curing
Staychip® 3090/ Cookson	Amine-cured epoxy underfill	N/A	Ceramics, laminates	Dispense, capillary underfill	Flip-chip in package or flip-chip on board
Staychip®81-3/ Cookson	Phenol-Anhydride no-flow, fluxing underfill	N/A	Devices and printed circuit boards	N/A	Flip-chip
NF260/Indiuim Corp.®	No-flow, epoxy resin	–	Flip-chip/BGA or CSP assembly and CSP bonding to PWBs (tin/lead eutectic solder bumped devices)	Automated dispensing (22-gauge needle)	CSP assembly and FCOB and CSP bonding to laminates
MC7883-UF/AI Technology	Quartz filled, ultra-low coefficient of thermal expansion	N/A	N/A	Dispense, capillary underfill at 40–80 °C and cure 90 s @150 °C	N/A
MC7885-UF/AI Technology	Alumina filled, ultra-low coefficient of thermal expansion	N/A	N/A	Dispense, capillary underfill at 40–80 °C and cure 90 s @150 °C	N/A

[a] Organic solderability preservative; N/A – Not Available; NA – Not Applicable.
Source: Compiled from suppliers' technical data sheets.

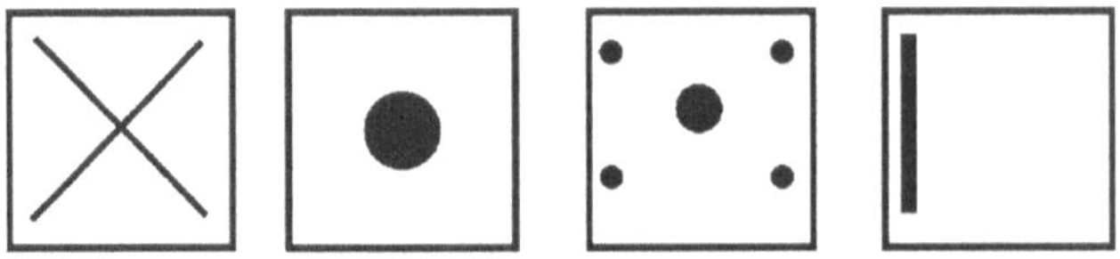

Figure 5.10 Dispense patterns for no-flow fluxing underfill.

approximately 50%, but flow-out completed the coverage.[14] Prebaking the printed-circuit board to remove moisture and limiting the exposure time of the adhesive to ambient moisture are recommended to reduce voids.[15]

5.1.3.3 Preforms

A third approach to underfilling involves applying an insulative thermoplastic preform (film) prior to attaching the device. Under pressure and heat, the preform softens and flows around the solder bumps, then solidifies quickly on cooling. The devices may then be solder reflowed to make the connections or the preform may be formulated to flow and encapsulate the solder during solder reflow. Table 5.6 lists some commercially available preforms used as underfills. Anisotropic film adhesives have also been used as underfill preforms, in which case z-direction conductive paths are formed beneath the solder bumps at the same time that the rest of the insulating film flows and cures around the bumps. Anisotropic paste adhesives can be similarly used (Figure 5.11).

5.1.3.4 Alternate underfill approaches

Alternate approaches that eliminate the capillary-flow underfill process currently used by electronics assemblers, involve pushing the problem back to the wafer stage. One process, *wafer-level underfill* (WLU) (Figure 5.12), consists of applying the underfill material to a bumped wafer and then B-staging the underfill. The wafer is then diced and the die are positioned on corresponding connection pads on an interconnect

Table 5.6 Examples of underfill preform adhesives

Adhesive/ Supplier	Type	Bonding conditions	Application
UF-511/Hitachi	Insulative film	180 °C, 145 psi, 20 seconds	Flip-chip to substrate, CTE mismatched substrates
UF-526/Hitachi	Insulative film	180 °C, 145 psi, 20 seconds	Flip-chip to substrate, CTE mismatched substrates
FC-212B/Hitachi	Anisotropic film	180 °C, 145 psi, 20 seconds	Flip-chip underfill and electrical connections

Source: Compiled from suppliers' technical data sheets.

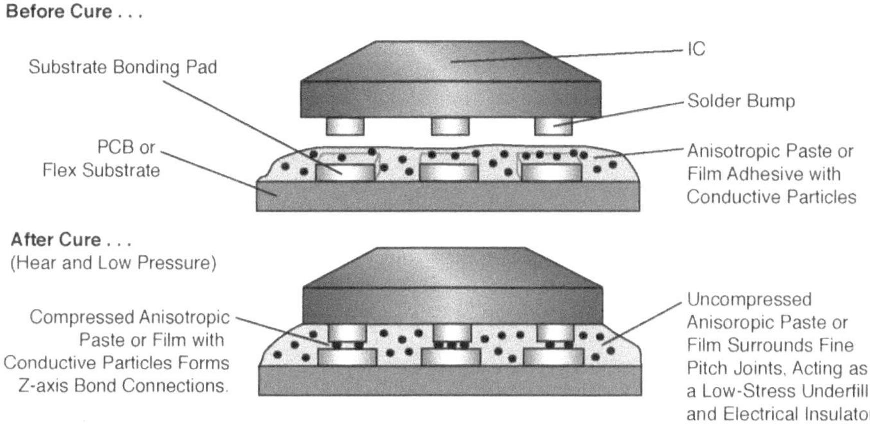

Figure 5.11 Anisotropic film or paste underfill.

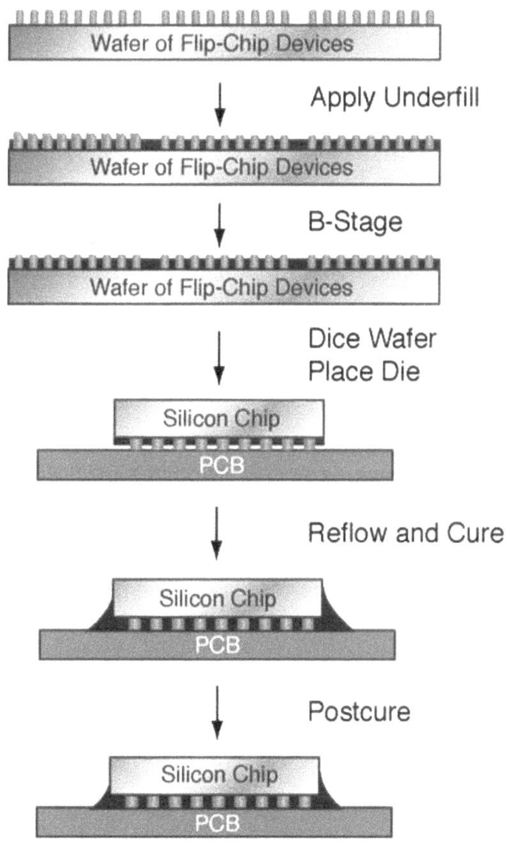

Figure 5.12 Process steps for underfill pre-applied at wafer stage.[10]

substrate and solder reflowed. During reflow, solder connections are made simultaneously with the flowing, filleting, and curing of the adhesive.

In a similar process, known as *polymer-film interconnect* (PFI), an insulative thermoplastic film is laminated over the devices at the wafer stage, and vias are opened over the bonding pads using a laser. At that point, either the normal solder bumps can be formed or a silver-filled conductive adhesive can be stencil printed into the vias to form polymer bumps. After printing, the epoxy is B-staged and the flip-chip devices are diced. In assembly, the devices are heated to a temperature that completes the cure of the B-staged bumps and simultaneously reflows the thermoplastic underfill material.[16,17]

Recently, Henkel has been investigating applying a coating around the solder balls of devices also at the wafer stage. The process, called *pre-applied underfill*, consists of stenciling a modified epoxy coating/adhesive to 50%–80% of the height of the bumps, heating to remove the solvent, and B-staging. The flip-chip devices are then singulated and ready for fluxing, attachment, and solder reflowing.[18]

A unique process, developed by K&S Flip Chip Division (Kulicke and Soffa), consists of forming a polymer collar around each solder bump at the wafer stage. The mini-collars reinforce and dissipate stresses from each solder bump. The collars are produced by a batch photolithographic process and are reported to increase the life-cycle performance of the solder connections up to 50% after temperature cycling.[19,20] Most of the problems associated with both capillary-flow and pre-applied underfills that are used with tin-lead solder connections have been resolved, and the materials and processes are fairly mature. However, with the trend toward replacing lead-alloy solders, a new set of underfill adhesives issues has arisen. Non-lead solders generally require reflow temperatures that are 10°C–30 °C higher than tin-lead solders and generate flux residues that are different. New underfill materials are needed that will survive these higher temperatures and be compatible with the flux residues. Pre-applied material should not gel or cure before the solder joints are formed.[10]

5.1.4 Conductive adhesives as solder replacements

Electrically conductive adhesives, primarily silver-filled epoxies, are finding uses as replacements for solder in surface-mounting components on printed-circuit boards and in flip-chip attachments. There are several driving forces for this application, a major one being the trend to eliminate lead and tin-lead solders because they may be health hazards. Also associated with the use of solder, is the need to eliminate ozone-depleting solvents presently used to clean and remove flux residues. Electrically conductive "polymer solders", as alternatives to tin-lead and other metal alloy solders, provide environmentally friendly materials and processes along with other benefits, such as the lower processing temperatures (often required in assembling heat-sensitive components), and finer pitch and tolerances for high-density interconnections. Cure temperatures of 120°C–150 °C can be used for most adhesives while reflow temperatures of 220°C–240 °C are required for solders. However, many adhesives have been formulated to cure during a typical solder-reflow cycle, thus are compatible with existing SMT production lines.

Two issues that must be addressed in substituting adhesives for solder are their compatibility with the variety of component-termination materials and PWB surfaces. Adhesion and electrical-contact resistance can vary widely depending on the nature of these surfaces. Terminations may consist of tin, tin-lead, nickel, silver-palladium, and gold. Board surfaces may consist of tin-lead solder, plated gold, or *organic-solderability preservative* (OSP) *copper*. OSP consists of a thin organic film that reacts selectively with the copper surface preventing it from oxidizing and guaranteeing solderability for many months. OSP provides a conductive surface that is compatible with fluxes and solder. The compatibilities of several solder-replacement adhesives with various metallizations, surface finishes, and printing resolutions are given in Table 5.7.

Table 5.7 Compatibility of various solder replacement adhesives with surfaces and printing resolutions

	Compatibility with surface finish							Printing resolution	
Solder replacement/ supplier	**Sn**	**Sn/ Pb**	**Ag**	**Ag-Pd**	**Cu**	**OSP-coated Cu**	**Au**	**20-mil pitch**	**12-mil pitch**
Polysolder LT/Cookson Electronics		X	X	X			X		
Ablebond®8175/Henkel				X	X			X	
Ablebond®8175A/ Henkel				X	X	X			X
Ablebond®8175Q/ Henkel					X				X
Emerson & Cuming™ 3103/Henkel					X			X (Au-Ni)	X
Emerson & Cuming™ 3103WLV/Henkel					X	X		X	
Emerson & Cuming™ 3103WXL/Henkel Labs.					X			X	
Epo-Tek® E 2101/Epoxy Technology			X	X	X	X	X		X
Epo-Tek® EE129-4/ Epoxy Technology			X	X	X		X		X
Epo-Tek® H20S5/Epoxy TTTechnology	N/A	N/A	N/A	N/A	N/A	N/A	N/A	N/A	N/A
ESS8459/AI Technology	N/A	N/A	N/A	N/A	N/A	N/A	N/A	N/A	N/A
ESS8450/AI Technology	N/A	N/A	N/A	N/A	N/A	N/A	N/A	N/A	N/A

N/A – Not Available.
[a] Organic Solderability Preservative.

Solder-replacement materials may be applied by stencil printing, screen printing, or automatic dispensing and can produce fine-pitch bonding pads for flip-chip devices. Emerson & Cuming CE-3100 and Ablebond 8175A from Henkel, for example, applied by stencil printing can produce 20-mil pitch pads, while Emerson & Cuming CE-3103 adhesive may be used to process 12-mil-pitch conductor pads. Many commercially available solder-replacement adhesives are also suitable for ultra-fine–pitch applications (<12 mils) and are compatible with a variety of surfaces including antioxidant-free copper, OSP-copper, and tin-lead soldered surfaces. The contact resistance of polymer bumps to aluminum bonding pads is higher than for solder bumps so that a more compatible metallization such as gold needs to be formed over the aluminum through barrier and adhesion layers such as titanium/tungsten and copper. The contact resistance of a silver-filled epoxy is reported to be approximately 25 micro-ohms compared with 10 micro-ohms for tin-lead solder.[16] The properties of some adhesives reported to be useful as solder replacements are given in Table 5.8.

5.2 Specific applications

For many applications, adhesives are formulated and tailored around specific products and environments. Among these are the ground communication and control equipment for the military that must withstand a wide spectrum of harsh terrestrial climates in addition to high levels of shock and vibration. Spacecraft electronics will be subjected to high vacuum, low temperatures, and particle bombardment from both radioactive and physical particles. Other specific applications include memory modules, automotive, flat-panel displays, optoelectronics, LEDs, MEMS, and EMI shielding.

Adhesives are also tailored to meet specific requirements of the packaging or assembly technology used (Table 5.9). Thus, the requirements for single-chip packaging in hermetically sealed ceramic packages such as CERDIPs will be somewhat different from those of plastic dual-in-line packages (PDIPs) and plastic-encapsulated microcircuits (PEMs).

5.2.1 Military applications

Adhesives are used extensively in military electronics. Communication systems, sensors, and navigation systems rely heavily on printed-wiring boards and assemblies because of their ruggedness and ability to survive shock and vibration, temperature cycling, and aggressive handling. Adhesives are also used to reinforce and protect other parts of an assembly, for example, to provide vibration damping, to seal enclosures from moisture, and to protect signals from electromagnetic interference.

Military electronics may be ground-based or airborne but, in either case, must be capable of withstanding the harsh extremes of terrestrial environments. Ground-based electronics must also be resistant to fungus and other microorganisms, salt spray (if

Table 5.8 Properties of adhesives used as solder replacements

Product/Supplier	Type	Viscosity cP (rpm)	Shelf Life (−40 °C)/ Pot Life (25 °C)	Cure Schedule	% Weight Loss @ 300 °C by TGA	Die-Shear Strength (psi/Surfaces)	Modulus of Elasticity (MPa /psi)	CTE (ppm/ °C) α_1	CTE (ppm/ °C) α_2	T_g (°C)	Volume Resistivity, ohm-cm	Thermal Conductivity (W/m K)	Ionic Impurities (ppm)
Polysolder LT/ Cookson Semiconductor Products	Silver-filled epoxy	150,000 (1 rpm)	6 mos/8–12 hours (on screen)	10 min @ 140 °C	N/A	>2,500/Sn/Pb surface-mount parts	N/A	54	N/A	90	$\leq 5.0 \times 10^{-4}$	3.5	N/A
Polysolder SE3001/ Cookson Semiconductor Products	Silver-filled epoxy	150,000–220,000 (1 rpm)	-/8–12 hours	15 min @ 130 °C	N/A	>2,300/Au to Au	N/A	54	N/A	90	$\leq 1 \times 10^{-4}$	4 - 5	N/A
Ablebond® 8175/ Henkel	Silver-filled epoxy	55,000 (5 rpm)	12 mos/2 weeks	30 min @ 150 °C	0.30	6,200/Cu to Ag	N/A	55	200	90	5×10^{-4}	3.2	N/A
Ablebond® 8175A/ Henkel	Silver-filled epoxy, fast curing	70,000 (5 rpm)	12 mos/ 24 hours	3 min @ 150 °C or 6 in @ 130 C	0.40	5,500/Ag to Pd/ Ag	N/A	65	250	80	5×10^{-4}	4.0	$Cl^- < 50$ $Na^+ < 10$ $K^+ < 5$
Ablebond® 8175Q/ Henkel	Silver-filled epoxy, low viscosity	35,000 (5 rpm)	12 mos/4 days	1 hour @ 150 °C	N/A	N/A	N/A	N/A	N/A	N/A	3×10^{-4}	N/A	Cl^- 10, Na^+ 5, K^+ 1
Emerson & Cuming™ CE3103/Henkel	Silver-filled epoxy	45,000–65,000 (N/A)	12 mos/ 24 hours	8 min @ 150 °C	N/A	N/A	5,600/ 812,000	45	146	109	7×10^{-4}	N/A	$Cl^- < 50$, $Na^+ < 10$, $K^+ < 5$
Emerson & Cuming™ CE3103WLV/ Henkel	Silver-filled epoxy, low viscosity	15,000–25,000 (N/A)	12 mos/3 days	3 min @ 150 °C or 10 min at 120 °C	N/A	N/A	4,500/ 652,500	45	225	114	8×10^{-4}	N/A	N/A
Emerson & Cuming™ CE 3103 WXL/ Henkel	Silver-filled epoxy, fine pitch printing	65.000 (N/A)	12 mos/3 days	8 min @ 150 °C or 10 min at 125 °C	N/A	N/A	4,500/ 652,500	N/A	N/A	100	7×10^{-4}	N/A	N/A

(Continued)

Table 5.8 Properties of adhesives used as solder replacements—*cont'd*

Product/Supplier	Type	Viscosity cP (rpm)	Shelf Life (−40 °C)/ Pot Life (25 °C)	Cure Schedule	% Weight Loss @ 300 °C by TGA	Die-Shear Strength (psi/Surfaces)	Modulus of Elasticity (MPa /psi)	CTE (ppm/°C) α_1	CTE (ppm/°C) α_2	T_g (°C)	Volume Resistivity, ohm-cm	Thermal Conductivity (W/m K)	Ionic Impurities (ppm)
Epo-Tek® E2101/ Epoxy Technology	Silver-filled epoxy	15,000–18,000 (20 rpm)	12 mos (25 °C)/5 days	1 hr @ 150 °C	N/A	≥1,700/none specified	N/A	56	131	≥90	5×10^{-4}	2.0	Cl^- 27, Na^+ 10, K^+ 1, NH_4^{+1} 80
Epo-Tek® E129-4/ Epoxy Technology	Silver-filled, low-temperature curing epoxy	2,000–4,000 (100 rpm)	12 mos/ 3 hours	15 min @ 100 °C	2.06	≥1,700/none specified	N/A	30	227	≥30	3×10^{-4}	1.6	Cl^- 223, Na^+ 26, K^+ 2, NH_4^{+1} 22
Epo-Tek® H20S/ Epoxy Technology	Silver-filled epoxy	1,800–2,800 (100 rpm)	12 mos/3 days	5 min @ 150 °C	0.60	≥1,700/none specified	N/A	31	120	≥80	5×10^{-4}	3.25	Cl^- 162, Na^+ 0, K^+ 4, NH_4^+ 282
DM 4030SM/F577/ Diemat	Silver-filled, lower temperature processing epoxy	44,000 (10 rpm)	N/A/6 mos (25 °C)	20–30 minute ramp to 150 °C and 15–30 min @ 150 °C	N/A	N/A	N/A	30	N/A	N/A	7×10^{-5}	10	<10 total (Cl^- F^- Na^+ K^+)
DM4130SM/F595/ Diemat	Silver-filled , lower temperature processing epoxy	41,000 (10 rpm)	N/A/6 mos (25 °C)	20–30 min. ramp to 150 °C and 15–30 min @ 150 °C	N/A	N/A	N/A	30	N/A	N/A	1×10^{-3}	5	<10 total (Cl^- F^- Na^+ K^+)
Solder Sub ESS8459/AI Technology	Flexible, gold-filled epoxy paste	N/A	1 year/N/A	5 min @150 °C	N/A	>2,000/None specified	N/A	120		−60	$\leq 5 \times 10^{-5}$	5.7	N/A
Solder Sub ESS8450/AI Technology	Flexible, silver-filled epoxy paste	N/A	1 year/N/A	5 min @150 °C	N/A	>2,000/None specified	N/A	120		−60	$\leq 4 \times 10^{-4}$	5.7	N/A

N/A – Not Available; ND – None Detected.
[a] Generally at 25 °C.
Source: Compiled from various suppliers' data sheets. Contact suppliers for specification values.

Table 5.9 Adhesives requirements for various packaging technologies

Packaging technology	Typical adhesive requirements	Examples of adhesives
Hermetic CERDIP, CBGA, CSP	High-adhesion strength, low-moisture absorption, stability of packages to processing temp. >350 °C (if silver-glass adhesives are used)	Silver-glass die attach, silver-filled cyanate ester, polyimide
Molded plastic, smaller die sizes (PDIP, QFP)	High strength, low-temperature processing (<250 °C), moisture resistance	Silver-filled epoxy, modified cyanate-ester and thermoplastic/thermoset blends
VLSIC in TQFP/thin outline plastic	Low modulus of elasticity (MOE), low-moisture absorption	Epoxy, cyanate-ester blends
PBGA, thermally enhanced BGA	Low MOE, low stress, high thermal dissipation	Epoxy blends, cyanate-ester blends, modified cyclo-olefin thermoset (MCOT), silicone materials, highly filled thermally conductive adhesives
CSP	Ultralow MOE, stress and moisture resistance, good mechanical integrity	Epoxy blends, cyanate-ester blends, modified cyclo-olefin thermoset (MCOT), silicone materials
Hybrid microcircuits and MCMs	MIL-STD-883, Method 5011-compliant, low outgassing, high-purity, reworkability	Silver-filled or insulative epoxies, thermoset or thermoplastic pastes or films
Chip-on board	Low MOE, reworkability	Epoxy pastes
Flip-chip bonding	Screen or stencil printable or *z*-axis (anisotropic conductive), or isotropic conductive bumps,	Anisotropic and isotropic conductive pastes or films, underfill epoxies or modified epoxies

close to an ocean), and high humidity. Moisture remains one of the major causes of electronic failures.

Thus, components that are surface mounted with adhesives must maintain their strength and electrical properties during and after these combined environments and after repeated stresses. To assure reliability, numerous accelerated tests have been developed and incorporated in military and industry specifications. Among specifications governing the mechanical, electrical, and qualification requirements for electronic components, modules, subsystems, and systems, most of which could not be met without using qualified and reliable adhesives, are: MIL-STD-883,[21] MIL-STD-810,[22] MIL-STD-202,[23] and MIL-PRF-38534.[24]

5.2.2 Space applications

Besides the mechanical and electrical requirements that adhesives must meet for terrestrial applications, a separate set of requirements apply to space applications. The space environment, depending on the orbit, consists of high vacuum (10^{-8} Torr or lower); extremes of temperature; proton, electron, and ionizing radiation; atomic oxygen; and debris and micrometeoroids. Adhesives and polymeric materials are a major concern in both unmanned and manned spacecraft. If directly exposed to the space environment, adhesives and other polymeric materials outgas and volatile products can condense and contaminate adjacent electronic, optical, or precision instrument parts. Even 100 angstroms of condensed organic material can affect the reflectance or absorbance properties of optical parts. In manned spacecraft, some outgassing products as low as parts per billion can be toxic to astronauts.

To assure safety and reliability, NASA and other government agencies require all polymeric materials to be qualified and, as a minimum, to pass outgassing tests defined in ASTM-E-595.[25] This specification defines two tests: the *total mass loss* (TML) and the *collected volatile condensable materials* (CVCM). The TML is the weight of material lost after exposure for 24 hours at 125 °C in a vacuum of less than 5×10^{-5} Torr and is specified as 1% or less.[26] The CVCM is the amount of volatiles that condense on a collector plate maintained at 25 °C during the same conditions and is required to be 0.1% or less. Through many years of testing, an extensive databank of materials that pass these requirements is available.[27] In some cases, an adhesive that fails when cured at one schedule will pass under other cure conditions. Outgassing data for some die- and substrate-attachment adhesives are given in Table 5.10.[27]

In direct exposure to the natural space environment, total radiation flux can also degrade the physical and electrical properties of the adhesive. Long-term radiation exposure embrittles some adhesives through continued crosslinking or causes polymer degradation with outgassing. The radiation resistance of polymeric materials varies depending primarily on their molecular structures. Aromatic and heterocyclic polymers and highly cross-linked polymers are more tolerant to high radiation doses than are aliphatic polymers. Thus epoxies cured with aromatic amines fare better than epoxies cured with aliphatic amines. Polyimides and phenyl silicones, because of their aromatic structures, are also highly resistant to total radiation doses up to 10^{8} rads. Where radiation data are not available, testing to the total ionizing dose expected should be performed according to MIL-STD-883, Method1019.5.[28] Fortunately, through metal shielding, encapsulation, or hermetic sealing of electronic assemblies, radiation exposures can be largely mitigated.

Another space-specific issue is the embrittlement or stiffening of an otherwise flexible adhesive at the low temperature extremes of space. Of the many polymer types, silicones best maintain their flexibility and physical properties at operating temperatures as low as −80 °C, or even lower for some formulations, thus are widely used in satellites and space vehicles.

Table 5.10 Outgassing data for some die- and substrate-attach adhesives[27]

Material information		Outgassing results (%)		
Adhesive/Supplier	**Cure schedule**	**TML**	**CVCM**	**WVR**
Ablebond® 8175 (silver-filled epoxy)/Henkel	2 hours @ 120 °C	0.18	0.01	0.05
Ablebond® 85-1 (gold-filled epoxy)/Henkel	2 hours @ 121 °C	0.20	0.01	0.08
Ablebond® 84-1 LMI (silver-filled epoxy)/Henkel	1 hour @ 150 °C	0.12	0.00	0.04
	2 hours @ 125 °C	0.12	0.01	0.09
Ablebond® 84-3 (electrically insulative epoxy)/Henkel	1 hour @ 150 °C	0.23	0.00	0.16
Ablefilm® 5020 K (epoxy film)/Henkel	1 hour @ 150 °C	0.24	0.02	0.18
Ablefilm® 5025E/Henkel	1 hour @ 150 °C	0.30	0.06	0.08
	2 hours @ 125 °C	0.32	0.08	0.08
Ablefilm® 561 K/Henkel	2 hours @ 125 °C	0.42	0.13	0.09
Ablefilm® 550/Henkel	3 hours @ 125 °C	0.42	0.12	0.24
Epo-Tek® H20E (silver-filled epoxy)/Epoxy Technology	1 hour @ 150 °C	0.62	0.01	0.09
Epo-Tek® H35-175MP/Epoxy Technology	1.5 hour @ 150 °C	0.30	0.02	0.18
Epo-Tek® H70E/Epoxy Technology	12 hours @ 60 °C	0.99	0.03	0.25
EG8050/Epoxy Technology	2 hours @ 121 °C	0.33	0.02	9.18
ME 7156/Epoxy Technology	1.25 hours @ 175 °C	0.5	0.15	0.1

Source: Outgassing Data for Selecting Spacecraft Materials, *NASA Publication 1124*.

Materials that are directly exposed to the space environment must also withstand or be protected from atomic oxygen. Atomic oxygen is prevalent in the low earth orbit (LEO), approximately 100 to 350 miles from Earth. At this distance, molecular oxygen reacts with ultraviolet light to produce monoatomic oxygen which is extremely chemically reactive and erosive. Most adhesives used in spacecraft, however, are sandwiched between two adherends, thus not directly exposed.

5.2.3 Automotive applications

Adhesives are used extensively in the manufacture of automobiles both for structural and non-structural assemblies, and the requirements they must meet are among the most severe in industry. Electronic assemblies situated under the hood or near the engine must withstand continuous temperatures of 125 °C to 150 °C, thousands of temperature cycles, vibration, and mechanical shock. In addition, when not operating, they must be capable of withstanding severe global climates and climatic changes ranging from tropical high humidity, high temperatures to frigid subzero temperatures. While the surrounding ambient may be as high as 150 °C, semiconductor device junctions must be kept cool. To this end, die-attachment and underfill adhesives must have the highest thermal conductivity possible in order to transfer heat from the devices. Low CTE, stress-absorbing adhesives are definitely required as underfills for flip-chip and BGA packages because of the added problem of solder ball deterioration expected under extended high-temperature and thermal excursions. Coupled with all this is the requirement for very low cost.

Among the better adhesives that meet automobile engine conditions are the aromatic anhydride-cured epoxies, polyimides, silicones, and silver-glass. A proposed qualification test requirement that adhesive-bonded assemblies survive 2,000 thermal cycles with 1% or less of failures has been difficult to meet.[29]

5.2.4 Medical applications

Many types of medical electronic devices, both internal to the body and external, rely on adhesives for their assembly. The requirements for devices in medical applications differ from, and are more severe than, those for other applications, especially where the devices will be implanted or come into contact with the body. The standards and regulations for the medical industry are far different from other industries, often requiring approval from the Food and Drug Administration. Medical electronic devices include pacemakers, defibrillators, drug-releasing pumps, hearing aids, and diagnostic equipment for measuring, monitoring, and recording body functions such as heartbeat and brain waves. Major requirements for medical devices are the following:

- Miniaturization
- Reliability (reliability is essential for implantable electronics, but also for external devices where accuracy is critical in diagnosing a patient's condition)
- Biocompatibility and non-toxicity
- Resistance to sterilization (parts should be able to be sterilized, often more than once, without degrading electrical or physical performance).

5.2.4.1 Miniaturization

Miniaturization, low weight, and low volume are as important in medical electronics as in space electronics and other applications. The small size and weight are benefits of hybrid microcircuits; they have been used for a long time in pacemakers and defibrillators. The attachment and packaging of bare-chip devices with conductive and insulative adhesives has provided major reductions in weight and space over printed-wiring board technologies. The hermetic sealing by welding has also assured protection of the electronics from moisture and other body fluids, as well as protecting the body from outgassing or other contaminants from the electronic modules. Miniaturization is necessary for the new generation of hearing aids. Some designs consist of a thinned silicon IC embedded and connected to a flexible polyimide circuit that is then folded to reduce volume.

5.2.4.2 Reliability

The performance of the electronic devices implanted in the body is extremely important. Malfunctioning can result in death. For hybrid microcircuits and multichip modules, the qualification tests and accelerated tests required of military and space applications also apply, although some shock and vibration tests are not realistic for humans. In addition, the Food and Drug Administration (FDA) becomes involved in testing and reviewing data for toxicity, compatibility with body fluids, and failure modes and mechanisms. Some of the early pacemakers failed after being implanted and had to be removed and replaced. Failures occurred due to corrosion within hermetically sealed hybrid circuits due to moisture outgassed from adhesives and other polymer materials and to contaminants that were not completely removed during cleaning.

For externally used diagnostic electronics, reliability may also be a life-or-death situation if instruments do not provide accurate information.

5.2.4.3 Toxicity and biocompatibility

Adhesives used in medical devices that are implanted or in contact with the body must be tested and shown to be non-toxic, biologically inert, and compatible with blood and body fluids. Compatibility with blood and other body fluids is especially critical. Surfaces in contact with blood must not serve as sites for coagulation and clotting of blood. Generally, qualification testing is performed to ISO-10993[30] or to U.S. Pharmacopoeia (USP) Class VI. The two standards specify slightly different tests.[31] The USP Class VI standard specifies acute systemic (over the tissue), intracutaneous (under the skin), and muscle implantation tests. The ISO-10993 standard is a set of 12 documents that is more universal and more extensive than the USP standard. It specifies:

- Intracutaneous injection tests to evaluate irritation potential
- Acute systemic injection to evaluate for toxic effects
- Cytotoxicity tests to evaluate the effect on monolayer cells
- Blood compatibility.

Some adhesives suppliers already have test data showing compliance of their adhesives to ISO-10993 and will provide a certificate of compliance for their products. However, passing the ISO or USP Class VI tests should be considered a starter in proving that an adhesive has a low level of toxicity; it does not guarantee FDA approval. The FDA often requires more extensive testing.

5.2.4.4 Sterilization

All devices implanted in the body or that come into contact with body fluids or blood must be sterilized and be resistant to the sterilization conditions. Adhesive-bonded parts are not used in direct contact with the body fluids. They are generally part of an electronic module that is hermetically sealed or otherwise encapsulated or shielded. However, the sterilization conditions for the module must not degrade the internal adhesive-attached components. Some adhesives such as cyanoacrylates and UV-cured acrylics degrade at the elevated temperatures and high humidity encountered in dry heat and autoclave sterilization. Thus, adhesives, even though not directly exposed, should be tested and qualified based on their resistance to the sterilization process that will be used. The main sterilization procedures are gaseous ethylene oxide, autoclave, dry heat, and gamma radiation.

5.2.4.4.1 Gaseous ethylene oxide (EtO)

Exposure to ethylene oxide, admixed with a carrier gas such as carbon dioxide or a chlorofluorocarbon, is the most widely used sterilization process because it has the least damaging effect on materials and devices and is low cost. EtO sterilization is a room-temperature process that affects only the external surfaces of a module.

5.2.4.4.2 Autoclave

Autoclave sterilization consists of exposing the parts to steam in a pressurized vessel. Temperatures of 121 °C to 132 °C at pressures up to 2 atmospheres are reached with saturated moisture. Under these conditions, some polymer materials hydrolyze and revert to monomers with loss of adhesive strength. Materials most susceptible to reversion include some polyurethanes, acrylics, and silicones. Epoxies are most resistant.

5.2.4.4.3 Dry heat

In dry-heat sterilization, the parts are exposed for 2–3 hours at 165°C–170 °C. Dry-heat exposure is the least effective sterilization method and most likely to degrade materials and electronic devices. Although adhesives may have been cured at 165 °C prior to hermetic packaging, the additional temperature exposure within the sealed package can result in further outgassing of moisture and corrosive volatiles causing electrical failures. Outgassing is worse if the adhesive had been cured at a temperature below 165 °C.

5.2.4.4.4 Gamma radiation

Exposure to gamma radiation or to other high-energy radiation source (beta, electron beam) is an effective sterilization process but requires expensive equipment and strict safety controls. Gamma rays are highly penetrating even affecting materials and devices within an enclosure. Some electronic components, especially bare semiconductor devices, may be temporarily or permanently damaged.

5.2.5 Chip-scale packages (CSP)

Bare die such as integrated circuits may be assembled in plastic or ceramic carriers whose dimensions are slightly larger than the chip. These are called *chip-scale packages* (CSP) and are defined as packages that are no larger than 1.5 times the area of the die or no more than 1.2 times the width or length of the die.[32,33] If the carrier is a BGA type, an alternate definition is that the solder-ball pitch be less than 1 mm.[34] The pitch of interconnects varies by I/O count. Parts having fewer than 100 I/Os have pitches of 1 mm decreasing to 0.5 mm as the number of I/Os increases to 500.[35] Chip-scale packages have many advantages over bare-chip attachment especially in their ease of handling, electrical testing, and automated soldering or epoxy attachment. They also have advantages over QFP and PGA packages in the higher levels of silicon density that are obtained. While the silicon density (ratio of silicon-to-package area) of a QFP is 10% to 60%,[36] that for a single chip in a CSP may be as high as 60% to 100%.[37] CSPs may be bumped for flip-chip connections or may be wire bonded. Adhesives are used to attach the chip to the carrier and the carriers to an interconnect substrate. In the past decade, there has been a proliferation of designs and configurations for CSPs, and they have become the preferred method of packaging ICs having up to 500 I/Os. Table 5.11 lists some examples of commercially available chip-scale packages, suppliers, and applications.

5.2.6 Vertical-stack, 3-D packaging

In the quest to produce ever-higher density memory modules, various stacked 3-D packaging configurations have been developed and are being used. Historically, focal-plane arrays with on-board processing and solid-state data recorders have driven the development of 3-D high-density memory modules. For example, the data storage unit of the Hubble space telescope consists of 3-D stacked memory chips. More recently, the driver for 3-D stacking has been cell phone handsets where increased functions are being integrated in smaller units. For these consumer products, separately packaged SRAM or flash memory chips are no longer sufficient to support the growing demand for new features such as voice, video, and data without substantially increasing the size and weight of the product.[38] Vertical-stack packaging designs may be of four types as illustrated in Figure 5.13.

5.2.6.1 Chip stacking

Bare ICs are stacked and interconnected to achieve the densest level of integration. The chips are held together by thin layers of adhesive. Interconnects are formed by several methods including vapor depositing and photodelineating thin-film metallization along the sides, by applying electrically conductive epoxy adhesive, using TAB leads or wire bonding, or through-silicon vias (TSV).

5.2.6.2 Package stacking

IC chips are first inserted into packages, then the packages are stacked and interconnected using either the package leads or extensions from the board. Package stacking is a better solution for some applications because the packaged die can be burned in and electrically tested prior to stacking, assuring known good devices

Table 5.11 Examples of Chip-Scale Packages, Suppliers, and Applications

Construction	Advantages	Examples (Supplier)	Types	Applications
Leadframe-Based CSP				
Copper leadframe, low-loop wire bonds or flip-chip interconnect, over-molded	50% size reduction compared to thin-shrink small-outline package (TSSOP) inductance (MicroLeadFrame®) and very low inductance flip-chip MicroLeadFrame	MicroLeadFrame® (wire bond) and flip-chip MicroLeadFrame® (*fc*MLF) (Amkor)	Quad flat pack no lead (QFN), small outline no lead (SON), Dual flat pack no lead (DFN)	Portable electronics (cell phones and personal data assistants) (PDAs), MicroLeadFrame® and high-performance applications in the 5–40 GHz range
Exposed die-attach pad	50% reduction in inductance/capacitance compared to leaded parts	MLP (MicroLeadFrame Package) (Carsem)	QFN, SON, MLPQ (quad format), and MLPD (Dual format)	PDAs, Global-Positioning Systems (GPS), RF product for Bluetooth™
Bumped Chip Carrier (BCC) and BCC+ based CSP				
Standard bumped chip carrier and electrically/thermally enhanced center-exposed-pad technology, over-molded.	40% less mounting area an 67% less mounting volume compared to Shrink Small Outline Package	BCC, BCC+ (Fujitsu)	Single-chip and stacked chip	RF devices, wide area networks, and Dense Wavelength Division Multiplexing systems
Rigid laminate substrate (with or without soldermask), wire bonds, over-molded	Small outline, low profile (1.0 mm), enhanced electrical performance over 2.4 GHz, JEDEC moisture sensitivity level (MSL) 3	Land grid array and fine-pitch ball-grid array (National)	Thin, fine-pitch QFN, Thin shrink SON	Portable applications, including cellular phone, wireless, lap-top computers, and DSP
FR-5 laminate substrate	Improved signal-to-noise ratio (> 1 GHz) compared to wire-bonded parts and low-inductance die interconnects	ChipArray® Ball Grid Array (Amkor)	fcCSP	Cell phones, handheld electronics, and high-performance workstations (notebooks and RF applications)

Laminate-Substrate-Based CSP				
Rigid laminate substrate (with or without soldermask), wire bonds, over-molded	Small outline, low profile (1.0 mm), enhanced electrical performance over 2.4 GHz, JEDEC moisture sensitivity level (MSL) 3	Land grid array and fine-pitch ball-grid array (National)	Thin, fine-pitch QFN, Thin shrink SON	Portable applications, including cellular phone, wireless, lap-top computers, and DSP
FR-5 laminate substrate	Improved signal-to-noise ratio (> 1 GHz) compared to wire-bonded parts and low-inductance die interconnects	ChipArray® Ball Grid Array (Amkor)	fcCSP	Cell phones, handheld electronics, and high-performance workstations (notebooks and RF applications)
Flexible Interposer CSP				
Polyimide flex substrate (single electrical layer), wire bonded and overmolded	Significant area reduction compared to quad flat packs Cost parity with QFPs	MicroStar™ BGA (Texas Instruments)	MicroStar™ and MicroStar™Junior BGA	Consumer-based end equipment using digital signal processor solutions (wireless communication, laptop computers, and hard-disk drives)
Silicon die wire-bonded to polyimide tape	N/A	Flex Tape Chip Scale BGA	CS144/CSG144	N/A
Wafer-level CSP (WLCSP)				
Repassivation and deposition of solder bumps	Eliminates need for underfill, standard SMT processes, smallest footprint per I/O with space savings on PWB	WLCSP (National)	Micro SMD	N/A
Polyimide layer for redistribution traces, over-molded	30% reduction in board area, 65% thinner profile	Super CSP (Fujitsu)	Ball-grid array, land-grid array	Flash memory and ASIC

BCC: Bumped Chip Carrier, PDA: Personal Data Assistant.
[a] Technology licensed to Amkor; NA – Not Available.

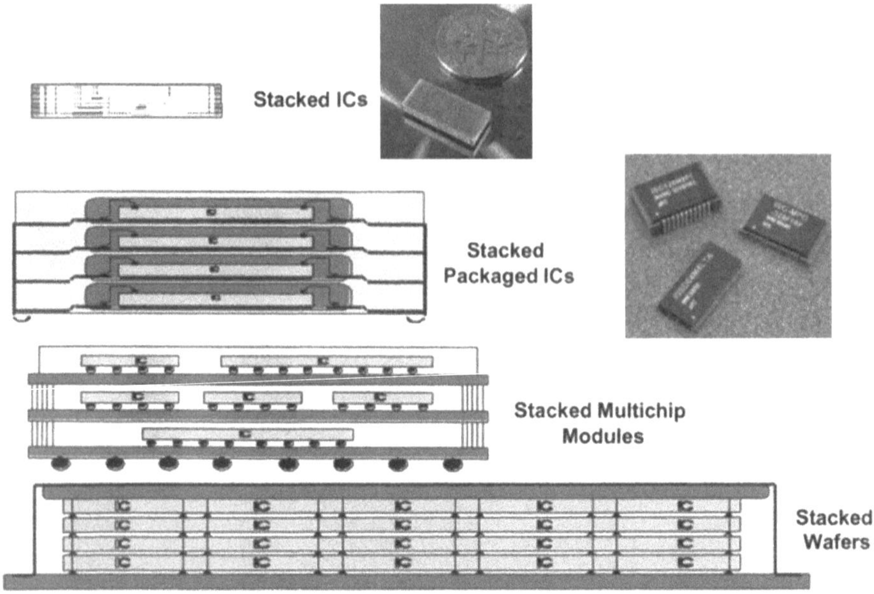

Figure 5.13 Vertical-stack packaging designs.
Source: Irvine Sensors.

5.2.6.3 *Multichip stacking*

Several IC chips are assembled using MCM techniques and the layers are stacked similarly to package stacking. Layer-to-layer interconnects are produced using an interposer such as fuzz button layers that differentiate this approach from the package-stacking approach.

5.2.6.4 *Wafer stacking*

In this approach, entire wafers or segments of wafers containing ICs are stacked. Interconnections are formed from vias in the silicon. In vertically stacking memory chips or other IC chips, the chips are first thinned to several mils (4 mils or less), then adhesive bonded, and electrically interconnected directly from chip-to-chip, either from the edges or through vias in the silicon. In these approaches, both electrically conductive and electrically insulative adhesives are used. Thermally conductive preform adhesives or thin thermoplastic films are used to bond and isolate the chips within the stack and to dissipate heat. Several processes are available to vertically stack and interconnect chips of the same size and function such as SRAM, flash, and DRAM memory chips. Other processes have been developed to vertically stack chips of different sizes and functions, or to horizontally interconnect different chips in one layer, then vertically stack the layers.[39,40]

In most cases, the vertical-stacking process begins at the wafer stage where the normal bond pads are first redistributed toward the edges of the die by several photolithographic process steps shown in Figures 5.14 and 5.15 and described as follows.

- Apply and cure a dielectric coating such as polyimide over the entire top surface of the fabricated memory wafer.
- Etch vias in the dielectric directly over the original bond pads using a photoresist process. Vias may be chemically etched, plasma etched, or laser ablated.

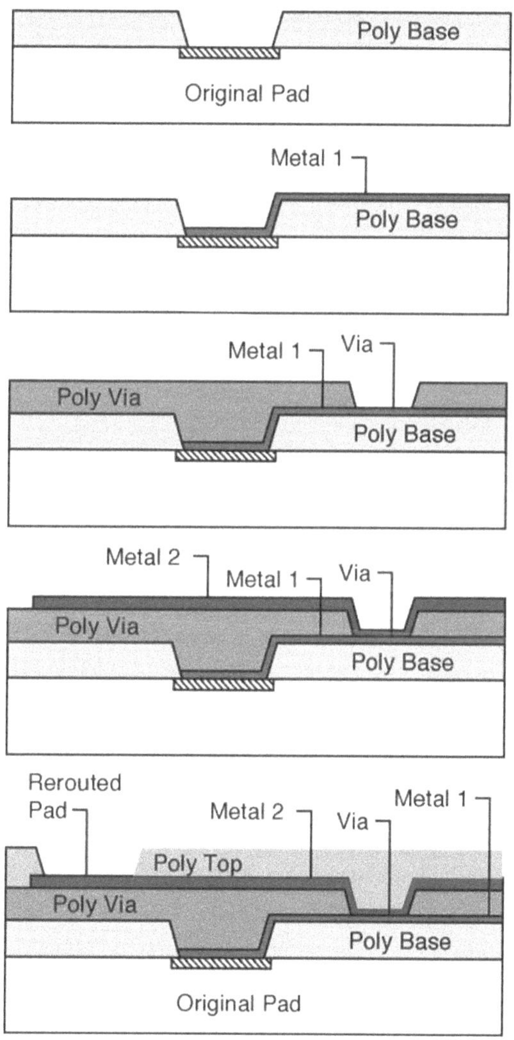

Figure 5.14 Redistribution of bond pads at wafer stage.[39]

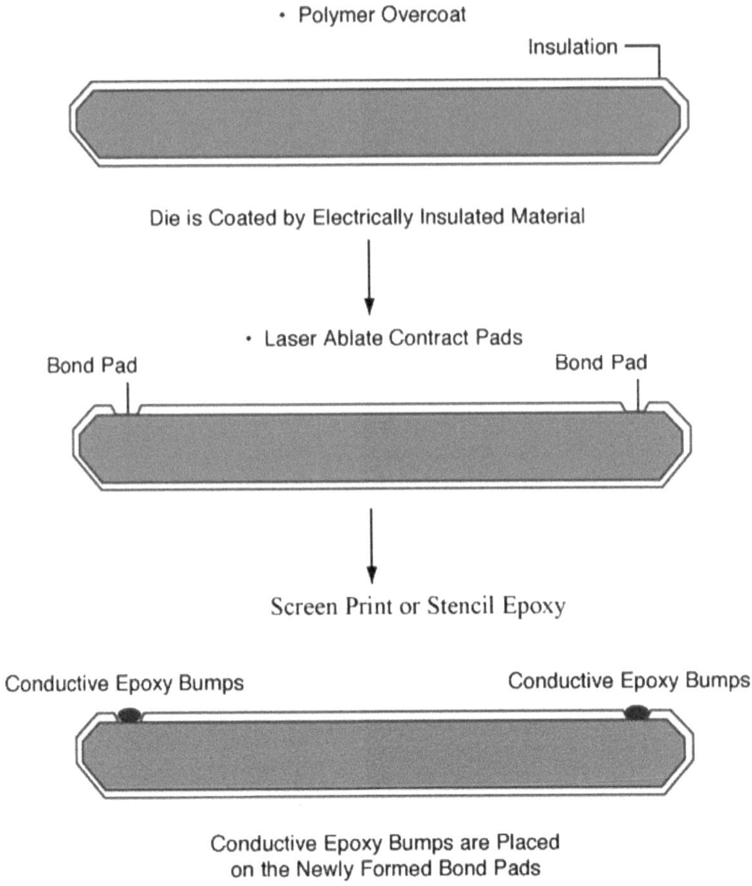

Figure 5.15 Conductive epoxy bumps on redistributed bonding pads.[39]

- Vapor deposit a thin layer of metal, such as gold, over the entire surface, covering all the dielectric and filling in the vias. The original aluminum bonding pads are usually transitioned to gold by first metallizing with barrier and adhesion metals such as titanium, tungsten, nickel, and copper.
- Delineate the metallization through another photoresist process to form conductor traces that extend to the new redistributed bonding pads.
- Coat the entire surface again with polyimide dielectric and etch vias over the new pads.
- Fill the vias by screen printing or stencil printing sufficient silver-filled epoxy to form bumps or leave the vias open for wire bonding.

Once the epoxy bumps have been formed, the wafer is thinned from the backside, the devices are singulated, and their edges beveled at 45° to expose gold pads at the periphery of die. The chips are then stacked and bonded with thermally conductive, electrically insulative epoxy (Figure 5.16). Paste adhesives are cost effective and can be dispensed with existing epoxy dispensing equipment. However, with

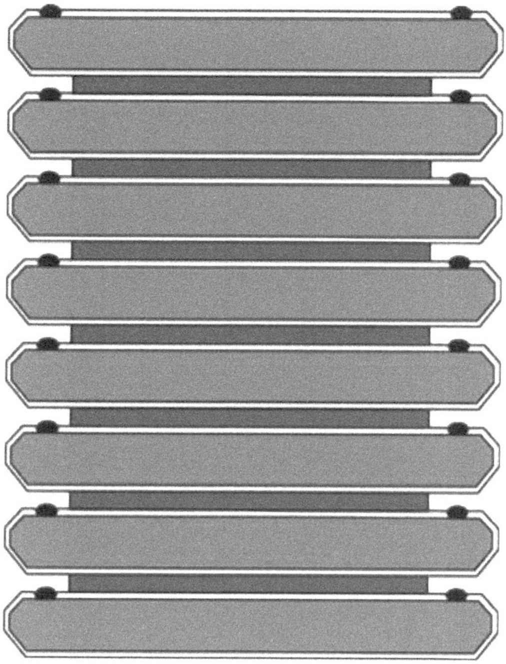

Figure 5.16 Chip stack built up using thermally conductive preform between die layers.[39]

increasing miniaturization, paste adhesives present problems in the control of voids, fillet coverage, bond line thickness, bleedout, and die tilting. Film adhesives have advantages over paste adhesives in all these areas and are being used to a greater extent in new designs. After stacking, the assemblies may be interconnected from the perimeter pads using fine wire or interconnected on the vertical edges by thin-film metallization or by applying electrically conductive epoxy stripes (Figure 5.17).

The assembled stacks are then attached to a leadframe or substrate again with conductive epoxy or wire bonds, and encapsulated with epoxy. Alternately, they may be packaged in thin, small-outline packages (TSOP) or BGA packages. If multiple die are used on each layer, as many as 64 die may be incorporated in one stacked module.[41]

By vertical stacking, eight 256-Mbit memory chips can be stacked and interconnected at their edges resulting in a 2-Gbit flash SDRAM module. Storage densities of more than 2.0 gigabytes per cubic inch using 16 Mbit DRAMs are reported.[41] Vertical stacking has resulted in numerous benefits, among which are reduced power consumption of 15%–20%, reduced capacitance, elimination of package parasitics, and, of course, smaller volume and higher density.[42] A 3-D memory stack is shown in Figure 5.18.

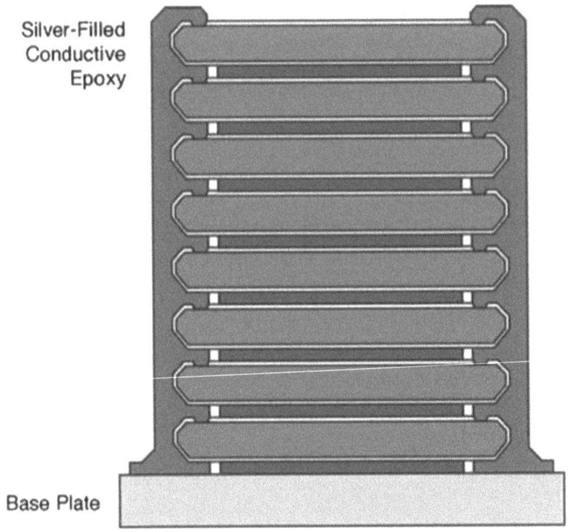

Figure 5.17 Interconnection of die from sides using conductive epoxy.[39]

Figure 5.18 3-D memory stack.
Source: Vertical Circuits, Inc.

5.2.6.5 Through-silicon vias

Through-silicon vias (TSV) are being widely investigated for vertical interconnections and 3-D stacking individual chips or multichip substrates. The TSV approach avoids the use of side edge connections such as wires, photoetched vapor deposited or plated metal conductors, or screen-printed electrically conductive adhesives, thus further reducing the size, weight and complexity of the modules.[43] TSV is also a direct approach, one that avoids the use of flip-chip interconnections or the use of interposers. A further benefit is the shortened electrical paths resulting in lower inductance, resistance and capacitance important for high-speed circuits. TSVs are an essential step in wafer scale packaging (WSP) where 3-D stacking of wafers processed using various types of devices can be interconnected and packaged at the wafer level then singulated to produce thousands of mini-modules at lower cost. Considerable interest exists in using TSV to further miniaturize and reduce the cost of cameras for mobile phones, video over the internet and security devices.

In TSV, vertical micro through holes (vias) are chemically etched, mechanically drilled or laser drilled in the silicon die at the wafer level, then made electrically conductive by plating or vapor depositing metal conductors. The sidewalls of the vias, however, must first be insulated from the silicon. A dielectric film overlaid with conductive metal may be applied to the sidewalls or the exposed silicon may first be oxidized to isolate it from subsequent metallization. Connections within the chips are generally made between bond pads on one face of the die to a land on the opposite side.

In one approach[44] thin-film aluminum is vapor deposited or sputtered onto the silicon wafer and photoetched to form a mask to etch the vias. The vias are plasma etched using fluorine or a gaseous fluoride compound such as carbon tetrafluoride but etching is stopped short of going completely through the silicon. The aluminum mask is then removed and silicon dioxide is deposited over the entire wafer to isolate the silicon from subsequent deposition of copper. A tie layer of Ti/W is generally deposited prior to copper deposition. The vias are then filled with either a conductive

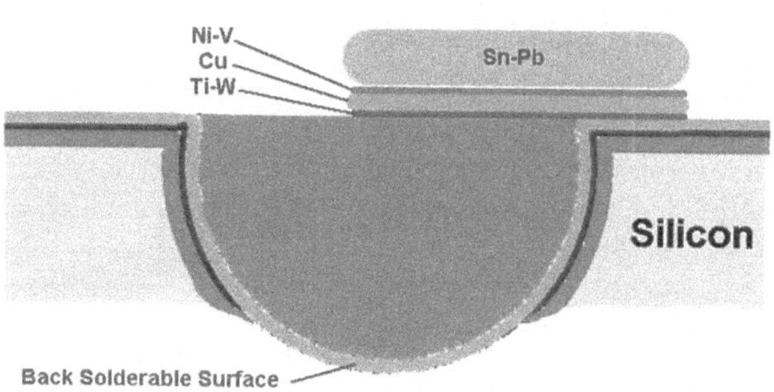

Figure 5.19 Stackable Interconnect Via.

or insulative adhesive and the backside of the wafer is thinned to the bottom copper contacts. Further processing of the top layer produces the top bonding pads thus allowing top-to-bottom via connections for subsequent stacking (Figure 5.19).

Stacking and interconnecting of wafers may be face-to-face or front-to-back, depending on the 3-D design. Interconnections between wafers may be made by flip-chip bonding or by using an anisotropic conductive adhesive film. The 3-D mini-stacks are then singulated.

In packaging imaging sensors at the wafer level (Wafer Level Packaging), insulative adhesive in the shape of window frames are screen printed or otherwise simultaneously applied around all the devices on the wafer, a glass wafer is then attached, and the individual packaged die are singulated.[45]

5.2.7 System-in-package (SIP)

At higher levels of integration, stacked IC chips of various functions may be combined with passive devices (capacitors, resistors) and connected as a functional block known as *system-in-package* (SIP). The SIP packaging approach provides a small form factor for multiple electronic functions important in mobile phones and other telecommunication products. The vertically stacked chips are interconnected either by wire bonding or by flip-chip attachment onto a BGA substrate, then epoxy molded to form a module that can be treated as a discrete standard component in board-level manufacturing. In maintaining a low-profile package, the die must be thinned before stacking. Thus, a 5-chip stacked package can be produced having a height of only 1.4 mm. An example of a SIP that uses chip stacking is shown in Figure 5.20. Carsem, Inc., offers a line of SIPs (Table 5.12).[46] As with memory stacks, both conductive and nonconductive paste and film adhesives may be used in processing SIPs.

5.2.8 Flat panel displays

Polymer adhesives are essential in the manufacture of flat-panel displays (Figures 5.21 and 5.22). Adhesives are primarily used in the back-end packaging as shown in Figure 5.22. Common uses of adhesives are in main and final seal application as well as fabricating the driver electronics. Anisotropic conductive adhesives are also used in the bonding of Indium Tin Oxide (ITO) electrodes to circuitry associated with the driver electronics. Other applications include bonding of the backlights and attaching polarizers and polarizing films. In particular, adhesives have played a critical role in the development of liquid-crystal displays (LCDs). The durability, reliability, and functionality of state-of-the art LCDs have been made possible by advancements in both film and paste adhesives.

5.2.8.1 Front and rear panel processing

The processes in Figure 5.21 are generally representative of those used in semiconductor fabrication.[47] Deposition and etching processes are used on both sides of

Figure 5.20 Stacked system-in-package.
Source: Strand Interconnect AB.

a glass substrate. The top side of the substrate is processed to produce a color filter. The backside of the substrate has conductive interconnects to energize and orient the crystals. [48] The LCD panel assembly process combines the color filter and the array substrate.

The front panel electrode is produced by sputter depositing indium tin oxide onto the glass. Then, a black matrix and red, green, and blue color filters are deposited onto the panel and patterned using standard photolithography processes. The black matrix creates a border around the color filters for contrast. Since the color filter produces a non-uniform surface, a polyimide coating is next deposited to planarize the surface. The last material to be added to the front panel is an alignment layer, which is also a polyimide applied by roll coating and then rubbed to the desired molecular orientation.

The rear panel is where transistors are formed. The fast response required for LCDs requires a switch at each pixel. The switch is a transistor that consists of a gate, source and drain, and channel. The gate metal is applied to the glass substrate and photodeliniated using standard processes. Next, a silicon-based dielectric, channel,

Table 5.12 System-in-package products[46]

Package Size, mm	I/Os	Substrate	Number of Passive Devices	Number of ICs	Number of Flip-Chip Devices	Moisture Sensitivity Level
1.42 × 1.8	8	BT	–	4		1
4.9 × 3.2	8	BT-epoxy	2	3		3
5 × 5	28	MLP	4	2		3
7 × 7	48	BP	6	1	1	3
11 × 9	24	BT-epoxy	1	1	1	3
10 × 14	34	Ceramic	49	1	1	3
11.4 × 10.2	36	FR4	1	1	1	3
11.5 × 7.5	56	BT-epoxy	13	3		3
11 × 11	133	BT-epoxy	3	6	2	3
14 × 14	219	BT-epoxy	16	6	1	3
15.5 × 9.25	161	BT	10	12		3

Source: Carsem, Inc.

and a doped silicon layer are applied and patterned. The pixel electrode is formed by sputtering indium tin oxide. The source/drain metal is sputtered and patterned, and a contact is formed between this layer and the doped silicon layer. The substrate is then cleaned, a thin polymer alignment layer applied to the surface, and then the layer is rubbed in the direction desired for liquid crystal orientation.

5.2.8.2 *Display cell assembly*

Figure 5.22 represents the actual module assembly. At this stage of the process, the color filter substrate (front glass) and the transistor substrate (rear glass) are joined with an adhesive seal material. Spacers are deposited and then a main seal (adhesive) is applied to mate the two glass plates and the adhesive is cured. Liquid crystal material is then injected between the two plates followed by an end seal and a cure cycle. Both the main and end seal materials are UV- or heat-curable epoxy resins. Low-viscosity adhesives that penetrate by capillary flow are required. Other requirements are that the adhesive must (1) selectively bond only the edges and not contaminate the liquid crystals, (2) cure at a low temperature to avoid degrading the liquid crystals, and (3) resist air and moisture permeation. Several examples are given in Table 5.13.[49,50] Among the materials used are silicones, acrylics, and epoxies. Examples of end seal materials are in Table 5.14.[50,51]

The final step in display cell assembly is applying polarizing films or polarizers to the outside of each panel. Every square inch of LCDs, both top and bottom, must be covered with a laminated polarizer and additional polymer films having unique optical birefringent properties. Polarizer films are needed to control the passage of light through the liquid-crystal cell. Polarizers and analyzers separate the image created by the electrically altered orientation of the birefringent liquid crystal film between the glass plates. The optical films are called "compensating films" and are used to enhance the viewing angle

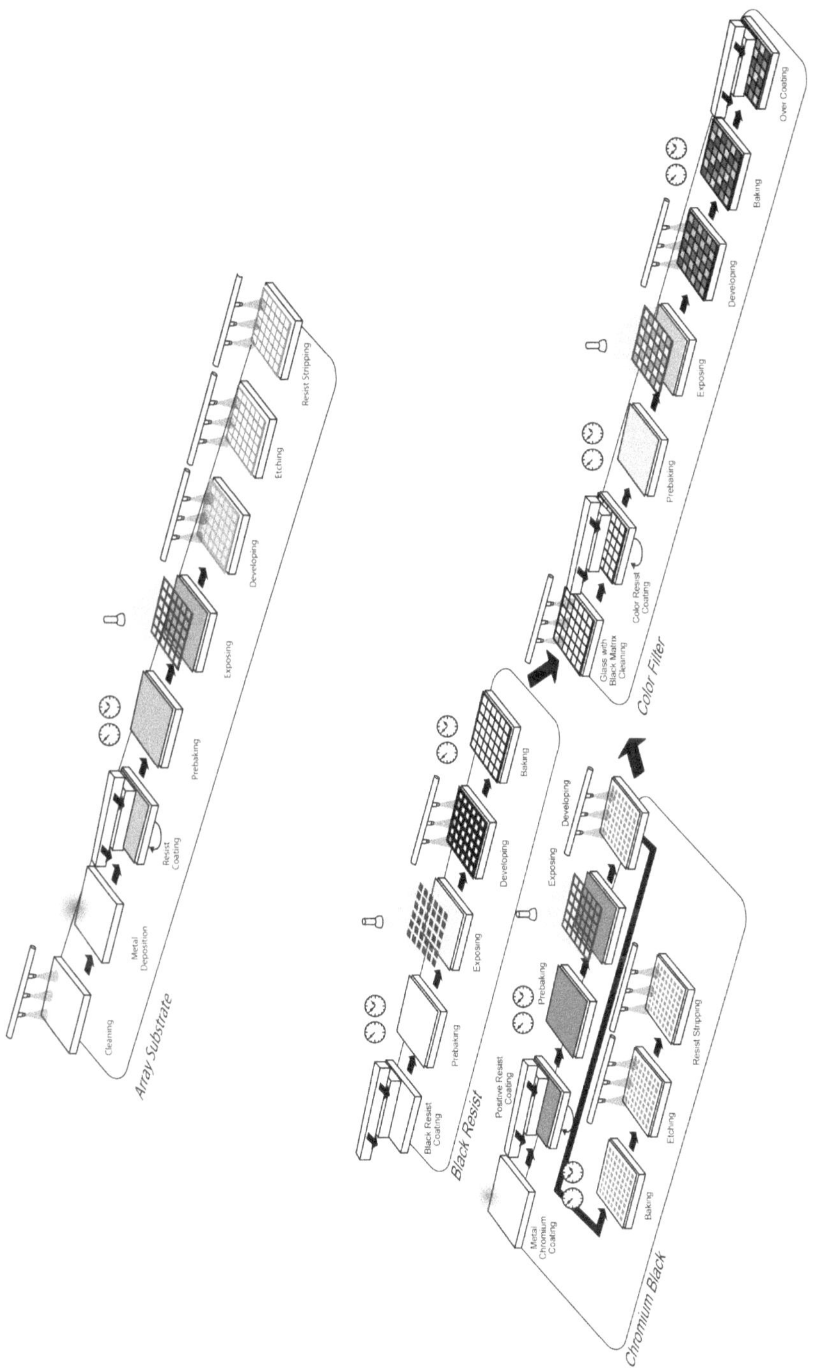

Figure 5.21 Front and rear flat panel process flows.[47]: Array substrate (Top); Color filter (Bottom).

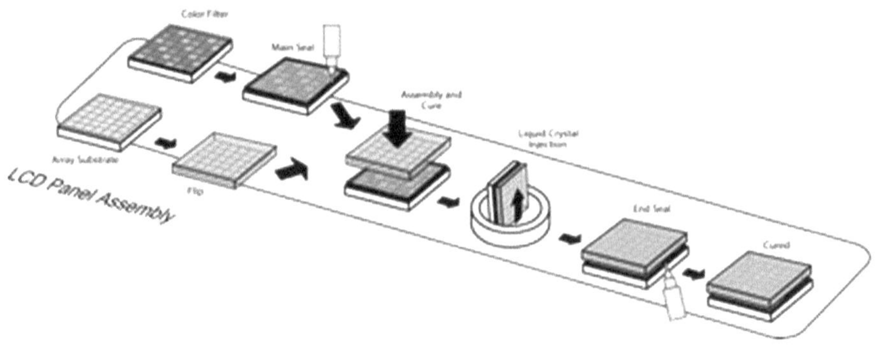

Figure 5.22 Flat panel assembly process flow.[47]

Table 5.13 Properties of main sealants[49,50]

Test Item	ThreeBond 3025	ThreeBond 3025 G
Type	UV-curing	UV-curing (Epoxy Based)
Cure condition	UV – 3,000 mJ/cm^2	UV – 4,000 mJ/cm^2
Appearance	Milky white liquid	Milky brown liquid
Viscosity, Pa·s (P)	80 (800)	45 (450)
Specific gravity, g/cm^3	1.42	1.38
Hardness (JIS* D)	90	90
Glass-transition temperature (°C)	95	140
Boiling water absorption rate (2 hour, %)	0.1	1.0
Moisture permeability (g/cm^2 /24 hours/ 25 μm)	7.0	5.0
Cure shrinkage (%)	6.5	3.0
Specific volume resistivity (ohm-cm)	8.0×10^{14}	N/A
Specific surface resistance (ohm)	5.0×10^{14}	N/A
Dielectric constant (20 °C, 1 MHz)	4.11	N/A
Dielectric constant (120 °C, 1 MHz)	4.38	N/A
Dielectric loss tangent (20 °C, 1 MHz)	0.032	N/A
Dielectric loss tangent (120 °C, 1 MHz)	0.030	N/A
Peeling adhesive strength (Corning 7059 glass, kg/cm^2)	2.0	6.4
Peeling adhesive strength (Indium tin oxide on Corning 7059 glass, kg/cm^2)	1.3	1.1
Total Ionic Contents (ppm)	Cl^- 10 Na^+ 3 K^+ 1	Cl^- 25 Na^+ 1 K^+ 1

* Japanese Industrial Standards.

with birefringent optical compensation in three axes. The polarizer film is pre-cut with adhesive layer, compensator layer, and second adhesive layer for lamination to the LCD glass. This laminate stack uses a carrier film for placement and protection. The film is packaged in large rolls for automatic application to the LCD using custom laminating

Table 5.14 Properties of end sealants[50,51]

Test Item	ThreeBond 3025	Loctite 3010
Appearance	Milky white liquid	Transparent amber liquid
Viscosity Pa·s (P)	15 (150)	(110)
Hardness (JIS[a] D)	90	87
Cure shrinkage (%)	6.1	7.4
Shear strength MPa (Kg/cm^2)	7.0 (71.4)	N/A
Glass-transition temperature (°C)	80	88
Thermal expansion coefficient (ppm/°C)	63	62
Moisture permeability (g/cm^2/24 hours/25 μm)	36	N/A
Ionic Contents (ppm)	Cl^- 100 Na^+ 3 K^+ 1	N/A

N/A – Not Available.
[a] Japanese Industrial Standards.

machinery. After application, the display is placed in an autoclave to cure the adhesive, typically at two atmospheres pressure and 60 °C. The cured adhesive layers must have unique optical and mechanical properties to be used in a LCD. The index of refraction of the polymer film must closely match that of the glass to assure minimum scattering and absorption. The most difficult properties to achieve are mechanical because the adhesive must maintain its bond strength for dissimilar materials over a wide temperature range from −40 °C to +80 °C, and preferably wider for industrial and aerospace applications. The polymer films must also be removable for repair of the LCD if the polarizer film becomes scratched or damaged. Selected adhesive materials and their properties are given in Table 5.15. Examples of adhesives used to bond these films are ARclear™ 8930-Strippable Acrylic Transfer Adhesive and ARclear™ 8932-Silicone Transfer Adhesive engineered for Adhesives Research Corporations's Optically Clear Pressure-Sensitive Adhesive Technologies for the display screen/flat panel display industry. ARclear 8930 is a 25-μm-thick strippable, acrylic optically clear transfer adhesive used in applications requiring high quality optical properties and preservation of expensive components. ARclear 8932 is a 50-μm-thick silicone-based transfer adhesive designed for low flammability and resistance to extreme temperatures.

5.2.8.3 Module assembly

The module assembly step includes attaching the backlight unit, which is the light source for the LCD. A typical desktop unit has four backlights that are placed around the edges of the display. The light projects across a diffuser screen to provide uniform illumination. The printed wiring boards, the power supply, and the plastic cover and stand are added to make an assembled monitor.

Generally, the electronics that drive the display have been assembled and packaged separately, then mechanically and electrically connected to the glass display portion with a flexible cable. The flex connector is mechanically and electrically bonded to the

Table 5.15 Pressure-sensitive adhesives for polarizer attachment

Product Name/ Supplier	**Construction**	**180° peel adhesion (12" min) (5'/SS)**	**Liner release**
ARclear 8930/ Adhesives Research Inc.	Polyester release liner, no carrier, 1-mil caliper	2.72 N/cm	N/A
ARclear 8932/ Adhesives Research Inc.	Polyester release liner, no carrier, 2-mil caliper	4.36 N/cm	N/A
ARclear 90176/ Adhesives Research Inc.	Polycoated craft liner: 6.5 mils ARclear Adhesive (AS-144): 1.5 mils Polyester carrier film: 2.0 mils ARclear Adhesive (AS-144): 1.5 mils Clear polyester liner: 2.0 mils	Liner 1 side: 50 oz/in Liner 2 side: 20 oz/in	Liner 1: 150 gm/2" width Liner 2: 20 gm/2" width
ARclear 90178/ Adhesives Research Inc.	Polycoated craft liner: 6.5 mils ARclear Adhesive (AS-144): 1.5 mils Polyester carrier film: 2.0 mils ARclear Adhesive (AS-144): 1.5 mils Clear polyester liner: 2.0 mils	30 oz/in	N/A

N/A – Not Available.
Source: Compiled from Supplier's Technical Data Sheets.

row-and-column ITO electrodes of the glass panel. The separate assembly and packaging of the electronics adds to the volume and weight of the final product and increases its cost. Thus, the focus in recent years has been to integrate the electronics with the panel by forming or assembling the driver electronics directly on or into the panel, reducing the overall thickness and assembly costs. The ultimate integration would be to extend the thin-film processes used to fabricate the active matrix circuits on the glass, that is, to integrate the circuits on the available area at the periphery of the glass display. However, the normally used solder attachments of devices or connectors to the LCD runs the risk of degrading the liquid crystals due to the high solder temperatures involved. Furthermore, the fine-pitch connections required for chip devices work poorly for solder due to solder bridging and smearing. Thus, polymer adhesives, as alternates to solder, because of their low temperature processing and rapid cures, have found a specific niche in the assembly of flat-panel displays.

Several approaches using electrically conductive adhesives instead of solder have been explored and are proving successful. Anisotropic adhesives (See Chapter 1), for example, have been used to connect flex circuits and cables from the separate driver circuits to the panel, avoiding the use of solder connections. More importantly, integrated-circuit chips can be bonded directly to the ITO conductor traces on the panel, a technology called chip-on-glass (COG). IC chips can be flip-chip bonded, then underfilled with a stress-free underfill adhesive if necessary. For protection, the chips may then be encapsulated with epoxy (glob-topped).

Alternately, flip-chip devices or TAB devices may be connected using an anisotropic film adhesive which, with moderate heat and pressure, forms electrical connections between the bumps or leads of the device and the conductor pads on the panel. In using an anisotropic film adhesive, the film is first attached to the glass area at 80 °C to 100 °C under pressure of 200–400 N/cm^2 for 20–30 seconds.[52] The protective layer is then removed from the film, components are placed and aligned, and the connections are made by bonding at 170 °C–180 °C under pressure of 50–100 N/cm^2. Anisotropic adhesives are used to connect (using the TAB process) large-scale integration (LSI) chip carrying polyimide film with traces to traces on the glass edge of the LCDs. Typical materials used are given in Table 5.16.[50,53,54,55]

The traces are at densities of over 200-line pair per inch. The adhesive must survive temperatures from −40 °C to +80 °C and be reparable. The resolution of the LCD is limited by the achievable line-pair density. If the lines are too closely spaced, the conductivity between the lines becomes too high. The LCD industry has been the prime motivator for improving anisotropic conductive adhesives with very low line-pair pitch.

UV-curing adhesives are also finding an application in flat-panel electronics. Because the panel is glass and transparent to UV light, devices can be attached with a UV-curing adhesive on the topside and, after accurate positioning, exposed to UV light from the backside of the panel and cured in a few seconds. Thermoplastic "hot melt" adhesives have also found use in flat-panel displays (FPDs) because of their rapid bonding and ease of reworking.

5.2.9 LED lighting and display applications

Light emitting diode (LED) applications are rapidly growing for both lighting and flat panel displays. Already in commercial use are LED traffic signal lights, signs, automobile lights, and light bulbs. Computers with LED monitor screens up to 17.3 inches are also available (Dell Studio 1749 Touch computer). Expected regulations for energy efficient home and industrial lighting and for large TV screens are already driving the development of low-cost high efficiency LED bulbs, lighting devices and flat panel displays.

Incandescent and fluorescent bulbs have been in use for lighting since their introduction in the late 19th and mid 20th centuries, respectively. Incandescent bulbs use heated tungsten filaments to produce visible light and fluorescent lights use an electric discharge into argon or mercury gas to produce UV light that results in visible light when striking a phosphor lining in the bulb. In comparison, light-emitting diodes (LEDs) produce photons directly from semiconductor junctions. A new transition

Table 5.16 Properties of anisotropic conductive adhesives[50,53–55]

Adhesive/Supplier	**ThreeBond 3370C/ThreeBond**	**Anisolm AC-2056/Hitachi**	**Anisotropic conductive Film 7393/3 M**
Type	Normal fine-pitch products	Tape adhesive	Fine (200-μm pitch)
Appearance	Gray, transparent film	Transparent	N/A
Composition	Thermoplastic resin	–	Thermoset-thermoplastic matrix
Conductive filler	Plated resin powder (8 μm)	Nickel and metal coated resin particles	Gold-coated polymer
Sheet structure	Three layers (with mold release film)	–	Polyester film liner (50 μm), 25 μm adhesive thickness
Film thickness	35+/−5 μm	35 and 45 μm	25 μm
Specifications	Width: 3 mm, 4 mm, and 5 mm Length: 30 m	Width: 1.5, 2.0, and 2.5 mm Length: 50 and 100 m	–
Connection resistance	1–5 ohm (FPC/PCB) (0.3 mm pitch, 3 mm width)	1 ohm (PWB.TCP) (2.0 mm width)	<10 ohm (Flex to ITO/PET) <0.1 ohm (Flex-to-Ag Ink/ ITO/PET)
Inter-line insulation resistance (0.2 mm-pitch comb10 V DC)	10^8 ohm or higher	10^{12} ohm	–
Minimum possible conductor width	0.1 m (0.2 mm pitch)	0.1 mm (0.2 mm pitch)	100 μm
Shear strength	500 N/m	500 N/m	–
Pressure bonding temperature	130–150 °C	170 °C	150–170 °C
Pressure bonding time	10–30 sec	20 seconds	10–15 seconds
Pressure bonding pressure	2.9 MPa	2 MPa	–
Peel strength (g/in width)	–	>2,540 (1.2 kN/m)	>1,524 (ITO/PET) >1,016 (silver ink/ITO/PET)

involving the use of digital lighting, also known as LEDs, promises to revolutionize commercial and residential lighting.

LED technology was first introduced in 1962. At present, LEDs account for less than 5 % of the lighting market, but by the end of the decade this technology is expected to dominate commercial and residential lighting.[56] This advance has been compared to the impact of digital cameras on the film business. The driving factors behind this change are reduced energy consumption, improved efficiencies, and longer lifetime of LEDs compared to more conventional methods of lighting.

Advantages of solid-state LEDs over conventional lighting are as follows:[56,57]

- Lower energy consumption. LEDs are 80 % more efficient than incandescent bulbs.
- Longer lifetime. The lifetime of a LED device is approximately 50,000 hours, far better than traditional light sources.
- Less wasted light. LEDs produce more uniform and directional lighting.
- Lower maintenance cost. The longer lifetime means fewer bulb replacements and cost savings.
- Higher mechanical integrity. LEDs are not affected by vibration or shock

Disadvantages of digital lighting are the following:

- Thermal sensitivity. Of the applied power, 75–85 % is dissipated as heat.[58]
- High material cost.

The overall configuration of a digital light bulb based on LED technology, also known as a *luminaire*, is shown in Figure 5.23. Basic elements in digital lights consist of an exterior lens case, a packaged semiconductor die on a heat sink, and a PWB with associated electronics.[56] The exterior lens and the lens case are made of high-impact plastic for ruggedization.

The major requirements for digital lighting are a long operational lifetime, high light output, and a fixed wavelength of light. Unlike incandescent and fluorescent lighting, LED light sources do not fail abruptly. Instead, the light output degrades over time[57] as shown in Figure 5.24 for a Phillips Lumiled White LXM3-PWx1.

Lifetime is affected by the junction temperature of the diode. The graph in Figure 5.25 refers to a 50 % failure rate and a lumen output maintenance of 70 %, the threshold at which the eye begins to detect a reduction in light output at various continuous forward currents. As junction temperature increases, lifetime decreases as indicated by a reduction in light output.[57]

In LEDs, a higher junction temperature also reduces light output. The light output is inversely proportional to junction temperature. In addition to the effect on lifetime and light output, the wavelength of the output light also increases at higher junction temperatures. This shift often makes orange light appear red or white light appear blue. In current leading-edge technologies the junction temperature is normally in the range of 60 to 75 °C.[57] Because of the high thermal sensitivity of LEDs, substrates, package materials, adhesives and interconnection methods must be carefully selected.

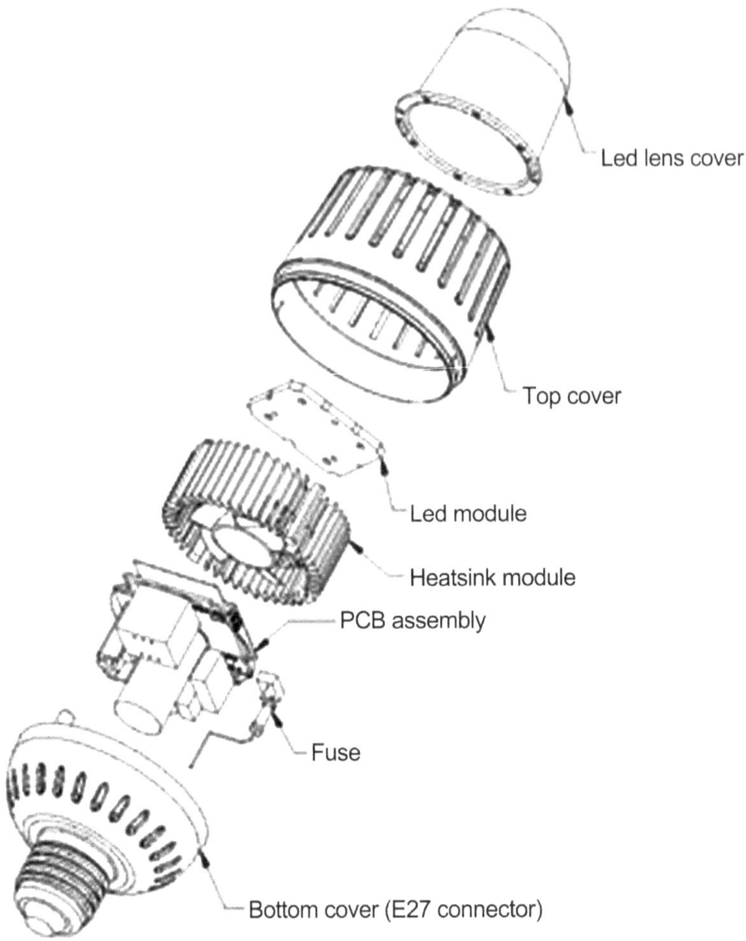

Figure 5.23 Packaging of digital lighting.
(Courtesy: Advanced Lumonics, EarthLED Div.)

LEDs are generally fabricated by growing compound semiconductors such as gallium nitride (GaN), indium gallium nitride (InGaN), aluminum gallium phosphide (AlGaP), or gallium phosphide (GaP) on wafer substrates such as silicon carbide, silicon, or sapphire. After singulation from the wafer, the LED chips are attached to a package or a heat spreader either with conventional solders or electrically conductive adhesives. As with many other electronic applications, adhesives play a vital role in the assembly and packaging of LEDs for lighting. Electrically conductive adhesives provide lower thermal resistance than solder since the thinner bond line possible with adhesives offsets their lower thermal conductivity (Table 5.17).[59] Electrically insulative adhesives, however, should be avoided due to their higher thermal resistance that can result in junction

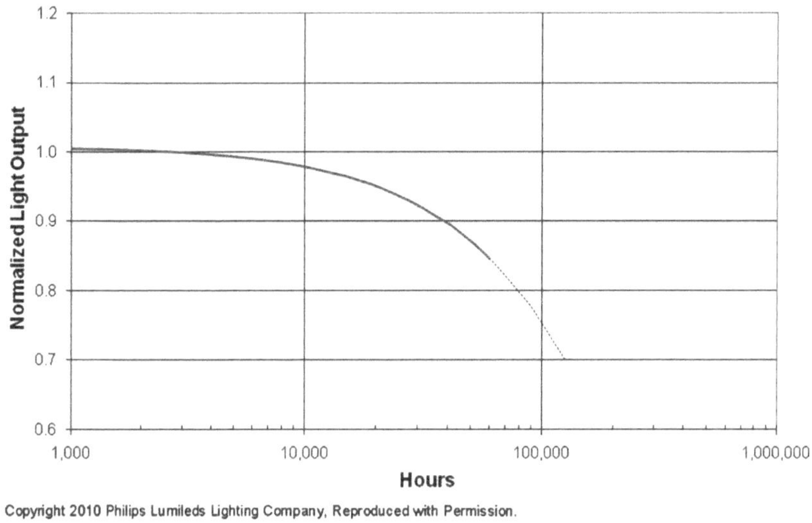

Figure 5.24 Degradation of light intensity with time.[57] (Lumen Maintenance Projection for White LXM3-PWx1 at 55 °C and 0.3A [$T_{junction}$ ~68 °C, Normalized to 1 at 24 hours]). Copyright(©) 2010 Philips Lumileds Lighting Company, reproduced with permission.

temperature rises of 20 °C for a 5 W LED. Among the best high thermally conductive adhesives used for LED attachments are the highly filled silver epoxies. Epoxies with up to 85% silver filler are reported to have thermal conductivities of 20 to 60 W/mK. The properties of DieMat® DM6030 adhesives used for LED attachment are given in Table 5.18.[60]

Besides high thermal conductivities, adhesives must combine low cost and relieve stresses from mismatches in the coefficients of expansion of the LED material and the substrates or packages to which they are attached.

Substrate materials often used for the packaged part include aluminum, copper, and composites of copper with diamond, molybdenum or tungsten. These materials and selected properties are shown in Table 5.19.[57]

Substrates such as diamond-copper combine excellent thermal conductivity and closely matched thermal-expansions but at a significantly higher cost. On the other hand, copper that has high conductivity and is low in cost, presents an expansion-coefficient mismatch with the LED materials. These problems may be resolved by formulating adhesives with modified epoxies, silicones, or flexibilizers to relieve stresses (see Chapter 2).

After attachment to the substrates, LED chips can be wire bonded or flip-chip bonded. If wire bonded (see section 1.2.1), the entire backside is available for heat dissipation. In flip-chip bonded formats, however, the thermal path is through the bumps only. Therefore, from a thermal dissipation viewpoint, direct die attachment to a heat spreader followed by wire bonding is preferred. Regardless of the interconnect approach, package materials having high thermal conductivities are required.

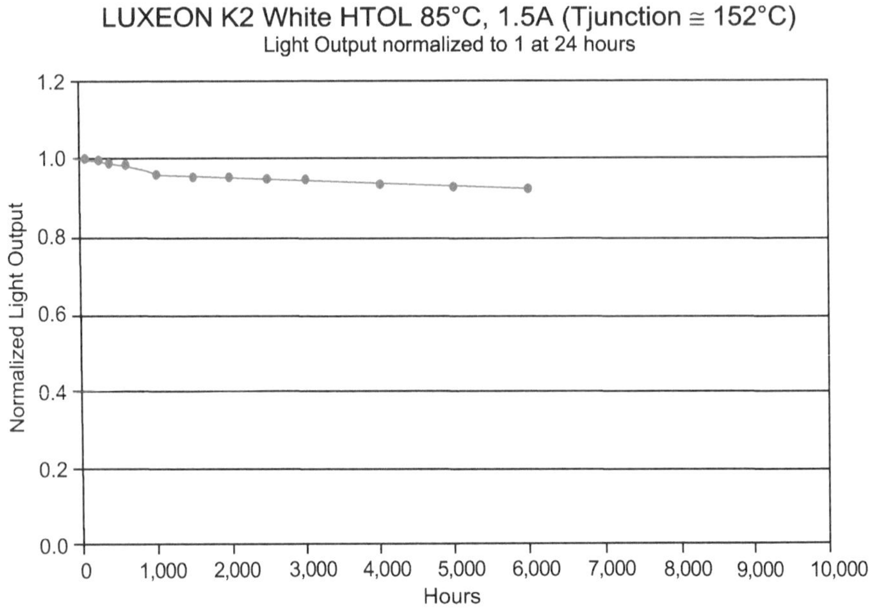

Figure 5.25 Effect of diode junction temperature on LED lifetime.

5.2.10 *Organic light emitting diodes (OLED)*

OLEDs are based on certain organic small molecules or polymers that behave as diode semiconductors with an applied current. The organic materials derive their semiconducting properties from their highly conjugated pi-bond molecular structures that allow electrons to flow within or across the molecular bonds. Whether

Table 5.17 Thermal characteristics of component attach materials[59]

Material	Thermal conductivity (W/m·K)	Thickness in package structure (μm)	Thermal resistance (°C/W)
Solder (63/37)	51	50	0.98
Conductive adhesive (DieMat® DM6030)	30	15	0.5
Non-conductive adhesive (Bond line 7258)	3	12	4

Table 5.18 Typical properties of DieMat® DM6030 adhesives[60]

Property	DM6030Hk	DM6030Hk Pt	DM6030SF
Brookfield Viscosity (cP, 25 °C, 10 rpm, Model RVT, Spindle T)	31,000	31,000	173,000
Thixotropic index (25 °C, 10/50 rpm)	2.4	2.1	2.8
Shelf life (−40 °C, mos)	12	12	12
Pot life (25 °C, mos)	6	6	6
Silver content (%, wet)	85	84	86
Percent solids (wet)	91	90	99
Density (wet, g/cm^3)	4.5	4.3	4.9
Thermal conductivity (W/m·K)	60	45	12
Volume resistivity (ohm-cm)	8.0×10^{-6}	9.0×10^{-6}	2.8×10^{-6}
Die-shear strength (250 mil die to ceramic, psi)	3,200	2,400	4,800
Coefficient of thermal expansion (ppm/°C)	26	26	35
Flexural modulus (psi)	600,000	500,000	800,000
Total[a] ionic impurities (ppm)	<30	<30	<30
Silver content (cured, %)	93	93	85
Density (cured, g/cm^3)	6.7	6.7	4.9

[a] Total of Na^+, K^+, Cl^-, and F^-.

Table 5.19 Thermal and mechanical characteristics of substrate materials[57]

Material	Thermal conductivity (W/m·K)	Expansion coefficient (ppm/°C)	Density (g/cm^3)	Young's modulus (GPa)
LED Materials				
Sapphire	30	8.0	4.0	372
Silicon Carbide	120	4.0	3.1	410
Gallium Nitride	1.3	3.2	6.2	204
Candidate substrate/heat spreader materials				
Al	216	23.5	2.7	69
Cu	394	17	8.9	180
Diamond-copper composite	1,200	6	5.5	255
Cu-Mo-Cu (13-74-13)	196 (in plane), 165 (through thickness)	5.9	9.9	195
W-Cu (90-10)	185	5.9	16.8	292

single molecules or polymers, they consist of aromatic structures having alternating single and double bonds. Examples of polymers include polyanilines, polyethylenedioxythiophene, and derivatives of polyphenylene vinylene.[61] Examples of small molecules are bis-biphenyl anthracene, pentacene, and metal complexes of aromatic compounds such as copper phthalocyanine.[62]

However, there are literally thousands of organic compounds and polymers that are possible in the world of organic chemistry that may resolve some existing problems and expand the applications for OLEDs. Further, the emitting properties and efficiencies of these materials may be augmented by adding phosphorescent dopants, metal ions, or salts such as lithium fluoride or cesium fluoride.

Advantages of OLEDs over inorganic LEDs are their potential lower cost, ability to be incorporated into flexible products and ability to be deposited in various sizes and shapes on a variety of substrates including glass and thin plastic flexible sheets. They are also produced in ultrathin layers where a multilayer structure of active emitting material may be only 100 to 200 nm thick.

Commercial applications to date have been largely for small displays for wireless devices, cell phones, cameras, night vision and medical imaging devices. Work continues on expanding their use for general lighting and for large area displays such as for TV. Among drawbacks has been the need to deposit thin uniform multilayers for large areas to produce white light. As many as 7 to 9 layers may be required; three separate emission layers each emitting a different color, electron and hole transport layers and electrode layers. Although the polymers may be deposited from solutions by low-cost automated coating processes, other layers (indium tin oxide for anodes and magnesium or lithium silver alloys for the cathodes) require sputtering or vapor deposition that significantly increase the cost. Improving the lighting efficiency at low operating voltages of 2–4 V is also an objective.

It is not clear yet how adhesives will be used for OLED lighting applications except that large panels need to be sealed off from atmospheric oxygen and moisture to prevent degradation of the organic materials. Sandwiching the active layers between glass panels that are adhesive bonded, much as is done with LCD displays, is one option.

An excellent overview and roadmap for the research and development of OLEDs was developed by the Optoelectronics Industry Development Association together with industry, government and academia personnel.[63]

5.2.11 Electromagnetic interference (EMI) shielding and grounding

Electrically conductive adhesives play a key role in shielding electronic circuits, modules, and subsystems from electromagnetic interference (EMI) and in grounding. EMI shielding generally involves surrounding or enclosing the part with an electrically conductive wire mesh, a metallic coating, or a metallized foil that is attached with pressure-sensitive adhesive tape. Conductive adhesives and sealants are also used to ground chasses and electronic boxes. Adhesives are used to fill in gaps and electrically bridge lids to chassis, to bond conductive gaskets in place, to attach conductive grounding pads, and to attach connector EMI gaskets. EMI shielding is important for the successful operation across all major electronics applications including military, space, and consumer communication systems, computers, servers, and medical instruments. With the proliferation of wireless electronic devices, EMI shielding becomes even more essential. Even though PWAs and other electronic modules can be designed to minimize or avoid EMI, it becomes increasingly difficult as frequencies increase beyond 800 MHz and into the high GHz regions where most wireless devices operate. Both conductive paste adhesives and pressure-sensitive tapes are widely used for EMI protection. Two-part epoxies or one-part silicones filled with silver or silver-plated copper particles are widely used as paste adhesives while acrylic conductive and nonconductive adhesives on aluminum or copper foil are used as pressure-sensitive tapes. EMI shielding tapes are suitable for temperatures ranging from −40 °C to 121 °C and are available in thicknesses from 1.5 mil to 5 mil.[64] Shielding effectiveness is measured as attenuation in dB at several frequencies, for example, from 1 MHz to 20 GHz. Values for several ARclad® shielding tapes are given in Table 5.20.[64]

The electromagnetic shielding effectiveness of a nickel-graphite filled silicone gasket with and without an EMI adhesive, measured over a wide frequency range, is shown in Figure 5.26.[65] Test methods and requirements for the control of EMI may be found in MIL-STD-461.[66]

5.2.12 Optoelectronics

Adhesives, especially those that cure on exposure to UV/visible light, are used in a myriad of optoelectronic (OE) applications among which are active and passive optical fiber alignment and coupling, charge-coupled device (CCD) attachment, lens bonding, fiber tacking, fiber strain relief, laser assembly, and light-emitting diode (LED) attachment. Adhesives are contributing to reducing the cost of assembling and packaging optoelectronics by facilitating the alignment and coupling of optical fibers with laser diodes, lenses, and other optical parts.

Precision alignment and attachment of optical fibers to couple them with lenses, transmitter and receiver components, and laser diodes is still largely a manual, labor-intensive process. Optical fibers must be aligned to insure that the optimum amount of light is transmitted between the fiber on the outside and the laser, photodiode, or other optical component on the inside of a package. Alignment may be active or passive.[67]

Table 5.20 Shielding effectiveness of EMI Tapes[64]

Attenuation at Frequency (in dB)	1 MHz	500 MHz	1 GHz	18 GHz
ARclad® 5795 (5-mil Al foil/acrylic adhesive)	55	71	85	49
ARclad® 6738 (3.5-mil Al foil/acrylic adhesive on Cu carrier)	93	65	82	52
Flexshield® 8269 (4-mil Cu foil/acrylic adhesive on Ni/Cu/polyester carrier)	102	90	90	80

Source: Technical data sheets, Adhesives Research, Inc.

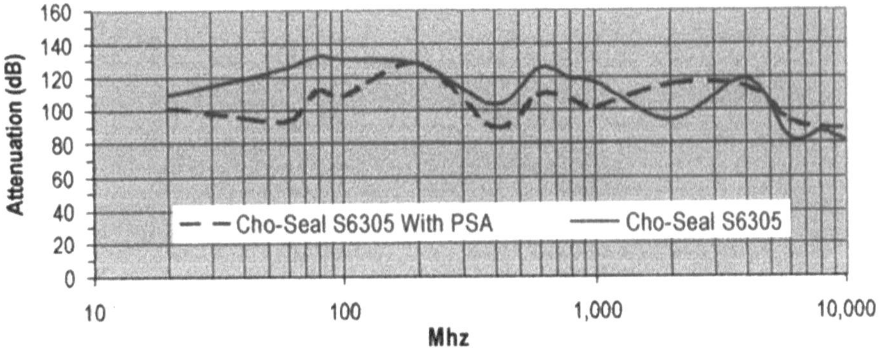

Figure 5.26 Electromagnetic shielding effectiveness of Cho-Seal S6305 gasket with and without CL-130 pressure sensitive adhesive.
Source: Chomerics, Div. of Parker Hannifin Corp.

In active alignment, the device is powered while a precision stage adjusts its position until maximum coupling is measured. The positioned components must then quickly and permanently be fixed in place, for example by adhesive bonding, soldering, or welding. Passive alignment relies on selecting materials and processes that have precise tolerances and known controllable dimensional changes after assembly.

Once in place and bonded, the bonding material must not change dimensionally, otherwise loss of coupling power occurs. Although metallurgical attachments (solder or laser welding) are used and are dimensionally stable, they require high temperatures or special, expensive equipment.[68]

Ultraviolet-light-curing adhesives that can reduce costs and be more easily manipulated than solder or other metallurgical attachment methods are being widely explored as alternates. The greatest benefit in using UV-curing polymer adhesives is that the adhesive will not cure until exposed to a UV source, thus curing is "on demand" allowing sufficient time for alignment and measurements. However, as with all polymeric materials, thermal expansion, shrinkage during cure, and continued dimensional changes occur during aging, thermal exposures, and humidity absorption. In spite of this, good results have been obtained with some UV-curing adhesives. In the assembly of 1×8 collimator array with a 1×8 lens array, Zhou et al.,[69] measured coupling losses as a function of adhesive thickness and displacement after curing. The coupling loss for three UV-curing adhesives evaluated was less than 1.5 dB for all channels, and the variation among all eight channels was less than 0.5 dB. A list of some commercially available UV-curing adhesives and their properties is given in Table 5.21. In active alignment processes, cyanoacrylate adhesives have been used for rapid tacking, but have had to be reinforced by laser welding because of the poor adhesive properties of cyanoacrylates in high humidity and temperature environments. Thermosetting adhesives such as epoxies provide excellent bond strengths, but require elevated temperatures and extended times to cure. The use of rapid setting UV-cured adhesives followed by a short heat cure holds promise of meeting all requirements.

Table 5.21 Examples of UV-curing adhesives

Adhesive/Supplier	Type	Viscosity cP (rpm)	Cure conditions	Depth of Cure (mm)	Tensile strength, MPa (psi)	Water absorption (%)	CTE (ppm/°C)		T_g (°C)	Refractive Index	Application
							$\alpha 1$	$\alpha 2$			
Dymax® OP-60/ Dymax	Urethane acrylate	150,000 (20)	UV (50–150 mW/ cm^2 (300–500 nm)	<3.2	27.6 (4,000)	2.8	58	N/A	N/A	N/A	Optical bonding where precision alignment and no movement after placement is required
Dymax® Ultra Light-Weld® 9-20269 Gel/Dymax	Solvent free, urethane methacrylate	25,000 (20)	2 minutes @ 1,000 $mWcm^2$ (365 nm)	6.4	17.9 (2,600)	5	N/A	N/A	N/A	N/A	Fast tacking of coils, LCD's, wires, and other components to PWBs
Dymax® Multicure® 997/Dymax	Modified acrylic	800,000 (N/A)	10–20 seconds (longwave, 365 nm)	N/A	27.6 (4,000)	N/A	70	N/A	60	N/A	Bonding surface-mounted components
Loctite 3525/Henkel	Modified acrylic	9,000–21,000 (20)	30 sec @ 50 mW/ cm^2(365 nm)	3.9 mm	24 (3,500)	N/A	N/A	N/A	N/A	1.51	General purpose bonding of a wide range of substrates
Loctite® 3104™/ Henkel	Acrylated urethane	90–210 (20)	< 5 sec @ 30 mW/ cm^2(365 nm)	3	19 (2,755)	3	N/A	N/A	N/A	1.51	General purpose bonding of a wide range of substrates
Loctite® 3336/ Henkel	Acrylated epoxy	14,500 (20)	270–300 sec @ 50 mW/ cm^2(365 nm)	N/A	41 (5,900)	N/A	480	N/A	43	1.575	Bonding and tacking applications in low-outgassing applications

(*Continued*)

Table 5.21 Examples of UV-curing adhesives—*cont'd*

Adhesive/Supplier	Type	Viscosity cP (rpm)	Cure conditions	Depth of Cure (mm)	Tensile strength, MPa (psi)	Water absorption (%)	CTE (ppm/°C) $\alpha 1$	CTE (ppm/°C) $\alpha 2$	T_g (°C)	Refractive Index	Application
OG147/Epoxy Technology	UV-curable epoxy	30,000–40,000 (10)	>2 minutes @ 100 mW/cm^2(320–350 nm)	N/A	N/A	N/A	55	174	130	N/A	Adhesive application in fiber optics and circuit assembly
OG134/Epoxy Technology	UV-curable epoxy 50 – 100 (100)	50–100 (100)	>4 minutes @ 100 mW/cm^2(320–350 nm)	N/A	N/A	N/A	156	208	≥15	<1.42	Adhesives for opto-electronics, fiber packaging, LCDs, LEDs, and general electrical assembly
OG115/Epoxy Technology	UV-curable epoxy	800–1,200 (100)	>2 minutes @ 100 mW/cm^2(320–350 nm)	N/A	N/A	N/A	28	223	≥90	N/A	Adhesive bonding in semiconductor, electro-optical, and fiber optical assembly
OG114-4/Epoxy Technology	UV-curable adhesive	150–250 (100)	1–2 minutes @ 100 mW/cm^2(320–350 nm)	N/A	N/A	N/A	73	209	≥50	1.5176	Adhesives for opto-electronics, fiber packaging, LCDs, LEDs, and general electrical assembly
OG172/Epoxy Technology	UV-curable adhesive	2,669 (50)	1–3 minutes @ 100 mW/cm^2(320–350 nm)	N/A	N/A	N/A	104	171	62	1.4727	Adhesives for fiber packaging, sensor development, and general electrical assembly

OG125/Epoxy Technology	UV-curable adhesive	2,669 (50)	1–3 minutes @ 100 mW/cm^2(320–350 nm)	N/A	N/A	N/A	86	199	≥55	1.4359	Adhesives for fiber packaging, sensor development, and general electrical assembly
OG116/Epoxy Technology	UV-curable adhesive	4,000 (N/A)	1–2 minutes @ 100 mW/cm^2(320–350 nm)	N/A	N/A	N/A	56	165	146	1.5756	Adhesives for fiber packaging, sensor development, and general electrical assembly
UVIBOND® 7100/ Aptek Labs	Semi-rigid polyurethane	4,000 (N/A)	1.30 Joules/cm^2 UV radiation	N/A	N/A	N/A	N/A	N/A	N/A	N/A	General electrical/ electronic applications (low ionic impurities)
UVIBOND® 7101/ Aptek Labs	Flexible polyurethane	5,500 (N/A)	2.20 Joules/cm^2 UV radiation	N/A	N/A	N/A	N/A	N/A	N/A	N/A	General electrical/ electronic applications (low ionic impurities, flexible adhesive)
UVISTAKE® 7207/ Aptek Labs	UV-curable polyurethane	N/A	2 passes, 300 W/in D-bulb lamp	N/A	N/A	N/A	N/A	N/A	N/A	N/A	General electrical/ electronic staking applications (low outgassing)
LC-1213/3 M Co.	UV/visible light-curable, low-viscosity adhesive	450 (N/A)	400–500 nm (visible) 250–380 nm (UV)	N/A	13.3 (1,930)	1.8	N/A	N/A	N/A	1.56	UV-visible deep-section curing through UV-opaque substrates, semi-rigid bond

N/A – Not available.
[a] Tensile strength, unless indicated.
Source: Compiled from suppliers' technical data sheets.

Passive alignment of fibers holds more potential for reducing costs than active alignment, but requires greater attention in selecting and processing materials so that they maintain tight dimensional tolerances and result in repeatable high efficiency signal propagation. Since adhesives and polymer materials in general, have high CTEs as well as CTEs and T_g that vary with degree of cure and small changes in composition, the choice and control of the adhesive is critical.

One study of a fiber pigtailed module illustrates the extensive use of adhesives in its assembly.[70] Optical fibers were assembled and coupled to laser diodes on both precision-machined silicon and injection-molded plastic substrates. V-grooves were formed in the substrates to contain and position the fibers. A second cavity was formed in-line with the V-groove to contain the laser-diode chip. Electrically conductive, electrically insulative, and thermally conductive adhesives and both heat-cured and UV-cured adhesives were all used at various stages in the assembly and alignment. First, the laser-diode chip was attached to the substrate with Epoxy Technology's Epo-Tek®H20E, a silver-filled epoxy, which also provided an ohmic contact between the n-portion of the diode and a conductor at the base of the cavity. The p-contact was made by bonding with a 25 μm-diameter gold wire. The wire-bonded contact was reinforced with a small drop of the conductive adhesive. The adhesive bond line thickness, combined with a 100 μm copper shim, provided the exact vertical alignment for the diode chip. The substrate was next mounted onto an L-shaped metal submount with a thermally conductive adhesive (Epoxy Technology's Epo-Tek 353ND T).

The fiber was aligned with the laser by inserting it in the groove and positioning it to within 30 μm of the laser. The fiber is supported by the sharp edges of the groove such that it does not touch the bottom of the groove. After positioning to obtain the maximum coupling efficiency, the fiber was bonded in place using several drops of a UV-curing adhesive (Loctite® 3525™) which cured in 10 seconds on exposure to a UV-light source. In dispensing the adhesive, spreading was controlled so it did not completely fill the groove. The assembled module is shown in (Figure 5.27).

5.2.13 Smart cards and RFIDs

Smart cards are among the fastest growing applications for electrically conductive adhesives. Smart cards are pocket-size plastic cards with thin embedded ICs that store, process and transfer data much in the manner of a mini-computer. Smart labels are also being developed where the substrate is paper instead of plastic. Flip-chip transponder chips and flexible antenna are incorporated producing, for example, low-cost radiofrequency identification (RFID) units.[71]

In the manufacture of smart cards, microprocessor chips, memory chips, EEPROMs or combinations in flip- chip configuration are electrically bonded onto the circuit card using anisotropic electrically conductive paste or film adhesive (see Chapter 1). Smart cards may be contact or contactless types. In contact cards, the electrical contacts of the card connect to a card reader when the card is inserted. Contactless cards transfer data in proximity to a reader (www.smartcard.gov) without physical contact.

Figure 5.27 Fiber pigtailed module using precision-machined substrate for laser-to-fiber passive alignment.[52]

Electrically conductive adhesives are being used to interconnect flip-chip devices in smart cards resulting in thinner and smaller structures. Flip-chip silicon devices that have been thinned to several mils may be connected to a substrate with silver-filled paste epoxy or with anisotropic film adhesive instead of solder, then embedded and laminated to form a card that is less than 40-mils thick. Requirements for adhesives used in smart cards, in many respects, are more severe than those for other commercial applications. Besides having to withstand high humidity and temperature extremes, smart cards must take the continued abuse of human handling, repeated bending, exposure to human sweat and salt residues, and exposure to ultraviolet radiation from sunlight.

In one study,[72] a series of smart cards was fabricated in which the microchip was attached and laminated with three anisotropic-conductive adhesives and one silver-filled paste adhesive. The anisotropic films were hot-melt types all filled with rigid silver-coated glass particles, but differed in the polymer matrix. Polymer matrices included thermoset nitrile rubber/phenolic resin, thermoplastic polyester, and thermoplastic polyetherimide. The paste adhesive was a silver-filled, one-part, 100% solids, modified acrylic that was used with an insulative thermoset nitrile/phenolic film. A series of accelerated tests was used as shown in Figure 5.28. Tests included functional testing before and after subjecting parts to torsion, bending (up to 1,000 cycles), temperature cycling, thermal shock, tropical conditions, low and high temperatures, chemicals, and UV irradiation. In addition, scanning-acoustic microscopy (SAM) was used to study the adhesion between the chip and antenna structure after the environmental tests. Functional tests were Answer to Reset (ATR) and Answer to Select (ATS). In general, parts assembled with the thermoset polymer binder passed all mechanical and functional tests while the parts that employed the thermoplastic anisotropic adhesives displayed many failures.

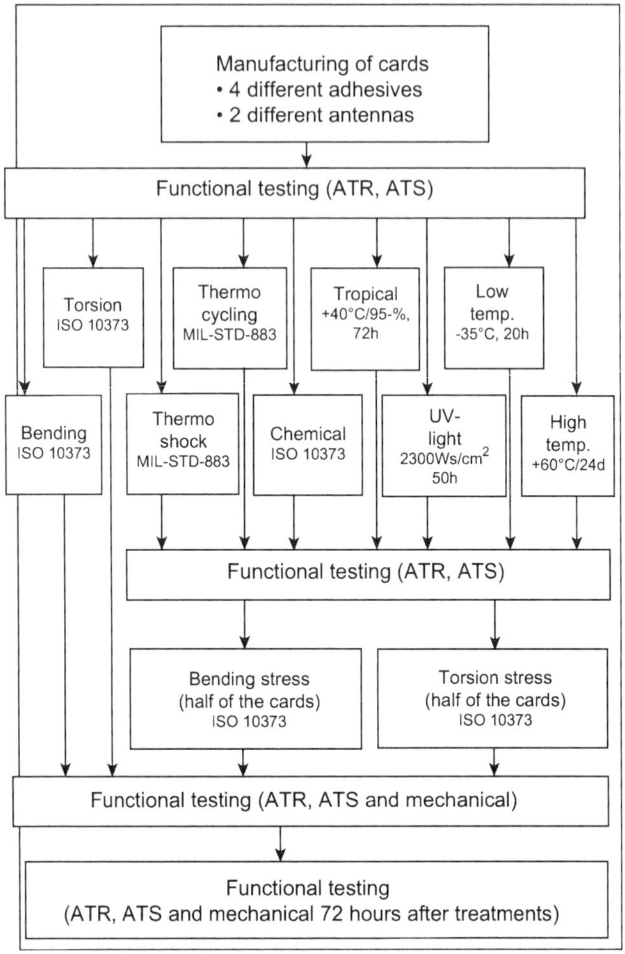

Figure 5.28 Accelerated reliability tests for smart cards.
Source: VTT Electronics.

The reliability of these cards has been reported to be quite good. VTT Electronics reported extended bend cycles up to 17,500 at which point the card plastic fractured, but electrical interconnects were still sound.

5.2.14 Lid sealing

Adhesives can be used to attach and seal lids to a variety of ceramic or metal packages in lieu of seam welding, soldering, or glass attachment (Figure 5.29). As with other applications, the main advantage of using polymer adhesives is low cost. In lid sealing with adhesives, there are the added advantages of ease of delidding for rework, avoidance of metal-particle contamination from solder or welding processes,

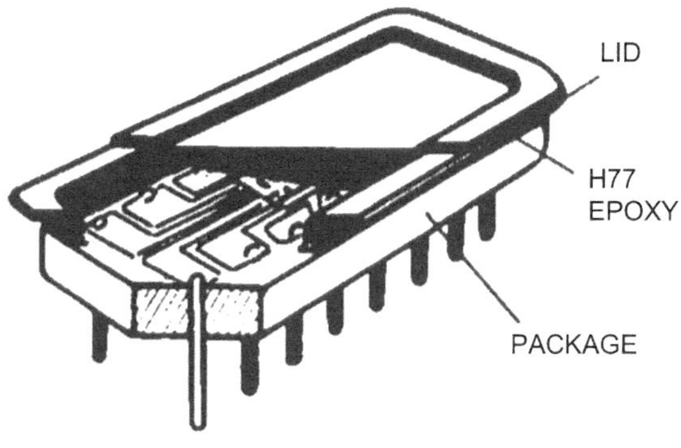

Figure 5.29 Cross-section of cavity package with epoxy-attached lid.
Source: Epoxy Technology.

low-processing temperatures (<165 °C for curing), and low-cost equipment. Both paste and preform insulative adhesives, typically epoxies, modified epoxies or silicones, have been used. B-staged epoxy preforms and thermoplastic films are convenient since the preform can be cut to size, tacked onto the lid, then stored at low temperature until ready for use. With preforms, there is also better thickness control than for paste adhesives.

The reliability of epoxy-sealed parts has been demonstrated based on helium leak rates for a rather benign commercial aircraft cabin environment. Packages were 2.54 × 2.54 cm, ceramic flat packs with ceramic lids. Hybrid microcircuits sealed with Ablefilm® 507 initially passed both the fine leak test ($<5 \times 10^{-7}$ atm-cc/sec) and the gross leak test according to MIL-STD-883, Method 1014, Condition C. Actual field data demonstrated reliable electrical performance. Although analysis of packages recovered from aircraft after three years of service showed the presence of moisture, some in amounts greater than the maximum 5,000 ppm specified, no failures occurred even in the presence of moisture. The internal condition of the circuit showed no corrosion even on the thin film, closely spaced aluminum conductors. It was concluded that the circuits were very clean and that moisture alone in the absence of ionic contaminants did not affect the reliability of the modules. However, a moisture requirement is still considered important since the cleanliness of circuits produced by different suppliers or by the same supplier for different batches cannot be guaranteed.

Sealing with adhesives, however, does not provide the long-term hermetic properties associated with solder sealing or seam welding. Some adhesives suppliers report passing a 1×10^{-8} cc/sec helium leak-rate after 100 hours exposure to 90% RH and 32 °C.[73] Others[74] showed higher water-vapor transmission rates for silicones (1.73–3.11 g mm/m^2 day) and polyurethanes (0.94–3.43 g mm/m^2 day) compared to epoxies (0.7–0.94 g mm/m^2 day). Epoxy sealing, therefore, will be more effective than either silicones or polyurethanes. For applications where humidities up to 100% and temperature extremes from sub-zero to 50 °C are encountered, there would be considerable risk in using adhesive-sealed packages. Sealed microcircuits must pass a series of mechanical and environmental tests for both qualification and acceptance prior to moisture analysis. In one study[75] the permeation of moisture through epoxy-sealed packages was measured by enclosing calibrated moisture sensors in epoxy-sealed packages and subjecting them to 98% RH and 60 °C. Although the moisture content remained low (less than 1,000 ppm for the first five days), an abrupt and linear increase occurred at the end of the fifth day, and continued to increase (Figure 5.30). The latent period of 4–5 days may be explained as the time that was necessary for moisture to saturate the epoxy adhesive, traverse the width of the bond, and be released at the other end into the package.

5.2.15 Power circuits and power devices

In power circuits or power devices, heat must be removed by convection by blowing air across the part, by conduction through several solid materials and interfaces to a heat sink or heat exchanger, or by a combination of conduction and convection. If heat transfer depends solely on conduction, the adhesive used for attachment is often the limiting factor providing the high thermal impedance compared to solder, metals, or ceramic interfaces. Generally, heat must flow through several materials and interfaces, each of which contributes its own thermal resistance. A simple thermal analysis can be performed based on the knowledge of the thermal conductivities of each of the materials, their thicknesses, and contact areas. The total thermal resistance is calculated by adding the individual resistances in a manner similar to conducting an electrical circuit analysis. Thus, if a power device dissipates 5 W/cm^2 and the difference between the device junction temperature and the backside temperature of a package must be kept at less than 20 °C, the total impedance must not exceed 4 °C/watt.

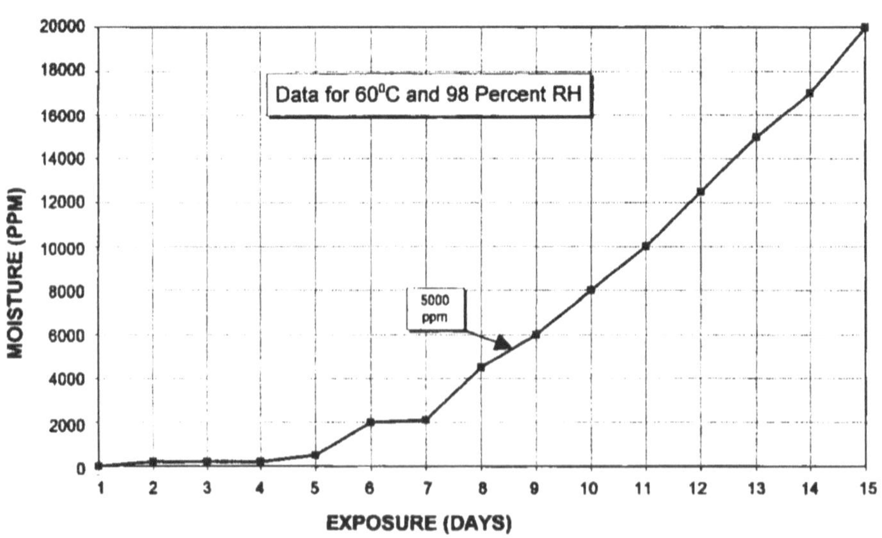

Figure 5.30 Moisture permeation through epoxy-sealed ceramic package.[75]

In a typical case where a silicon IC is attached to an alumina ceramic substrate that, in turn, is attached to the inside of a metal or ceramic package, the two epoxy interfaces can easily contribute 2.5 °C/watt to the total resistance. However, some silver-filled epoxies are reported to have high thermal conductivities, thus contributing 0.6 °C to 1 °C/watt.[76] Actual measurements may differ considerably from calculated values because of reported thermal conductivities that differ from the actual, differences in the thicknesses of bond lines, voids in the adhesive, and incomplete mating of surfaces. Further in the analysis, the effects of lateral flow of heat and interactions of heat flow among adjacent components are often neglected.

In practice, several techniques can be used to maximize the dissipation of heat. Of course, solder and other metallurgical attachments provide the lowest thermal resistance but, where adhesives must be used, those having the highest thermal conductivities should be selected. Some silver-filled adhesives, formulated with unique suspensions of silver, resin, and carrier fluid are reported to have thermal conductivities of 20 $W/m \cdot K$[77] approaching those of ceramics and some metals. An adhesive having a high silver loading is produced once the carrier fluid has been removed, and the adhesive is fully cured. This adhesives was developed for thermal management applications requiring high heat extraction from the die. Once the material is fully cured, the adhesive provides very low thermal resistance between the chip and the case, nearing solder and eutectic type materials.

To assure optimum conductance, the adhesive must be applied as thinly and uniformly as possible. To control the thicknesses bond lines, thermally conductive paste adhesives have been formulated with collapsible spacers. The spacers are reported to control bond line thickness to 30 μm. Bond line thicknesses and uniformity may also be achieved by using film or preform tape adhesives and controlling the applied pressure and heat during cure.

Lastly, a technique that has been used for many years involves first eutectic or solder attaching high-power devices to small heat spreaders such as *molytabs* (gold-plated molybdenum tabs that are the same size or slightly larger than the die). Thermally conductive adhesive is then used to attached the die-tab parts to a ceramic or laminate interconnect substrate.

References

1. Marinelli C, Burns B. Adhesives/epoxies and dispensing. *SMT Special Supplement.* July 2001.
2. Fisher E, Hinerman JB, Dixon D. SMT Adhesive deposition: the line to success. *SMT.* Dec. 2000.
3. Gossling M. The properties of SMT adhesives, *Circuits Assembly.* Feb. 1994.
4. Adhesives for SMD Technology, SN 59651, Edition 1997–09. Berlin: *Siemens AG.* Sept. 1997.
5. Burns B, Fisher E. Adhesives, epoxies and dispensing. *SMT.* May 2001.
6. Beckman S, et al. SMT Adhesives—evaluating the materials critical parameters. *SMT.* Jul. 1997.
7. Van den Bosch A. Behavior of adhesives before curing. *SMT.* Feb. 1997.

8. Epibond 7275-Series Surface Mount Adhesives Technical Data Sheet, Cookson Electronics (Revised Mar.29,2007).
9. Wong CP, Vincent MB, Shi S. Fast-flow underfill encapsulant: flow rate and coefficient of thermal expansion, *IEEE Transactions on Components. Packaging, and Manufacturing Technology – Part A* June 1998;Vol. 21(No. 2).
10. Tonapi S, Reitz B. Flip-chip underfill: design and material challenges. *Chip Scale Review.* Jan./Feb. 2004.
11. Adamson SJ. Review of CSP and flip-chip underfill processes and when to use the right dispensing tools for efficient manufacturing, GlobalRONICS Technology Conference 2002, Singapore.
12. Li Z, Lee S, Lewis B, Houston P, Baldwin D, Stout E, Tessier T, Evans J. No flow underfill assembly process development for fine pitch flip chip silicon to silicon wafer level integration, Advancing Microelectronics. July/August 2010; 20–25.
13. Katze D. No-flow fluxing underfill material and process considerations. *Proc. HDI Conf.* 2000.
14. DeBarros T, Katze D. Achieving SMT compatible flip-chip assembly with no-flow underfills. *Proc. NEPCON West Conf.* Anaheim, CA: 2000.
15. Lim R, Duck A. Controlling process parameters for flip-chip underfill. *Electronics Engineer.* Jul. 2000.
16. Patterson T, Esfandiar F. Bump types, metallization and substrate considerations. *HDI.* Oct. 1999.
17. Patterson T, Esfandiar F. Interconnect options for small precision assemblies. *Proc. IPC National Conf.* Mar. 1998.
18. Shi G, Tang H, Crane L, Todd M. Preapplied underfills—technology and applications. *MAPS SoCal '04 Symp. and Exhibition.* Pasadena, CA: May 2004.
19. Korczynski E. Wafer-level processing, variations on a theme. *Semiconductor Magazine.* Sep. 2002;5(9).
20. Patterson DS. Continued improvements in the use of alloys and polymers to enhance wafer-level IC performance. *Chip Scale Review.* Oct. 2002.
21. *Test Method Standard Microcircuits.* MIL-STD-883; Jun. 2004.
22. *Environmental Test Methods and Engineering Guidelines*, MIL-STD-810.
23. *Electronic and Electrical Component Parts* Test Method Standard. MIL-STD-202; Feb. 2002.
24. Hybrid microcircuits, performance specification, General Specification, MIL-PRF-38534.
25. Total Mass Loss and Collected Volatile Condensable Materials from Outgassing in a Vacuum Environment, ASTM E595, American Society for Testing and Materials. 1999.
26. Selection of Spacecraft Materials & Supporting Vacuum Outgassing Data, NASA Practice No. PT-TE-1410 (undated), klabs.org/DEI/References/design_guidelines/test_series/1410.pdf.
27. Outgassing Data for Selecting Spacecraft Materials, NASA Reference Publication 1124, Rev. 4, Goddard Space Flight Center (Jun. 1997) and http://outgassing.nasa.gov.
28. Ionizing Radiation (Total Dose) Test Procedure, Method 1019.6 MIL-STD-883F. Mar. 2003.
29. Johnson RW. MEPTEC Symp., Sunnyvale, CA: Feb. 2003.
30. Guidance on Selection of Tests, ISO 10993, International Organization for Standardization.
31. Petrie EM. Adhesives for medical devices, www.specialchem4adhesives.com. Jun. 2, 2004.

32. DiStefano TH, Fjelstad J. Chip-Scale packaging meets future design needs. *Solid State Technology.* Apr. 1996.
33. Bauer CE. Let's talk chip scale. *Circuits Assembly.* Apr. 1997. IPC Supplement.
34. Ngulty TA, Ekere NN. Chip scale vs flip chip: issues to consider. *Circuits Assembly.* Nov. 1999.
35. Schueller RD, Portable CSP. *Advanced Packaging.* May 1998.
36. Baliga J. High-density packaging: the next interconnect challenge. *Semiconductor Intl.* Feb. 2000:91-4.
37. Karnezas M. Advantages of system on a package and system on a chip. *Electronic Packaging and Production.* Jul. 2001.
38. Miles B, et al. 3-D Packaging for wireless applications. *Semiconductor Intl.* Feb. 2004.
39. McCray E. What's New in Packaging. *HDI.* May 2001.
40. Ozguz V, Carson J. SPIE *Optoelectronic critical review on heterogeneous integration.* In: Towe E, ed. SPIE Press; 2000.
41. Entering a new era, cubic memory brochure, cubic memory (now Vertical Circuits, Inc.).
42. Al-Sarawi SF, Abbott D, Franzon PD. A review of 3-D packaging technology. *IEEE Trans. Components, Packaging, and Manufacturing Tech.* Feb. 1998;21(1).
43. Savastiouk S, Siniaguine O, Korczynski E. 3-D Stacked Wafer-Level Packaging. *Advanced Packaging.* Mar. 2000.
44. Reche J. Tru-Si Technologies IEEE/CPMT Meeting, Santa Clara. CA: May 10, 2000.
45. Humpston G. Setting a new standard for through-silicon via reliability. *Electronic Design.* Oct. 22, 2009.
46. Carsem system in package, Technical Data Sheet, Revision B. May 2005.
47. Henkel Solutions for Flat Panel Displays. 2005.
48. Question 2: which computer displays were investigated during the project? life-cycle assessment of desktop computer displays: summary of results, http://www.epa.gov/dfe/pubs/comp-dic/lca-sum.
49. *ThreeBond Technical News.* July 1, 1997;(Issue 49).
50. *ThreeBond Technical News.* May 1, 1994;(Issue 43).
51. *Henkel Data Sheet for Loctite Product. 3010* May 2001.
52. Kristiansen H, Liu J. Overview of conductive adhesive interconnection technologies for display applications, Conductive Adhesives For Electronics Packaging. In: Liu J, ed. Electrochemical Publications; 1999.
53. Hitachi Anisotropic Conductive Film ANISOLM AC-2056, Technical Data Sheet. July 3, 2001.
54. Shiiki M, Imaizumi J, Miyata T, Chinda A. Materials and components for flat panel display applications. *Hitachi Review.* 2006;Vol. 55(No. 1).
55. 3M Product Data Sheet 3M™ Anisotropic Conductive Film Adhesive 7393. May, 2009.
56. The Light Bulb Goes Digital. *Fortune.* February 8, 2010:33-8.
57. Chanchani R. An overview – Advanced Packaging for Solid State Light Emitting Diode (SSLED) Lighting. *Advancing Microelectronics.* January/February 2010; 6-12.
58. http://www1.eere.energy.gov/buildings/ssl/printable_versions/comparing_lights.html.
59. Chanchani R. An Overview – Advanced Packaging for Solid State Light Emitting Diode (SSLED) Lighting, 42nd International Symposium on Microelectronics. (IMAPS 2009) Proceedings, pp. 492–499, San Jose McEnery Convention Center, San Jose, CA. ISBN 0-930815-89-0 (Nov. 1–5, 2009)
60. DieMat DM 6030 Series, High thermal conductivity silver epoxy adhesive, DieMat Product Data Sheet. March 19, 2008.

61. Rupprecht L, ed. *Conductive polymers and plastics, Plastics Design Library.* William Andrew Publishing; 1999.
62. Brütting W. *Physics of Organic Semiconductors.* Wiley-VCH; 2005.
63. Organic Light Emitting Diodes (OLEDs) for General Illumination Update 2002OIDA Technology Roadmap, OptoElectronics Industry Development Association, http://lighting.sandia.gov/lightingdocs/OIDA_SSL_OLED_Roadmap_Full.pdf). Aug. 2002.
64. ARclad®, Adhesives Research, Inc.
65. CL-130 Adhesive on Cho-Seal S6305, Test Report # CHO-TR CL130, Chomerics, Div. *of Parker Hannifin Corp.* Aug. 2001.
66. *Requirements for the control of electromagnetic interference characteristics of subsystems and equipment.* MIL-STD-461, Defense Information Systems Agency; Aug. 1999.
67. Tomaiuolo P. EMS Opto success: another testimonial. *Electronic Packaging and Production.* Mar. 2003.
68. Liu J, et al. *ECTC Proc.* 2000;975-80.
69. Zhou H, et al. Packaging of fiber collimators. *Advanced Packaging.* Jan. 2002.
70. Karioja P, et al. Comparison of active and passive fiber alignment techniques for multimode laser pigtailing. *Proc. 50th Electronic Components and Technology Conf.* Las Vegas, NV; May 2000.
71. Rasul JS. Chip on paper technology utilizing anisotropically conductive adhesive for smart label applications. *Microelectronics Reliability.* Jan 2004;Vol. 44(Issue 1).
72. Holmberg M, Lenkkeri J, Lahti M, Wiik B. Reliability of Adhesive Joints in Dual Interface Smart Cards. *Proc. 2nd Intl. IEEE Conf. on Polymers and Adhesives in Microelectronics and Photonics*, Zalaegerszeg, Hungary. Jun. 2002;23–26.
73. Lid Sealing Epoxy Technical Data Sheet, Epo Tek H77, Epoxy Technology, Billerica, MA.
74. Nagy M. The effectiveness of water vapor sealing agents when used in application with thermoelectric cooling modules, Undated TE Technology, Inc., Publication, www.tetech.com/publications/pubs/ICT97MJN.pdf, (accessed July 13, 2010).
75. Perkins KL, Licari JJ. Development of low-cost, high-reliability sealing techniques for hybrid microcircuit packages. *NAS-831992*, http://ntrs.nasa.gov/details.jsp?R=311623; Aug. 1977.
76. Hodgin MJ. Epoxies for optoelectronic packaging: applications and material properties. *J. Microelectronics and Electronic Packaging.* Second Quarter 2004;1(2). IMAPS.
77. Abletherm® 2600BT, Technical Data Sheet. Ablestik Laboratories. Sep. 2003.

6 Reliability

Chapter Outline

Adhesives Technology for Electronic Applications. DOI: 10.1016/B978-1-4377-7889-2.10006-3

The reliability of an adhesive and its impact on the performance of an electronic assembly should be considered in the initial selection of the adhesive and the design of the system. The function that the adhesive must perform for a specific application, the environment it is expected to encounter, and its duration are all important. Various approaches may be used to predict and assure reliability. Key among these approaches is a basic understanding of possible failure modes and mechanisms. Most failure modes attributed to adhesives are now well understood and documented so that they can be avoided in the initial selection and qualification of the adhesive and in its processing.

Secondly, accelerated tests can be used to qualify adhesives for specific applications, and the data obtained can sometimes be extrapolated to long-term, real-time performance. Accelerated tests are also valuable as screen tests and as acceptance tests and are specified in material-procurement documents or in hardware-acceptance specifications. Adherence to material and process specifications and their quality-control provisions is an essential element in assuring reliability. Life prediction through modeling is yet another approach, but is outside the scope of this chapter.

Lastly, there are extensive databanks of real-time performance for adhesives that have been used in numerous applications, various environments, and for extended periods of time, the so-called heritage data.

6.1 Physics of failure methodology

The physics of failure (PoF) approach to determine the mode and mechanism of failure of an electronic part or assembly has been used successfully for many years. PoF is a step-by-step approach that assures that information is not irretrievably lost. Thus, in performing a failure analysis, it is important to perform all the tests that are nondestructive first. These tests may or may not give a clue about the cause of failure but, if omitted, either can no longer be performed or are compromised by the subsequent destructive tests.

6.1.1 Physics of failure step-by-step approach

As a first step, before a part is removed from an assembly, subjected to extensive handling, or uncontrolled storage, an external visual and microscopic examination should be performed and any anomalies recorded and photographed. These include corrosion products, discolorations, deformed package leads, or poor solder connections.

If corrosion products or other extraneous materials are present, samples should be taken and subjected to Fourier Transform Infrared (FTIR) spectroscopy, X-ray, or other analytical tests.

A second step involves complete electrical testing of required parameters to determine those that are out-of-specification. Controlled overstressing may be used, if applicable.

If package cracks are suspected, a dye penetrant test should be performed. In this test, a fluorescent dye is applied to the package surface and then wiped clean. Examination of the part under UV-light reveals residual dye in package cracks. This test is applicable both to hermetic parts and plastic-encapsulated microcircuits (PEMs).

For hermetically sealed microcircuits several screen tests should be performed before delidding the package. A particle impact noise detection test (PIND [MIL-STD-883 Method 2020 or MIL STD-750D, Method 2052-2]) is recommended. This "shake" test combines a series of shocks and vibrations and, through an attached transponder, acoustically detects any loose particles (for example, chips of cured adhesive, solder, wire, ceramic, or loose plating) that have detached and can cause shorted or open circuits and damage or breakage of thin wires and connections.

Another test, constant acceleration, is useful in determining the reliability of chips; adhesive attached to the substrate and substrates; adhesive attached to the inside floor of the package. The acceleration force pulls on the wires and causes any marginal bonds to detach. According to this test, the parts are centrifuged up to 5,000 g in a Y_1 direction. The packages are mounted with the lid against the chamber wall so that the centrifugal force is pulling away from the attached devices, wire bonds, and lid (MIL-STD-883, Method 2001). The centrifugal force may be increased until the part fails in order to determine the failure limits but this, of course, becomes a destructive test.

Fine and gross helium leak testing will detect any marginal or bad seals through which moisture and other contaminants can enter the package.

As a final test, prior to delidding a sealed cavity package, a residual-gas analysis (RGA) should be performed to measure the presence, types, and amounts of moisture and other gases within the package. Moisture and organic contaminants such as absorbed water, entrapped cleaning solvents, or gaseous decomposition products from adhesives are most often involved in electrical failures. Sources of moisture may be from poor vacuum baking prior to seal, absorbed moisture in the adhesives, decomposition of the adhesive, desorption during screen testing involving heat exposures, or the penetration of ambient moisture through marginal seals.

After RGA data are collected and analyzed, the package is delidded to examine the inside components, interconnections, materials, and surfaces. Because removing the lid is a destructive process, it should be done in a lid-down position with special equipment to vacuum off any debris. After delidding, the components, surfaces, and connections are optically examined followed by analytical testing such surface analysis for contaminants, especially chloride and sodium ions. Scanning acoustical microscopy (SAM) is very useful in detecting voids, cracks or delamination within the adhesive or at the adhesive bond line. An excellent overview of SAM and its variations may be found in Ardebili and Pecht.[1]

6.1.2 Physics of failure for plastic-encapsulated microcircuits (PEMs)

The failure analysis approach for plastic-encapsulated microcircuits differs somewhat from that of hermetically sealed cavity packages. Because PEMs have no cavity, internal moisture and gas analyses, PIND tests, and gross and fine leak testing are not applicable. For a PEM, a Shore Durometer hardness test (ASTM D2240) should be performed first. The durometer test is a simple indentation test that measures softness or hardness of the plastic encapsulant and provides an indication of the degree of cure of the plastic. A small amount (10–20 mg) of the plastic is then scraped off and examined by FTIR spectroscopy. FTIR gives considerable information on the type and chemistry of encapsulant, degree of cure, and changes in the spectra when compared with a "good" sample (see also Chapter 4). Differential scanning calorimetry (DSC), thermogravimetric analysis (TGA) and changes in the glass-transition temperature (T_g) may also be used to establish the degree of cure (see also Chapter 4).

To conduct tests on the embedded devices and circuitry, the molding compound must first be removed (decapsulation) either by wet or dry etching. Both processes are rather aggressive involving exposure to strong concentrated acids or laser ablation and must be performed carefully to avoid destroying the underlying devices and circuitry. Once exposed, the integrity of the adhesive and adhesive bonds can be determined visually (excessive bleedout, metal migration, fracture, delamination, discoloration) or tested using a variety of analytical techniques. Key among these tests are SAM for voids, cracks or delamination, FTIR spectroscopy for organic surface contaminants or residues, and atomic absorption spectroscopy or water extract resistivity tests for traces of sodium, potassium, chloride, ammonium, or other ions.

Besides FTIR, other test methods such as thermogravimetric analysis (TGA) and differential thermal analysis (DTA) are useful in assessing the condition of the adhesive, for example, whether it is undercured, overcured, or decomposed.

Finally, once a failure site or failure mode had been found, a mechanism based on knowledge of electrical, chemical, and physical reactions can be postulated, simulated, and verified and corrective action taken. Figure 6.1 is a flow diagram of some tests used in a PoF study for electronic modules containing adhesive-attached components.

6.2 Failure modes and mechanisms

The major failure modes of adhesives include loss of adhesion, high thermal impedance, loss of electrical contact, and corrosion. To a lesser extent, failures may occur because of metal migration causing high-leakage currents and even electrical shorting. Sloughing of particles from the adhesive during vibration is also known to damage wire bonds, fine wires, and thin-film metallization in sealed packages, causing opens or shorts, depending on the type of particle.

6.2.1 Loss of adhesion (delamination)

Loss of adhesion or loss of mechanical strength may result from cohesive failure within the adhesive or adhesive failure at the bond line interfaces, such as between the

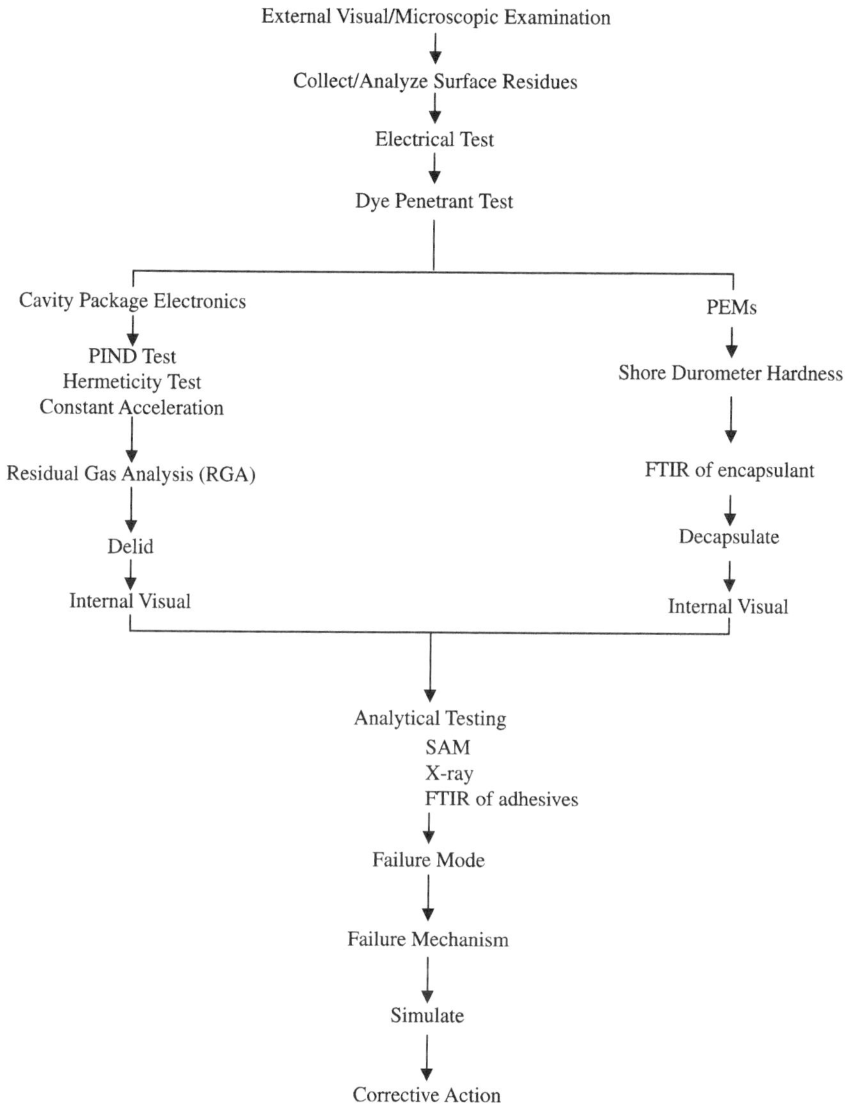

Figure 6.1 Flow diagram for PoF of electronic modules containing adhesive-attached components.

adhesive and the substrate or the adhesive and the device. Adhesion failures result in delamination or separation of ICs, substrates, packaged devices, heat sinks, or other attached parts. It is important that the mechanical strength of adhesives be maintained over the operational lifetimes of the bonded parts, at elevated temperatures, and under other environmental conditions.

Reliability requirements may be classified as short term or long term. Thus, delamination may occur shortly after curing due to contaminated surfaces, improper ratios of adhesive-curing agents, or inadequate cure or later, during environmental stress testing. Short-term requirements include maintaining mechanical strength under conditions encountered during the assembly, processing, and testing of electronic modules. Long-term reliability requires maintaining adhesion strength after repeated thermal cycling, elevated-temperature exposure, or temperature–humidity exposure. Mechanical reliability for die or substrate attachment over time may be determined by comparing die-shear strength or lap-shear strength before and after aging or after exposure to actual- or accelerated-stress conditions.

Delamination may be complete or partial. Complete delamination results in gross separation of the device, substrate, or other adherend. Partial debonding can result in device or substrate cracking with complete detachment occurring in time. Numerous factors and mechanisms contribute to delamination, some of which are as follows:

- Selection of an adhesive whose adhesive or cohesive strength is initially marginal or poor. In the case of two-part systems, poor strength may be due to using non-stoichiometric ratios of resin to hardener.
- Insufficient cure time or temperature, leaving unreacted resin or hardener; or excessive cure conditions, resulting in embrittlement.
- Insufficient cleaning leaving surface contaminants that reduce wetting (resin bleedout or volatile outgassing that redeposits on surfaces).
- Insufficient or nonuniform adhesive coverage (generally, a minimum thickness of 2 mils is required).
- Nonuniform or insufficient pressure in applying film adhesives.
- Stresses due to large mismatches in the CTEs of the adherends or residual stresses due to shrinkage during cure.
- Adhesives having too high or too low a modulus for the size of the device.
- Voids in the bond line due to entrapped air, moisture, solvents, or low-molecular-weight volatile compounds.
- Large amounts of absorbed moisture in the die-attach or underfill adhesive.

Of these failure mechanisms, the major ones associated with delamination include contaminated surfaces, inadequate adhesive coverage, stresses, voids, and moisture absorption. Many of these mechanisms are also responsible for other failure modes besides delamination.

6.2.1.1 Contaminated surfaces and poor wettability

A major cause of delamination is poor wettability of liquid or paste adhesives, or insufficient contact and surface wetting for film adhesives because of low or nonuniform pressure applied during cure. In either case, inadequate cleaning is a prime cause of partial or complete separation of the adhesive from the device, the substrate, or both. Besides the normal contaminants from processing or handling that may remain because of inadequate cleaning, contaminants may be introduced during cure. Contaminants may consist of solvents, epoxy diluents, silicone fragments, or other low-molecular-weight compounds that volatilize and deposit onto die-bonding pads, leadframe fingers, top of the die, or bottom of the die pad in single-chip packaging or onto substrates in multichip packaging. In the case of leadframe attachment, redeposited volatile materials reduce adhesion of the subsequently applied molding compound, effectively acting as a release agent. Reduced adhesion occurs because residues as thin as 70–500 Å are enough to "coat" the highly polar and reactive surface groups such as free hydroxyl or oxide groups that normally form strong bonds with the adhesive. Depositions on the bottom of leadframes can occur when parts are stacked in magazines and cured in an oven. In comparison, parts cured on a hotplate showed excellent resistance to delamination.[2] Volatile diluents from epoxy adhesives may also deposit on the walls of curing ovens and then redeposit onto other parts being cured in the same oven.

6.2.1.2 Adhesive coverage

Poor adhesion and delamination can also result from dispensing insufficient adhesive to cover the entire bond line (Fig. 6.2). Generally, a minimum thickness of 2 mils is required. A visible fillet around two or more sides of an attached die or substrate provides assurance that sufficient adhesive has been used and has spread and covered the bond line. In fact, the appearance and extent of fillet are inspection criteria for the acceptance of high-reliability circuits as defined in MIL-STD-883, Method 2010 for monolithic ICs and Method 2017 for hybrid microcircuits. According to these methods, at least 75% of the fillet should be visible along each side of the device. Applying a uniform thickness is also important. Nonuniform thicknesses result in tilted devices with the potential for cracking and separation.

6.2.1.3 Stresses

Besides inadequate cleaning, the second most common cause of delamination is stress buildup in adhesive joints due to residual stresses generated during thermal curing and cooling or to those resulting from large mismatches in the CTEs of the adherends. It should be noted, however, that although adhesives can contribute to stresses and

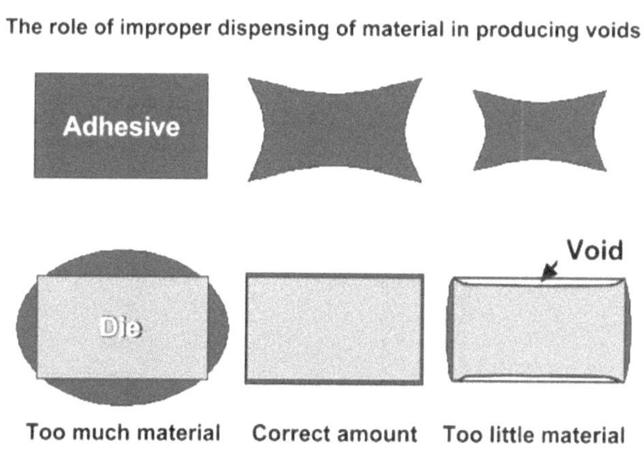

Figure 6.2 Accept/reject criteria for adhesive coverage.
(Courtesy CALCE Electronic Products and System Center, Univ. of Maryland.)

failures, they may also absorb, dissipate, and reduce stresses. An example is their wide and indispensable use as underfills for flip-chip devices. Figure 6.3 compares the reliability of flip-chip devices with and without underfill adhesives showing the dramatic advantage of the underfills in dissipating stresses generated during temperature cycling.[3]

6.2.1.3.1 Residual stress

Residual stresses may be measured from the radius of curvature (ROC) of the warpage that occurs on a substrate when the adhesive is cured on that substrate. The bending stress, S_b, is inversely proportional to the radius of curvature, R, and directly proportional to the modulus, E, and the thickness, h, of the adhesive according to:

$$S_b = Eh/2\gamma R \tag{6.1}$$

where γ is Poisson's ratio.

High-modulus adhesives are likely to have high bending stresses, high deflections, and low radii of curvature. A comparison of typical and ultra-low-stress conductive adhesives was reported using 250-mil square silicon die (20 mils thick) bonded to 10-mil-thick copper leadframes.[4] A conventional silver-filled adhesive had a deflection of 3 mm and an ROC of 1,070 mm while a low-stress adhesive showed a deflection of 0.01 mm and an ROC >125,000 mm.[4]

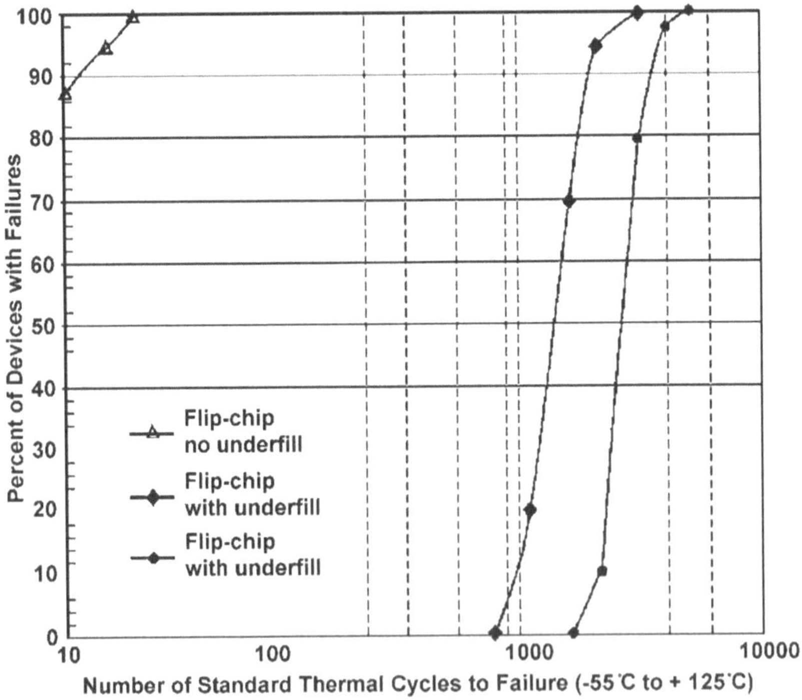

Figure 6.3 Reliability of flip-chip devices with and without underfill.

In another evaluation,[5] silver-glass, polyimide, and silicone adhesives were analyzed for residual stress. The silver-glass adhesive showed the greatest deflection; the silicone, the lowest and the polyimide was intermediate. This order is explained by the high modulus of the silver-glass and the low modulus of the silicone relative to the polyimide.[5]

Stress buildup in bonded devices may also be characterized over a temperature range as a function of the coefficients of expansion of the adhesive and of the substrate according to the following one-dimensional model.[6]

$$\sigma_{Si} = \varepsilon_a \times E_a = (\alpha_a - \alpha_{Si}) \times E_a \times \Delta T \tag{6.2}$$

where σ_{Si} = stress on the silicon, ε_a = strain of the adhesive, E_a = modulus of the adhesive, α_a = CTE of the adhesive, α_{Si} = the CTE of silicon, and ΔT = temperature difference.

Yet, a third equation defines maximum stress for rectangular devices taking into account the size of the device, bond line thickness, and cure and exposure temperatures, in addition to modulus and CTEs.[7]

$$S_{max} = K(\alpha_{sub} - \alpha_{Si})\,(T_o - T)\sqrt{\frac{E_a E_{sub} L}{x}} \tag{6.3}$$

where S_{max} = maximum stress, K = geometric constant related to shape and filleting, α_{sub} = CTE for the substrate, α_{Si} = CTE for the silicon die, E_a = modulus for the adhesive, E_{sub} = modulus for the substrate, L = length of the die, x = bond line thickness, T_o = cure temperature, and T = application (exposure) temperature.

Based on these equations, stresses that can result in failures are a function of many parameters including modulus of elasticity, expansion coefficients of the adhesive and adherends, glass-transition temperature, cure temperature, operating or exposure temperatures, and bond line thickness.

6.2.1.3.2 Modulus of elasticity

The modulus of elasticity, also known as tensile elastic modulus or Young's modulus, is the stress required to produce unit strain. The modulus is measured by pulling a sample of the cured material in a tensile-testing machine while measuring the change in length by attaching an extensometer to the sample. The stress (force per unit area) divided by the *strain* (change in length normalized by the initial length) is the *elastic modulus*. In flexural modulus, a bending stress instead of a tensile stress is applied, the deflection produced is recorded and the modulus calculated from the ratio of stress to strain. The method applies only to materials within the elastic region, that is, where applied forces do not irreversibly stretch the material. Thus, modulus is a measure of the rigidity or flexibility of adhesives—an important parameter in selecting the type of adhesive to use for various semiconductor packages. Moduli can range from less than 30,000 psi to greater than 1 million psi at 25 °C. In all cases, the moduli decrease with increasing temperature. Those that decrease the most are classified as low-modulus types (Fig. 6.4). Adhesives having the lowest moduli at both room temperature and at elevated temperature generate the lowest stresses and

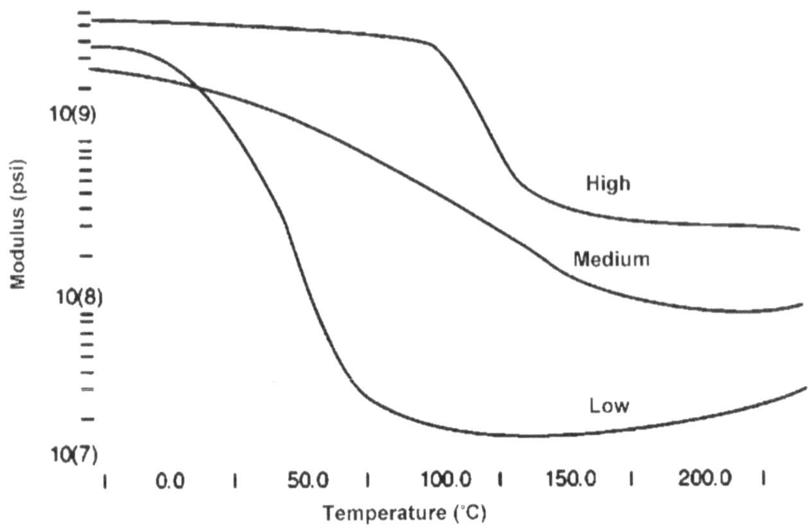

Figure 6.4 High, low, and intermediate-modulus curves as a function of temperature.[6]

are especially useful in preventing delamination and cracking in thin plastic packages such as thin quad-flat packages (TQFP).

Stresses in adhesive-attached die become critical as the size of the die increases and as dissimilar adherends are used. Silicon die, smaller than 100-mil square, attached with high-modulus adhesives to leadframes or to other substrates have been quite reliable even when large differences in the CTEs of the adherends exist. However, for large devices (greater than 300-mil square) low-modulus adhesives are necessary to absorb or dissipate stresses.[8]

In one study, adhesives having various moduli ranging from high to low were evaluated for bonding ferrite cores and magnetic devices to low-temperature, cofired-ceramic (LTCC) substrates. Two adhesives having high moduli of 1.3×10^6 psi at 25 °C (35,000–40,000 psi at 125 °C) displayed the highest stresses at 25 °C and produced cracks in the substrates. An intermediate-modulus adhesive had a lower stress while the lowest-modulus adhesive (19,000 psi) displayed the lowest stress and produced no failures (Table 6.1).[9,10]

Low-modulus adhesives produce low bending stresses and have higher radii of curvature than intermediate- or high-modulus adhesives. The radius of curvature of parts bonded with low-modulus adhesives increases with the size of the die. Post-cure-temperature exposures such as those subsequently encountered during wire bonding or molding also decrease the warpage, indicating stress relaxation in the adhesive by an annealing mechanism.

6.2.1.3.3 CTE mismatches

Large mismatches between the expansion coefficients of the adhesive and the two adherends have been the primary cause of numerous failures resulting in high

Table 6.1 Effect of modulus, expansion coefficient, and temperature on adhesive stress[9,10]

Property	Adhesive A (high modulus)			Adhesive B (high modulus)			Adhesive C (intermediate modulus)			Adhesive D (low modulus)		
Temperature (°C)	**−55°**	**25°**	**125°**	**−55°**	**25°**	**125°**	**−55°**	**25°**	**125°**	**−55°**	**25°**	**125°**
ΔT (150 $-T_{use}$)	205	125	25	205	125	25	205	125	25	205	125	25
CTE (α_1 or α_2)	40.3	40.3	119.9	25.9	36.1	117.0	37.8	37.8	116.1	33.2	167.4	159.3
Modulus (psi)	1.61×10^6	1.31×10^6	35,000	1.73×10^6	1.29×10^6	40,000	621,000	432,000	15,000	2.59×10^6	19,000	12,000
Stress[a] (psi)	10,000	4,962	96	5,639	4,209	107	3,539	1,501	40	12,318	373	45
% Reduction from adhesive A value	–	–	–	43.6	15.2	11.5 (increase)	64.6	69.8	58.3	23.2 (increase)	92.5	53.1
Lowest possible cure stress (psi) (25 °C only)	–	3,030 (125 °C cure)	–	–	0 (25 °C cure)	–	–	1,201 (125 °C cure)	–	–	373 (150 °C)	–

[a] Stress = $(\alpha_{dhs} - \alpha_{sub}) \times 10^{-6} \times (T_{cure} - T_{use}) \times$ modulus (−55°, +25°, +125 °C), where $\alpha_{sub} = 10$ ppm/°C for ferrite cores and α_{dhs} (above or below T_g, respectively). (Copyright ©1999 IMAPS – International Microelectronics and Packaging Society, and 2001 by Elsevier Press. Reproduced by permission.)

stresses and cracking or delamination. Unfortunately, the CTEs of most polymeric materials are much higher than those of the ceramic substrates or silicon devices that are used. Further contributing to stresses is the often abrupt increase in the CTEs of adhesives and other polymers that occurs at and above the T_g. Since the thermal expansion coefficients of adhesives below the T_g may range from 25 to 100 ppm/°C (that is, greater than the CTEs of the device or substrate materials) stresses in bonded parts are compressive on the die and substrate and tensile in the adhesive.

Where the adherends have closely matched CTEs as for silicon die (CTE = approx. 3 ppm/°C) attached to Kovar® leadframes (CTE = approx. 5 ppm/°C), a high-modulus adhesive could be used. As Kovar leadframes were replaced by copper leadframes (CTE = 16–17 ppm/°C), low-modulus, stress-free adhesives were required to compensate for the large mismatches in CTEs. Large CTE mismatches also occur when silicon devices are directly attached to epoxy printed-circuit boards as for COBs or CSPs.

6.2.1.3.4 Voids and moisture absorption

Two other related factors that can cause delamination are voids and moisture absorption. Voids can be generated during the attachment process and may result from dewetting, entrapment of air, or volatilization of absorbed gases during curing. Voids may also be produced during cool-down from heat curing if shrinkage occurs with separation from the substrate. The distribution and size of voids across the bond layer determine the effects they might have on device performance (Fig. 6.5). Generally, small random voids have a minor effect on both mechanical strength and thermal conductance. Large, contiguous voids are detrimental to both mechanical strength and thermal performance.

Acoustic microscopy is a widely used method for analyzing the integrity of adhesive bond lines. Acoustic microscopy is based on the fact that each material has an acoustic impedance designated as:

$$Z = \sqrt{K\rho}$$

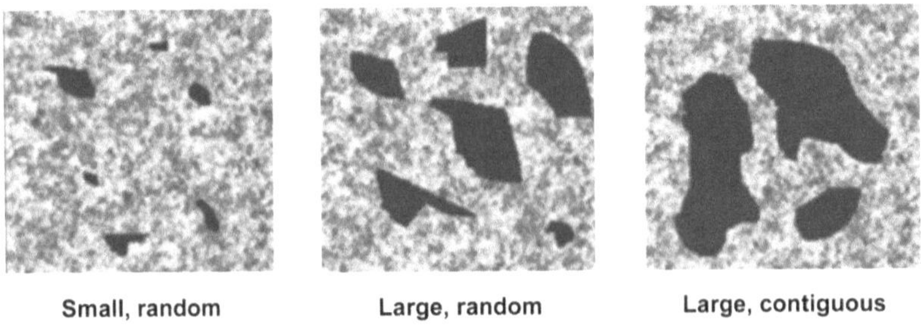

Figure 6.5 Types of die-attach voids.
(Courtesy CALCE Electronic Products and System Center, Univ. of Maryland.)

where K is the bulk modulus and ρ is the density.[11] Acoustic reflection at interfaces is based on the following:

- Ultrasound propagating from a high acoustic impedance material to a lower impedance material that produces a negative reflection.
- Ultrasound propagating from a material of low acoustic impedance on top of a high impedance material, thus generating a positive reflection.

When voids are present in a bond line or when delamination occurs, the air that fills the voids has an acoustic impedance of 0 kg/m^2 second and a large amplitude and total negative reflectance occurs. Most polymeric materials have acoustic impedances of $2.0–4.5 \times 10^6$ kg/m^2 second compared to copper (42×10^6 kg/m^2 second) and silicon (20×10^6 kg/m^2 second). Examples of acoustic micrographs showing voids and delamination in an adhesive bond line are pictured in Figs. 6.6 and 6.7.

Moisture is absorbed by adhesives and other plastic materials to various degrees and can accumulate in voids within the bond line during assembly, testing, and operation. The absorption of moisture in adhesives as well as in other polymeric materials, such as molding compounds, has been proven to cause failures when parts are subsequently solder reflowed and exposed to the high solder melt temperatures of 200 °C and above. The rapid evaporation and expulsion of the moisture result in stresses that cause cracking and delamination, a phenomenon referred to as *popcorning*.

The absorption of moisture in underfill adhesives induces a tensile hygrothermal stress on the solder-ball connections causing electrical opens in the connections and cracking in the adhesive. These tensile stresses offset the compressive stresses that underfill adhesives provide in improving the reliability of flip-chip and ball-grid-array devices.

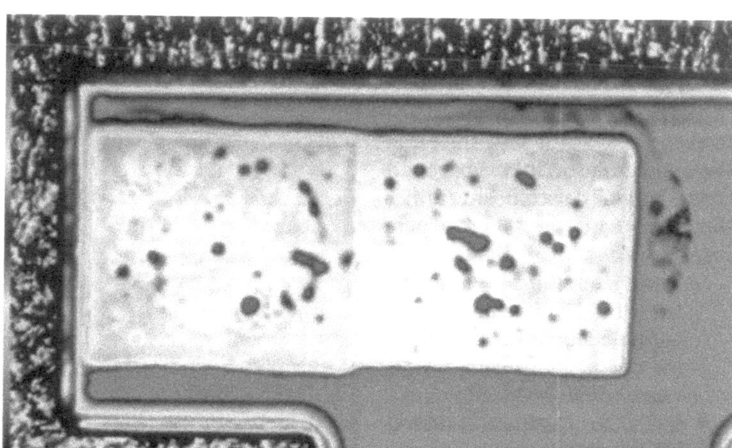

Figure 6.6 Acoustic micrograph of voids.[11]

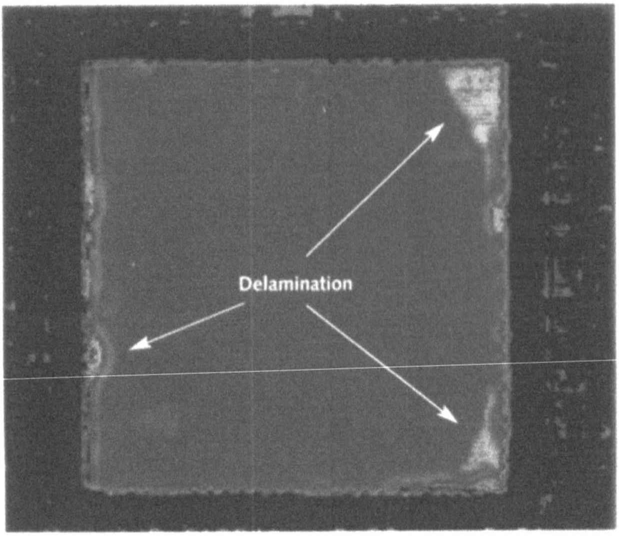

Figure 6.7 Acoustic micrograph of delamination.[11]

6.2.2 Die fracture

Besides delamination, actual cracking of silicon die is a failure mode that can occur from excessive adhesive stresses, voids, and moisture absorption. Residual stresses in adhesive-attached single-chip devices become critical as the size of the device increases and dissimilar die and leadframe materials are used. Several types of fractures that can occur within the die or within the adhesive are shown in Fig. 6.8. A hairline crack in an IC chip that resulted from adhesive stress is shown in Fig. 6.9.

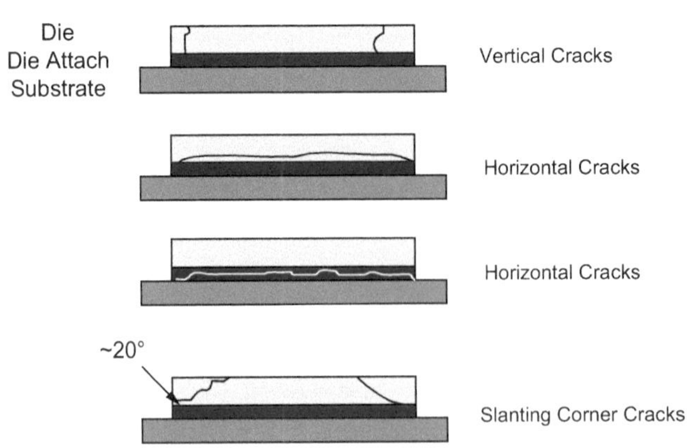

Figure 6.8 Types of fractures in die and die-attach adhesives.
(Courtesy CALCE Electronic Products and System Center, Univ. of Maryland.)

Figure 6.9 Fracture in silicon IC device resulting from adhesive stress.
(Courtesy CALCE Electronic Products and System Center, Univ. of Maryland.)

6.2.3 Adhesive fracture/fracture mechanics*

The failure of an adhesive joint can be considered to involve the initiation and propagation of naturally occurring (intrinsic) flaws or defects. Fracture mechanics is the field of mechanics concerned with the study of the formation and propagation of cracks in materials. The objective of using fracture mechanics is to determine the bond durability of a specific adhesive and provide a basis for estimating the fracture, fatigue, and service life of the adhesive joints. It uses methods of analytical solid mechanics to calculate the driving force on a crack and those of experimental solid mechanics to characterize the resistance of a material to fracture.

Fracture mechanics applies the physics of stress and strain, and the theories of elasticity and plasticity to microscopic defects found in real materials in order to predict their macroscopic mechanical failure. Fractography is widely used along with fracture mechanics to understand the causes of failures and also to verify the theoretical failure predictions with real life failures.

Although important for structural adhesive bonds, fracture mechanics is not as critical for non-structural low load-bearing adhesives as used in most electronic modules. For the most part, passing minimum specification requirements for peel and tensile strengths both at ambient conditions and accelerated test conditions are sufficient. However, computer-simulated modeling and reliability analysis have been used for evaluating electrically conductive adhesives as used in electronics assembly.[12]

* This section was based in part on material found at www.adhesivestoolkit.com.[13]

The main advantage of using fracture mechanics in design analysis is that geometry independent results are obtained from the laboratory testing. In one type of testing, strain energy release rate (SERR) is measured from beam test specimens during cyclic fatigue testing together with the associated crack growth rates. Crack growth testing is performed over a range of SEER values including the threshold value below which negligible crack growth is expected, typically associated with rates between 10^{-6} and 10^{-7} mm per cycle. The threshold value is used during joint design as a limiting criterion and the designer may limit the maximum SEER value within the joint at or below the measured threshold value (G_{th}). This would be classified as a damage intolerant design (i.e., no cracking permitted). If a damage-tolerant system (SEER) value greater than G_{th} is acceptable, then the designer can predict the expected life of the component by applying the measured crack growth model.

6.2.3.1 Fracture mechanics design approach

Two parameters have been used as the basis to quantify crack growth behavior:

- Stress-intensity factor, which considers crack tip stresses.
- Strain energy release rate, which considers the energy required to create new crack surfaces.

The stress-intensity factor K was developed for bulk materials. For elastic materials, the stress field around a crack tip is considered to be singular. The stress-intensity factor represents the strength of the singularity. The general form of the stress-intensity equation for a material is:

$$\sigma_{ij} = \frac{K}{(2\pi r)^{1/2}} f_{ij}(\theta) \tag{6.4}$$

where

$$\sigma_{ij} = \sigma_0 \left(\frac{a}{2r}\right)^{1/2} f_{ij}(\theta) \tag{6.5}$$

and

- σ_0 is the applied stress,
- $2a$ is the crack length,
- r and q are the coordinates of a point,
- σ_{ij} are the components of the stress tensor at that point.

By combining these equations the stress intensity can be expressed as:

$$K = Q\sigma_0 a^{1/2} \tag{6.6}$$

where Q is introduced as a geometry factor.[14]

With the stress-intensity method K is used as the basis for assessing failure; for fracture to occur the stress intensity must exceed the critical fracture toughness K_c. Examples of critical fracture toughness values for some materials are shown in Table 6.2[15].

Table 6.2 Fracture toughness values of common materials[15]

Material	Critical fracture toughness (K_c, MPa-m$^{1/2}$)
Aluminum alloy 2014	18–31
High strength steel	50–154
Silicon nitride	4–5
Silicon carbide	3
Alumina	3–5
Epoxy	0.3–0.5
Polyester	0.5
Polyethylene	1–2
Polymethylmethacrylate	0.5–1.75

In the strain energy release rate method, the parameter considered to determine failure is the strain energy release rate, which is the energy released per unit area of crack growth, commonly identified as G. For a uniformly loaded joint, the strain energy release rate, is derived from energy balance considerations as:

$$G = \frac{P^2}{2b}\left(\frac{\partial C}{\partial a}\right) \tag{6.7}$$

where P is the load, b is the width of the joint, C is the compliance of the joint, and a is the crack length.

For a crack to grow under monotonic loading, the strain energy released must exceed the energy required to create the new crack surface. For cracks to occur, the strain energy release rate must exceed a critical fracture energy value (G_c). Contributions to G_c include the breakage of fundamental bonds in the adhesive and energy dissipating mechanisms around the crack tip. In general, G_c depends on loading rate, temperature, thickness of the adhesive layer, and type of loading at the crack tip such as a combination of peel and shear.

Although K and G can be related, the stress-intensity method is less useful for adhesives because K values are more difficult to calculate and also because this method is limited to linear-elastic behavior only. With the strain energy release rate method, the nonlinear effects of viscoelasticity and plasticity can be more readily accounted for in the analysis.

In design analysis, G can be used to assess the onset of fracture under extreme loads and the onset of crack growth under static loads (creep failure), but more often the rate of crack growth under cyclic loads (fatigue failure).

The failure of adhesive joints depends on the combination of peel and shear loading. In fracture mechanics, three basic 'modes' of loading (Fig. 6.10) are the following:

- Mode I – tension.
- Mode II – shearing.
- Mode III – tearing.

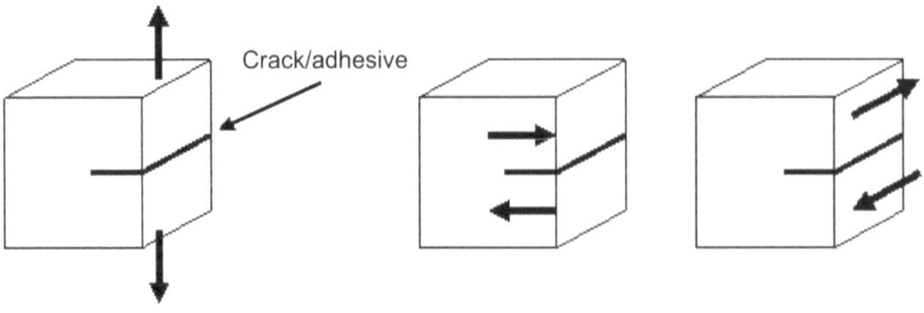

Figure 6.10 Basic modes of loading on cracks[13] (left to right: Mode I, Mode II, Mode III).

Mode I consists of tension across the crack leading to crack opening or widening. Mode II is a shearing action and Mode III, a tearing action. Mode I is generally considered to be the weakest mode. However, this may not always be the case because the area of failure within the bond line may be different for mixed-mode conditions compared to Mode I. It is, therefore, necessary to determine the contributions from the individual modes (G_I, G_{II}, and G_{III}) to determine the total SERR. In reality, most cracks will experience mixed-mode or combinations of loading which contribute to the total strain energy release rate G_{tot}. The most common combination is Mode I and Mode II.

6.2.3.2 Fracture mechanics design methodology

In fracture mechanics design, the growth of a crack from an intrinsic flaw or pre-existing crack is considered using strain energy release rate (SERR) as the basis. The relationship between the rate of crack growth and *G* is shown schematically in Fig. 6.11.

Two extreme levels of SERR are identified: a high catastrophic or critical level G_c, at which rapid fracture occurs and a low threshold level G_{th}, below which crack

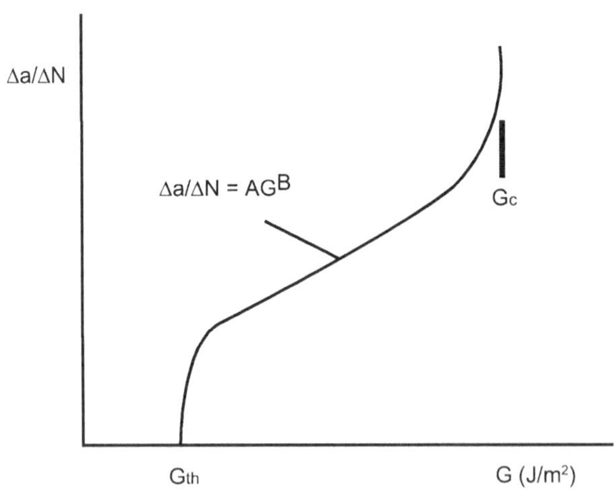

Figure 6.11 Schematic plot of G versus $\Delta a/\Delta N$.[13]

growth does not occur. Between these extremes, the crack growth behavior tends to follow a power law and $\Delta a/\Delta N = AG^B$, where A and B are constants. A materials' model that represents the rate of joint failure as a function of strain energy release rate is required. In the case of fracture, the critical strain energy release rate dominates the model and crack propagation is very fast. For fatigue, the rate of crack growth per cycle is the major attribute. This is an incremental approach to calculate the total number of cycles to failure. This is shown by the rate of crack growth per cycle, $\Delta a/\Delta N$ as a function of G and G_{th} (Fig. 6.11). The fatigue material model is used to predict crack growth based on the number of cycles and $\Delta a/\Delta N$.

6.2.3.3 Fracture mechanics design data

The design data required for the fracture mechanics approach are the model parameters that define failure in terms of strain energy release rate. Most commonly, test pieces are required to obtain data for Mode I loading, Mode II loading and Mixed-Mode I/Mode II loading. Various test pieces have been developed for measurement of joint properties, generally using beam type geometries. A double cantilevered beam specimen is used for Mode I testing, and single-cantilevered test specimens used for Mode II and mixed-Mode I/II. A four-layer 'sandwich' containing metal plates 175 mm × 110 mm is constructed with adhesive applied between each layer. The starter crack length, a_0, and the central (test) bond thickness are controlled using a mask. After curing, the four individual reinforced double cantilevered beam test pieces are carefully machined from the molded panel.

Testing should include the effects of adhesive layer thickness, mode ratio, and any relevant environmental effects that need to be accounted for in the analysis such as the rate of loading (i.e., Modes I, II, and mixed I/II), temperature, humidity, and long-term aging. The loading conditions, including gripping of the test piece, determine the mode of loading imposed at the crack tip. The test coupon is cycled under displacement control to generate crack propagation within the central adhesive layer. Typically, testing is started at high strain energy release rates, and as the crack grows SEER is progressively reduced. This allows a single test to generate the entire plot of strain energy release rate or G as a function of crack growth rate or $\Delta a/\Delta N$.

Standards such as the British Standards Institute BS 7991, "Determination of Mode I Adhesive Fracture Energy, GIC, of Structural Adhesives Using The Double Cantilevered Beam (DCB) and Tapered Double Cantilevered Beam (TDCB)" apply to these tests.

6.2.4 Thermal impedance

Overheating of devices resulting in electrical malfunctioning or reduction in the life expectancy of the device and circuit can be the direct result of poor conductivity and high thermal impedance of the adhesive. Thermal conductivity of adhesives becomes critical as devices are operated at higher speeds, consume more power, and dissipate more heat. Semiconductor junction temperatures must remain within normal limits, particularly for gallium-arsenide devices that operate at high frequencies and whose thermal conductivity is much lower than that of silicon devices (approximately 60% lower). In such cases, high thermal conductivity materials throughout the conduction path from the junction of a semiconductor device through the die-attach adhesive to the package case are required to maintain the junction-to-case temperature (θ_{JC}).[16] The flow of heat through materials is directly proportional to the cross-sectional area, the temperature difference across the material, and the time, but inversely proportional to the thickness (see Chapter 2). Some mechanisms for increased thermal impedance attributed to adhesives are as follows:

- Poor initial thermal conductivity of the adhesive chosen.
- Voids in the adhesive bond line generated during curing from entrapment of solvents, air, or moisture.
- Air pockets and delaminated areas due to poor wetting.
- Too thick or nonuniform thickness adhesive layer.

6.2.4.1 *Initial thermal conductivity*

Thermal conductivity is defined as the proportionality constant between the rate of heat flow and the cross-sectional area of known thickness and temperature drop. *Thermal impedance* is the temperature drop, from junction to case, divided by the power dissipated by the device and is reported in °C/W. Among the important steps in assuring adhesive reliability is the selection of an adhesive having the highest thermal conductivity. If electrical insulation is not a factor, the silver-filled epoxies are among the best thermal conductors. The thermal conductivities of silver-filled epoxies range from 0.5 W/m K to 8 W/m K. However, some silver-filled epoxies formulated with unique solvent blends are reported to have values as high as 20 W/m K. The highest values (>40 W/m K) are found with silver-glass adhesives.

Thermal-transfer adhesives that are electrically insulating also exhibit wide ranges of thermal conductivities, depending on the filler type and amount. The thermal conductivities of epoxies filled with boron nitride or diamond are approximately 4 W/m K and 12 W/m K, respectively, while those of the more common aluminum-oxide-filled adhesives range from 1 to 2 W/m K.

Thermal conductivity requirements are specified in both military and industry specifications generally as 1.5 W/m K or greater for electrically conductive adhesives and 0.15 W/m K or greater for electrically insulative adhesives.

The selection of highly conductive adhesives is a necessary factor, but not the only one in minimizing thermal impedance. Applying as thin a bond line as possible and avoiding the introduction or creation of voids are also essential in reducing thermal impedance.

6.2.4.2 *Bond line thickness*

The control of bond line thickness is essential for high-power devices. A thin bond line is required since thickness is directly proportional to the junction-to-case thermal resistance, θ_{JC} (see Chapter 2). Specially formulated solvent-based hybrid adhesives can produce uniform bond line thicknesses of 1 mil or less.[17] Small hard spacers, acting as shims, added to paste adhesive formulations, have also been used to control thickness.[18,19] The use of preforms instead of paste adhesives also assures reproducible thicknesses provided uniform pressure and temperature are applied during curing.

6.2.4.3 Voids and outgassing

Besides contributing to delamination, voids and outgassing in adhesive bond lines lower thermal conductance due to entrapped air. Voids may be introduced during the application of the adhesive because of poor wetting or improper dispensing conditions. Other sources of voids include volatilization of moisture or solvents. Entrapment of moisture, solvents, or low-molecular-weight polymer additives during curing also reduces thermal conductance. Steps that are often used to prevent voids include use of low-moisture-absorption adhesives, degassing adhesives prior to cure, prebaking the components to remove adsorbed moisture, and step curing at low temperatures to allow air and volatiles to escape before full cure.

6.2.5 Electrical stability

Silver-filled epoxies and other electrically conductive adhesives are widely used to electrically connect chip devices or packaged components to interconnect substrates or printed-circuit boards. Chip capacitors, resistors, transistors, diodes, and magnetic components may be attached with silver-filled epoxies whose volume resistivities range from 1×10^{-4} to 3×10^{-4} ohm-cm or with gold-filled epoxies whose volume resistivities are approximately 8×10^{-4} ohm-cm. Conductive adhesives are also finding use as replacements for solder balls in flip-chip devices. In all cases, to achieve reliable connections, initially low-contact resistances or volume resistivities must remain low on aging and on exposure to operational stress conditions, such as humidity, temperature, vibration, shock, and power.

6.2.5.1 Effect of cure and current on contact resistance

The reliability of electrically conductive adhesives (ECAs) used for chip attachment, flip-chip bonding, and replacements for solder depends on numerous factors. In addition to the type, amount, and morphology of the filler and exposure conditions, notably elevated temperatures and moisture, the contact resistance is affected by the degree of cure and current density. An issue with ECAs is the increasing resistance of contacts with high current densities that are encountered with small-area microelectronic connections. Kim and Shi,[20] for example, showed that a high degree of cure as measured by differential scanning calorimetry did not necessarily result in the most stable contact resistance. Contact resistances were measured of test samples bonded with a silver/silver-copper filled epoxy cured at 150 °C for three periods of time (5.95 minutes, 7.63 minutes, and 11.88 minutes). As expected, resistance decreased as the cure progressed and stabilized at 11.88 minutes (Fig. 6.12). However, when subjected to current densities of 10, 25, and 30 amp the high current densities of 25 and 30 amp resulted in the worst degradation while the best were those cured at an

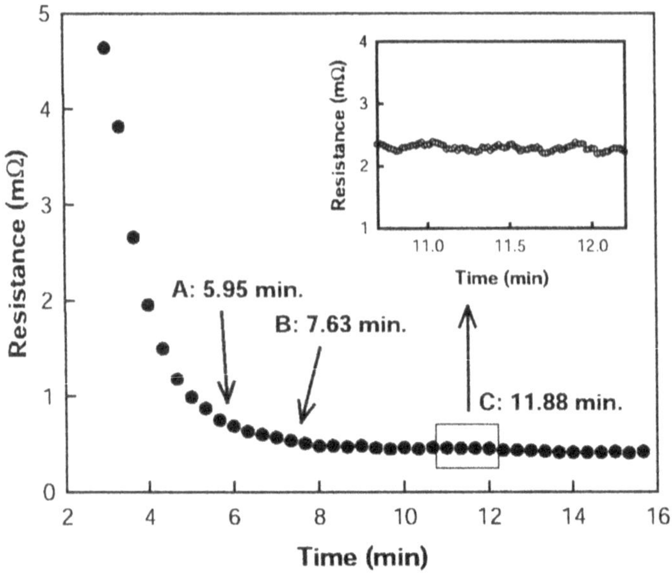

Figure 6.12 Resistance change during curing of conductive adhesive.[20]

intermediate time (7.95 minutes) (Fig. 6.13). Very little degradation occurred at 10 A except for the least cured sample.

Increases in electrical resistance or complete loss of electrical contact (electrical opens) can result from several failure mechanisms, such as:

- Loss of backside electrical contact on transistors and other semiconductor devices due to redistribution of filler at the interface or to oxidation/corrosion of the metallization.
- Separation of the adhesive and electrical opens due to one or more of the factors already discussed for delamination, such as poor wettability, shrinkage stresses, and large CTE mismatches.
- Electrical instability of silver-filled epoxies on non-noble metal surfaces when used as replacements for solder.
- Instability of silver-filled epoxies on aluminum surfaces due to oxidation and corrosion processes.

6.2.5.2 Loss of backside electrical contact on semiconductor devices

A reliability concern with semiconductor devices, such as chip transistors in which one of the connections is made through the backside metallization by attaching with either eutectic alloy or conductive epoxy, is the loss of backside ohmic contact. Loss of ohmic contact may be due either to mechanical/physical or chemical mechanisms.[21] Mechanical failures occur from partial or complete delamination of the adhesive interface. The smooth surface of the die metallization contributes to the problem whereas roughening the surface improves results. Loss of contact is evident from increases in resistance during initial electrical testing, but may better be detected

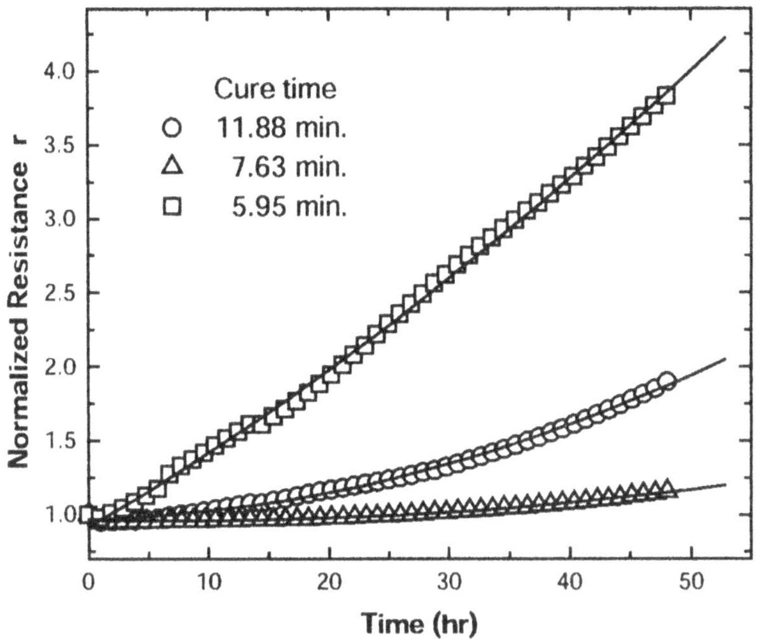

Figure 6.13 Contact-resistance changes for three different cure degrees versus time at 30 A.[20]

after temperature cycling or mechanical-shock screen tests, for example with hybrid microcircuits, before the product is shipped.

Failures may be related to the adhesive formulation as well as the contact metallization. The first generation of 100%-solids conductive epoxies contained 70% by weight of silver flake, and the final cured adhesive contained 21% silver by volume. Settling or disproportioning of the filler during cure in these early adhesives resulted in a resin-rich layer at the die metallization interface and a high contact resistance. Subsequent adhesives were formulated with the same amount of filler by weight but with a higher volume (34% after cure). The risk of conductivity loss due to filler redistribution for these formulations was thus minimized.[22] Increases in the "effective filler loading" is also possible from the shrinkage that occurs when some solvents in the formulation evaporate. The shrinkage prevents separation effects at the interfaces resulting in more stable interconnect resistances. However, as previously discussed, that excessive shrinkage during cure produces high stresses.

Compared with mechanical and physical mechanisms, chemical mechanisms are slow and may not be detectable until well after the product has been in use, at which point the failures may be catastrophic. One chemical mechanism involves the formation of an oxide layer at the die-epoxy interface. A poorly alloyed gold-backed surface has been identified as one cause of degraded interfacial resistance. Although this is one cause, adhesives formulations also may play a role. For example, adhesives having excess (unreacted) curing agent absorb large amounts of water, thereby accelerating the oxidation process. Adhesives having less than the stoichiometric amount of curing agents may improve performance. Adhesives having a high chloride-ion content (greater than 200 ppm) are also known to cause corrosion and increase the contact resistance, especially when used to attach die having less than 125 μm of gold thickness.

6.2.5.3 Stability of solder replacements on non-noble metal surfaces

A new area of concern for electrical stability arises because of the increasing use of conductive adhesives as replacements for solder. Some conductive adhesives show unstable electrical-contact resistance when used on non-noble metal surfaces such as copper or tin–lead solder. Although stable on gold, palladium, platinum, and silver surfaces, the same adhesives were found to be unstable on tin, tin–lead, copper, and nickel surfaces.[23,24] The unstable resistance and increase in resistance in temperature–humidity exposures have been attributed to the growth of an oxide layer separating the filler particles from the substrate at the interface, a mechanism similar to that for the loss of backside contact in die-attach materials.

The unstable behavior of some solder-replacement adhesives has been attributed to galvanic corrosion. Similar to most corrosion mechanisms, condensed or absorbed moisture on the surface and dissimilar metals are required to form a galvanic cell. The silver filler acts as a cathode while the substrate metallization acts as an anode and is oxidized. In the case of tin–lead solder surfaces, the solder, which has a lower electrochemical potential (0.13 V) than silver (0.79 V), becomes the anode at which corrosion and oxidation occur. A smaller potential difference between a copper surface and silver accounts for some improvement in contact resistance over the solder-silver couple.

The reliability of several silver-filled epoxies as solder replacements has been studied and reported by several investigators.[25,26] In one study,[27] adhesives formulated with an epoxy resin (Epon 862®) and cured with several anhydrides were evaluated for their electrical stabilities on non-noble metal surfaces and extended exposure to 85/85 conditions (Table 6.3). The anhydride-curing agents evaluated were hexahydrophthalic anhydride (HHPA), methyl hexahydrophthalic anhydride (MHHPA), and 4-methylphthalic anhydride (MPA). In all cases, resistance changes were negative up to 300–600 hours after which they became less negative and finally positive after 1,000–1,500 hours exposure. The initial decreases in contact resistances were attributed to further leaching out of the lubricant from the silver-flake filler or to fillers penetrating the initial oxide layer present on the non-noble metal surface during postcuring of the adhesive. Results were promising based on the acceptance criterion set by the National Center of Manufacturing Science (NCMS) that contact-resistance change should be less than 20% after 500 hours aging at 85/85 conditions.[28]

Although some results are promising, the extent of resistance change and direction depend largely on the formulation used and should be evaluated for each application, especially on non-noble metal surfaces.

To minimize the corrosion reactions, oxygen scavengers and corrosion inhibitors may be used as additives in solder-replacement adhesives. The percent changes in

Table 6.3 Behavior of solder-replacement adhesives on non-noble metal surfaces[27]

Adhesive no.	Curing agent	T_g (°C)	Moisture absorption (%, 350 hours of 85/85)	Percent resistance change (85 °C/85% RH exposure) (relative to initial reading)								
				100 hours			300 hours			500 hours		
				Sn	Sn/Pb	Cu	Sn	Sn/Pb	Cu	Sn	Sn/Pb	Cu
1	HHPA[a]	114.76	0.9	−30	−15	−50	−80	−40	−70	−90	−40	−75
2	MHHPA[b]	97.89	1.1	−40	−10	−70	−60	−30	−75	−70	−20	−90
3	MPA[c]	136.93	1.0	0	−30	−50	−35	−60	−55	−40	−60	−60

Adhesive no.	Curing agent	T_g (°C)	Moisture absorption (%, 350 hours of 85/85)	Percent resistance change (85 °C/85% RH exposure) (relative to initial reading)					
				1,000 hours			1,500 hours[d]		
				Sn	Sn/Pb	Cu	Sn	Sn/Pb	Cu
1	HHPA[a]	114.76	0.9	−90	−40	−70	−70	−10	−70
2	MHHPA[b]	97.89	1.1	−45	+50	−75	0	+100	−80
3	MPA[c]	136.93	1.0	−40	−35	−50	0	+10	−55

[a] Minimum at 400 hours exposure.
[b] Minimum at 300 hours exposure.
[c] Minimum at 600 hours exposure.
[d] Exception for Adhesive #1 (1,400 hours).
Table compiled from Lu and Wong.[27]

contact resistances for cured formulations without scavengers or inhibitors were found to be 2–3 times greater than those containing scavengers.[24] Contact resistances after 500 hours of 85/85 testing were significantly reduced for those adhesives containing oxygen scavengers and less than 10% and 20% for those with and without the inhibitors, respectively.

6.2.5.4 Stability of epoxies on aluminum surfaces

Erratic and high-resistance changes have also been reported when silver-filled epoxies are used to attach devices to thin-film-aluminum metallization and exposed to temperature–humidity environments.[29] In practice, however, aluminum-bonding pads and aluminum conductors on the top surface of an interconnect substrate are protected with barrier coatings of titanium–tungsten or chromium followed by plated gold as the top bonding surface.

6.2.6 Corrosion

Corrosion of conductor metallizations, devices, wire, wire bonds and other connections, and platings occur by direct or indirect-chemical reactions. The metals most susceptible to chemical attack are the non-noble metals among which are thin-film aluminum used for conductors or bonding pads, thin-film nichrome used for precision resistors, tin–lead solders used for surface-mounting components, and copper used for conductor traces on PWBs or as heat sinks. All these metals are susceptible to chemical attack and corrosion, ultimately leading to high-resistance paths, electrical opens, or high-leakage currents. Corrosion may result from constituents from the curing agent, catalyst, or resin portion of an adhesive that has not been fully cured or has been used in excess of the optimum ratios. These constituents may be reactive enough to directly react with the metal or may do so after decomposing or first reacting with moisture. Other constituents may outgas and, in the presence of moisture, produce highly acidic or alkaline solutions that attack the metals.

6.2.6.1 Corrosion from curing agents

In two-part systems, the curing agents and catalysts are generally much more corrosive than the basic resin. Early epoxy-adhesive formulations used boron trihalide complexes such as boron trifluoride ethanolamine to achieve rapid room-temperature cures. Unreacted trihalides dissociated forming hydrogen chloride or hydrogen fluoride that, in the presence of small amounts of moisture, generated hydrochloric or hydrofluoric acids, both highly corrosive to aluminum.[22]

Unreacted excess amines, such as dicyandiamide (Dicy), were also a cause of corrosion in some early epoxy formulations. Dicy decomposes producing ammonia and other reactive amine compounds. The primary and secondary amines, being quite alkaline, react rapidly with aluminum. Analysis of sealed hybrid microcircuits that used Dicy-cured epoxies showed quantities of ammonia as high as 30,000–100,000 ppm_v and extensive corrosion in the presence of moisture.[22] Other investigators corroborated the presence of ammonia and low-molecular-weight amino compounds. Ammonia and amines were generated from a Dicy-cured epoxy (Ablefilm® 550) substrate-attachment film adhesive after post-cure aging 24–300 hours at 125 °C.[30] Ammonia, in the presence of moisture, forms a highly alkaline ammonium hydroxide solution that reacts with and converts aluminum to aluminum hydroxide as follows:

$$\begin{aligned} &NH_3 + H_2O \rightarrow NH_4^+ + OH^- \\ &Al + 3OH^- \rightarrow Al(OH)_3 \end{aligned} \tag{6.8}$$

Corrosion due to alkaline condensates may be controlled by selecting curing agents that do not generate ammonia and by vacuum baking assembled circuits at 125–150 °C for 16–96 hours to remove moisture prior to sealing.[31] A related failure mechanism involving alkaline solutions is that of "vanishing" nichrome thin-film resistors. Nichrome resistors were found to disappear in the presence of moderate-to-high amounts of moisture and alkaline or ionic contaminants, especially during the cold cycle of temperature-cycle testing where the moisture condenses as a liquid. Residual-gas analysis showed that no failures occurred if the moisture content was less than 1,000 ppm_v. However, moisture ranging from 7,000 to 10,000 ppm_v having dewpoints close to room temperature produced failures during the cold cycle. In quantities above 17,000 ppm_v, liquid water and contaminants exist at room temperature and failures may occur early in testing.[21]

6.2.6.2 Ionic impurities

Ionic contaminants from adhesives or from other sources have been a serious problem, not only in contributing to corrosion, but also in causing electrical-parameter changes in semiconductor devices and high-leakage currents across closely spaced conductors. Chloride ions are particularly detrimental since they act as catalysts and are continuously regenerated while etching away aluminum and other metals. Increasing the chloride-ion content in adhesives from 10 ppm to over 200 ppm was reported to reduce the time to failure by a factor of 4.[32] The mechanism for chloride-ion etching of aluminum is shown by the following series of equations.[33]

$$\begin{aligned} &4Al + 3O_2 \rightarrow 2Al_2O_3 \\ &Al_2O_3 + 3H_2O \rightarrow 2Al(OH)_3 \\ &Al(OH)_3 + Cl^- \rightarrow Al(OH)_2Cl + OH^- \\ &Al + 4Cl^- \rightarrow AlCl_4^- + 3e^- \\ &AlCl_4^- + 3H_2O \rightarrow Al(OH)_3 + 3H^+ + 4Cl^- \end{aligned} \tag{6.9}$$

High levels of chloride ions, above 600 ppm, and sodium ion levels above 100 ppm were reported for the early "first generation" epoxy adhesives. The epoxy resin itself is a source of both chloride and sodium ions if not properly purified. Most epoxy resins are synthesized from epichlorohydrin and bisphenol-A in the presence of sodium hydroxide and produce sodium chloride as a byproduct (see Chapter 3).

Continuous extraction of the ions with deionized water and distillation of the resin are two methods used to purify the resin.

Besides corrosion, the presence and mobility of ions, together with moisture, contribute to the phenomenon of metal migration resulting in high-leakage currents between conductor lines and eventual electrical shorting. Ions and other contaminants, together with moisture, form an electrolytic medium and generate an electrochemical cell between two metal conductors. A small applied bias of several volts causes oxidation of the positively charged metal electrode (anodic conductor), migration of the positively charged metal ions towards the cathode, reduction back to pure metal at the cathode, and a metal buildup that eventually bridges the gap, known as dendritic growth.[34–36] Although most non-noble metal conductors are prone to metal migration, silver has been the prime cause of most circuit failures.

Perkins and Licari[37] showed the importance of moisture in initiating and accelerating silver migration from silver-filled epoxy adhesives. Test specimens consisted of pairs of thick-film gold pads separated by a 10-mil gap and formed on ceramic substrates. One pad of each pair was overcoated and cured with a silver-filled epoxy adhesive. Tests were run on samples that were cooled to just below the room-temperature dew point at which point condensation of moisture from the air occurred. Other samples were cooled using dry ice so that moisture condensed as frost. With 2 V/mil applied, water droplets condensed and coalesced. Dendritic growth of silver occurred when continuous water paths formed across the gap. Current increased from initially low microamps to high milliamps until shorting occurred in less than 1 hour. On the other hand, the frost-covered samples showed no silver migration and low microamp current for over 5 hours, but rapid migration as soon as the frost was melted.

6.2.7 Outgassing and moisture

Outgassing from adhesives that are used in hermetically sealed circuits, especially the outgassing of moisture, has been and still is a major cause of circuit failures and a reliability concern. The amount of moisture released depends on the adhesive used, its cure schedule, the extent of vacuum baking prior to sealing, and the temperature and exposure conditions after sealing. Even with optimum-curing and vacuum-baking conditions, if the temperature of the sealed units during thermal cycling or burn-in exceeds the cure or vacuum-bake temperatures, an extra surge of outgassing occurs. If exposed to very high temperature, major decomposition of the adhesive occurs as indicated by large amounts of moisture, carbon dioxide, and hydrocarbons. Semiconductor device failures due to moisture are not as prevalent now as during the 1970s and 1980s when the silicon oxide or silicon nitride passivation layers on semiconductor devices were porous and not yet fully perfected. Yet today, even with improvements in highly-reliable devices, the presence of moisture and ionic contaminants leads to corrosion of thin-film metallizations and wire bonds. Generally, moisture alone without contaminants is not damaging to electronics but, because it is difficult to determine how clean a circuit is and because ions can be generated after cleaning, it has been found necessary to place a limit on the amount of moisture allowable. Over the years, the limits for moisture that started out at 3,000 ppm_v for space applications (Class K) and 5,000 ppm_v for military applications (Class H) were, for a period of time, tightened to 1,000 ppm_v and 3,000 ppm_v, respectively. Today, the moisture limit for both classes and for all sealed electronic devices used in aerospace and other high-reliability applications has been established at no greater than 5,000 ppm_v, measured after a series of screen tests including thermal cycling according to MIL-PRF-38534.

Residual-gas analysis (RGA), a mass-spectrometric method, is the most accurate method for measuring moisture and other low-molecular-weight volatile compounds released from polymeric materials. Interpretation of RGA data has at times proven difficult since small differences in the processing of samples intended to be identical resulted in major differences in the results and analyses of the same circuits by different laboratories. Most of these differences have been resolved by using laboratories that are certified by government agencies, such as Defense Supply Center at Columbus (DSCC), using their own standards.

Analysis of RGA results provides a good indication of the sources and mechanisms for the generation of the outgassed materials. Table 6.4 shows RGA data of sealed packages containing various die-attachment materials and processing conditions.[38]

In Table 6.4, Case 1 shows an analysis of dry nitrogen used as an inert ambient during sealing. The analysis showed almost 100% nitrogen, no detectable oxygen, and insignificant amounts of moisture and carbon dioxide. The small amount of argon in the absence of oxygen could not be explained.

Control packages in which the die was attached with eutectic alloy (no adhesive) and sealed in air indicated the approximate composition of air (Case 2), but, in a second analysis, the high carbon dioxide and moisture content were attributed to an organic contaminant on a new lot of ceramic lids. In Case 3, the die was again attached with eutectic alloy, but was overcoated with polyimide. Analysis indicates extensive outgassing from the polyimide (apparently not completely cured) but also the possibility of a package leak due to the presence and amounts of oxygen and argon. The ambient analyses for Cases 4 and 5 are quite good.

Cases 6a and 6b show the importance of properly curing (processing) the silver-glass adhesive. The high carbon dioxide and moisture content of Case 6a indicates that the polymer binder or other organic constituents of the silver-glass were not completely burned off. After a second cure cycle was used, good RGA results were obtained (Case 6b).

In Case 7, a thermoplastic die-attach adhesive was used. The large amount of carbon dioxide indicates decomposition, but the small amount of moisture indicates that moisture may have been generated, but gettered by oxides or metals in the package. Often when moisture is expected to be high, but is low, and hydrogen is present, the moisture is believed to have reacted with oxides or metals to produce hydrogen.

Case 8 shows the evolution of ammonia that was common with some of the early epoxy formulations that were cured with Dicy. Once the Dicy-curing agent was replaced, results showed no ammonia (Case 9) although some alcohol and ketone were present; the source was residual solvents used in the new formulations.

Table 6.4 Examples of residual gas analysis (RGA) data[38]

Case no.	Description	Nitrogen (%)	Oxygen (%)	Argon (ppm_v)	Carbon dioxide (ppm_v)	Moisture (ppm_v)	Helium (ppm_v)	Hydrogen (ppm_v)	Other (ppm_v)
1	Common sealing atmosphere (dry nitrogen)	99.9	ND	125	$<$100	$<$100	ND	N/R	N/R
2	Glass-frit sealed ceramic packages/eutectic die attach	81.0	17.9	9,221	1,100	541	N/R	N/R	N/R
		78.9	15.6	9,211	36,700	8,795	N/R	N/R	N/R
3	Glass-frit sealed ceramic packages/eutectic die attach with polyimide barrier coating	86.1	5.09	9,808	54,900	2,207	N/R	12,900	Methane: 8,745
		94.8	1.55	724	30,500	642	N/R	N/D	Methane: 4,448
4	Solder-sealed device/eutectic die attach	99.4	N/R	N/R	154	473	N/R	5,845	N/R
5	Solder-sealed device/silver-glass die attach	99.0	N/R	N/R	3,420	439	N/R	5,556	N/R
		99.8	N/R	N/R	3,150	6,240	N/R	450	N/R
6a	Solder-sealed, low-temp. silver-glass adhesive, single cycle fired	97.1	N/R	N/R	14,300	12,600	N/R	1,985	N/R

6b	Solder-sealed, low-temp. silver-glass adhesive, double cycle fired	99.2	N/R	N/R	4,100	2,900	N/R	1,252	N/R
7	Solder-sealed packages/ thermoplastic die attach	89.6	N/R	N/R	92,700	511	N/R	4,587	Hydrocarbons: 5,900
8	Seam-welded packages/ epoxy cured with Dicy	91.0	N/R	N/R	2,091	1,680	8,340	N/R	Ammonia 3,161
9	Seam-welded packages/ epoxy cured without Dicy	85.1	N/R	N/R	5,792	4,720	13,300	N/R	MEK 1,446 ppm_v, Methanol 1,826, and THF 862
10	Seam-welded packages/ epoxy and poor pre-seal bakeout	95.3	N/R	N/R	1,220	10,500	N/R	350	MEK 1.80%, methanol 1.32%, and hydrocarbons 3,520 ppm_v
11	Seam-welded packages/ thermally overstressed epoxies	95.6	N/R	N/R	23,500	19,300	N/R	1,220	N/R
		92.0	N/R	N/R	20,700	18,900	N/R	9,780	N/R
12	Seam-welded packages/ intermittent leak phenomenon	96.1	0.12	274	1,210	3,600	24,400	897	Fluorocarbons: 0.76%
		90.5	6.07	2,940	1,575	30,400	N/R	425	N/R
		87.7	4.87	2,415	2,750	27,200	30,700	890	Fluorocarbons 1.02%

ND – none detected; N/R – none reported.

Insufficient pre-seal vacuum bake is indicated in Case 10 where large amounts of moisture were detected. Extended vacuum baking reduced the moisture content.

The presence of large amounts of carbon dioxide, water, and some hydrocarbons, as in Case 11, indicates decomposition of the epoxy adhesive possibly due to overheating.

In some cases, packages develop leaks because they were improperly sealed or subsequently stressed. These are evident from the presence of large amounts of oxygen together with argon. Since argon is present only in air, the detection of argon along with oxygen indicates the presence of air. In leaky packages, fluorocarbons may also be detected. The presence of fluorocarbons indicates an intermittent or a gross leak in which the fluorocarbon liquid used in testing for gross leaks has entered the package (Case 12). Fluorocarbon species may also originate from solvents used in cleaning that did not completely evaporate or somehow became tenaciously absorbed in the package.

6.2.8 Detachment of adhesive particles

Particles of adhesive and filler can be generated from aggressive handling, during rework, or during vibration or mechanical-shock testing. Particles can dislodge from the adhesive fillets around devices or substrates. In sealed packages, loose conductive particles can become wedged between closely spaced conductor lines causing electrical shorting. Nonconductive adhesive particles and abrasive-filler particles, such as aluminum-oxide, can collide with and damage thin wires, wire bonds, and thin-film metallization. Abrasive particles have been known to completely sever wires and wire bonds during vibration.

To assure the absence of particles in hermetically sealed single-chip or multichip modules, the Particle Impact Noise Detection (PIND) test[39] is employed. For high-reliability circuits such as Class K (space qualified) hybrid circuits, the PIND test is required of all circuits. To assure that particles are not generated subsequently to passing the PIND test, for example during vibration or mechanical shock, a particle-immobilizing coating such as parylene may be used. An alternate approach has been to apply a particle getter, usually a tacky silicone adhesive, to the inside center of the lid prior to sealing.[40,41]

6.2.9 Adhesives quality

FTIR spectra have been used for many years to monitor the quality of adhesives both prior to and after curing. The infrared spectra of the basic adhesive ingredients: resins, hardeners, and additives and of the cured adhesive can be measured and retained as standards with which to compare subsequent lots. Changes in the absorption peaks are noted if a supplier has changed the formulation, for example, by introducing a new additive or by substituting one resin or hardener for another.

FTIR may also be used as a quality-control test to determine whether an adhesive was under- or overcured,[42] or it can be used as a failure analysis tool. In one failure analysis study, adhesive failure of a lid attached to a flip-chip package occurred. FTIR

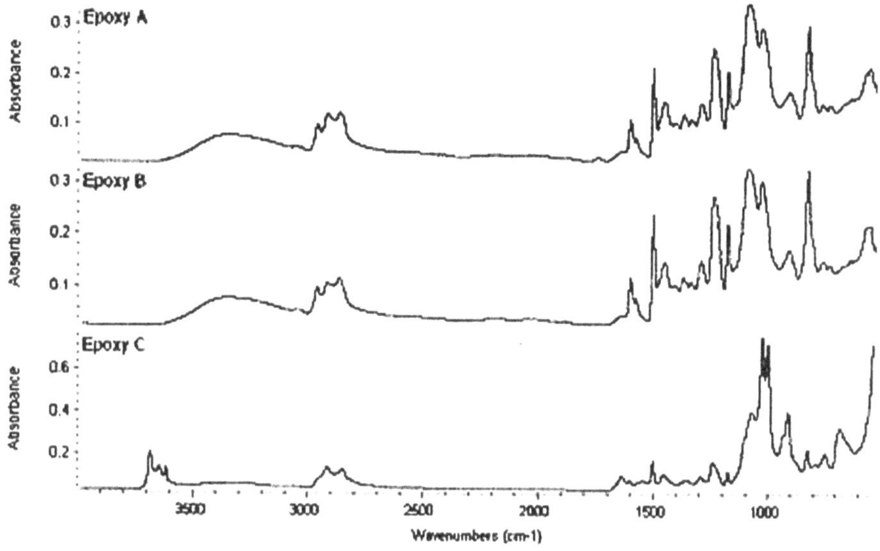

Figure 6.14 FTIR spectra for cured epoxy samples from three different manufacturers.[42]

analysis showed residual contaminants that had not been completely removed that affected the cure of the adhesive and its adhesion to the substrate.[43] In another study, the infrared spectra of cured samples of epoxy adhesives from three suppliers were examined (Fig. 6.14). Epoxy C differed from Epoxies A and B in the extra absorbance bands at the range of 3,600–3,700 cm^{-1}.

An examination of the far infrared region, the so-called "fingerprint region" (approx. 1,800–500 cm^{-1}) shows how similar or different epoxies from two vendors can be. In Fig. 6.15, Epoxies A and B appear to be identical but expanded views showed differences in the concentrations of the components and in an absorption band at 1,743 cm^{-1} for Epoxy A that is not present in Epoxy B, indicating the presence of a unique material in A (www.piketech.com).[42]

Lastly, FTIR can be used to study the kinetics of degradation of cured adhesives on exposure to various conditions such as temperature, time, radiation, humidity, or exposure to chemicals.

Test methods include ASTM E1252 (Standard Practice for General Techniques for Obtaining Infrared Spectra for Qualitative Analysis), ASTM E1790 (Standard Practice for Near Infrared Qualitative Analysis) and ASTM E168 (Practices for General Techniques of Infrared Quantitative Analysis) (see also Chapter 7).

6.3 Reliability prediction and modeling

The definition of reliability is the probability that a component part, circuit board, equipment, or system will perform its intended function under environmental

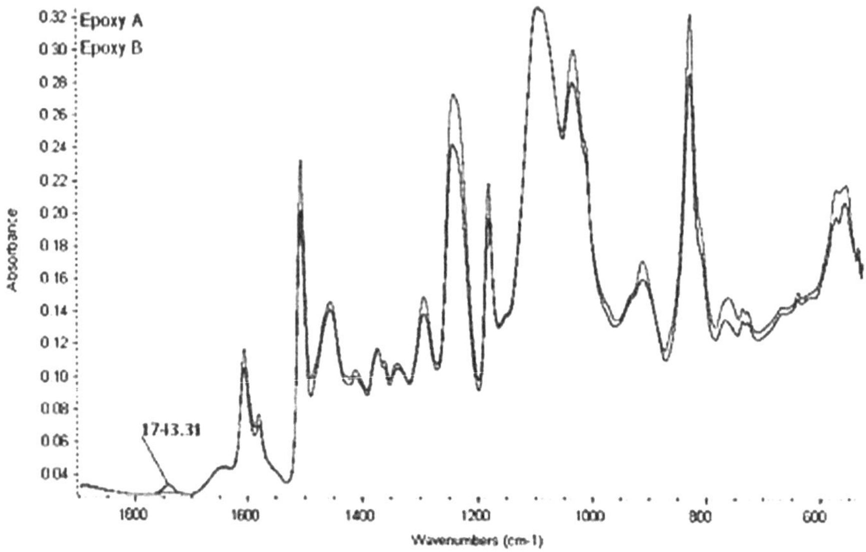

Figure 6.15 Expanded fingerprint region for Epoxies A and B.[42]

conditions, operating time, frequency, and thoroughness of maintenance for a specified lifetime.

The lifetime of a population of units at the component, board, box, or system level can be divided into three distinct periods. This is most often defined by the so-called reliability "bathtub" curve (Fig. 6.16). The bathtub curve describes the "cradle-to-grave" failure rates or frequency of failures as a function of time. The curve is divided into three distinct areas; early failure rate (also known as infant mortality), the useful life period, and the wearout failure period. The infant-mortality portion of the curve, also known as the early life period, is the initial steep slope from the start to

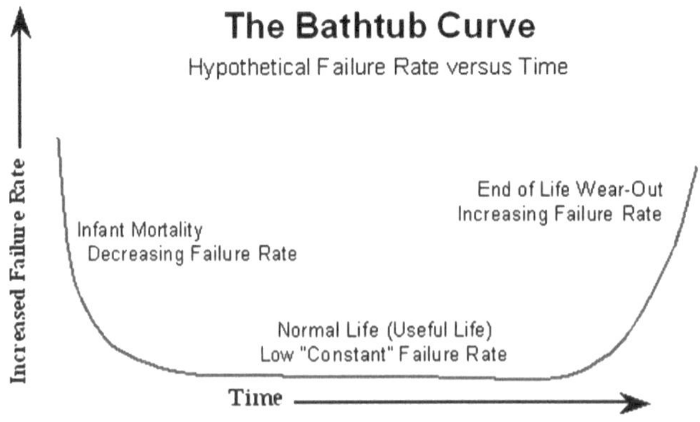

Figure 6.16 Bathtub curve.[45]

where the curve levels out or begins to flatten. The next distinct region in the curve is the flat portion of the graph and is called the useful life period. Failures may occur randomly, but at a mainly constant rate. The third period or wearout or failure period starts at the point where the slope begins to increase and continues to the end of the graph. Increasing rates of failure occur in this region.

Infant-mortality period. In the early period of failures, accelerated stress tests such as burn-in, power cycling, temperature cycling, and vibration as well as highly accelerated stress and life testing are used. These methods are used to screen out parts that are inherently not reliable and prevent the delivery of "dead on arrival" parts to a customer. While this consumes some early life of an assembly by stress screening, the remaining population begins the useful life period or the flat area of the bathtub.

Useful life period. After the weaker units die off in the infant-mortality period, the failure rate becomes nearly constant and the assemblies have entered the normal or useful life period. This period is characterized as a relatively constant failure rate and is also referred to as the system life of a product. Mean time between failures is calculated in this section of the curve. Mean time between failures (MTBF) is the predicted elapsed time between inherent failures of a system during operation.[44] MTBF can be calculated as the arithmetic mean (average) time between failures of a system. Failure rates calculated from MIL-HDBK-217 and Telcordia-332 apply only to this period.

Wearout period. Near the end of the curve, failures occur at increasing rates as components begin to wearout or fatigue. In this final area of the curve, the mean time between failures calculated in the flat area of the curve no longer applies. Products with longer MTBFs can still wearout in just a few years.

6.3.1 Weibull analysis

The Weibull distribution is a life distribution model that is flexible enough to be used to characterize failure distributions in all areas of the bathtub curve. The basic Weibull function is:

$$f(t) = (\beta/\eta)\,(T/\eta)^{(\beta-1)}e^{-(T/\eta)\beta} \quad (6.10)$$

The Weibull function has two parameters. The first is β or a shape parameter and the second is η a scale parameter. The scale parameter determines when, in time, a given portion of the population will fail (say 75%) at a given time *f*(t). The shape parameter enables the Weibull distribution to be applied to any portion of the bathtub curve as follows:

- $\beta < 1$: Infant-mortality period. Typical infant-mortality distributions for state-of-the art semiconductor chips follow a Weibull model with a β in the range of 0.2–0.6.[45]
- $\beta = 1$: Constant failure rate or useful life period.
- $\beta > 1$: Wearout period. A β value of 3.4 found in an evaluation of an automotive component indicated a wearout mechanism.[46]

The graphical results of various βs on the Weibull analysis can be seen in Fig. 6.17. If all curves are taken together, the result is the bathtub curve.

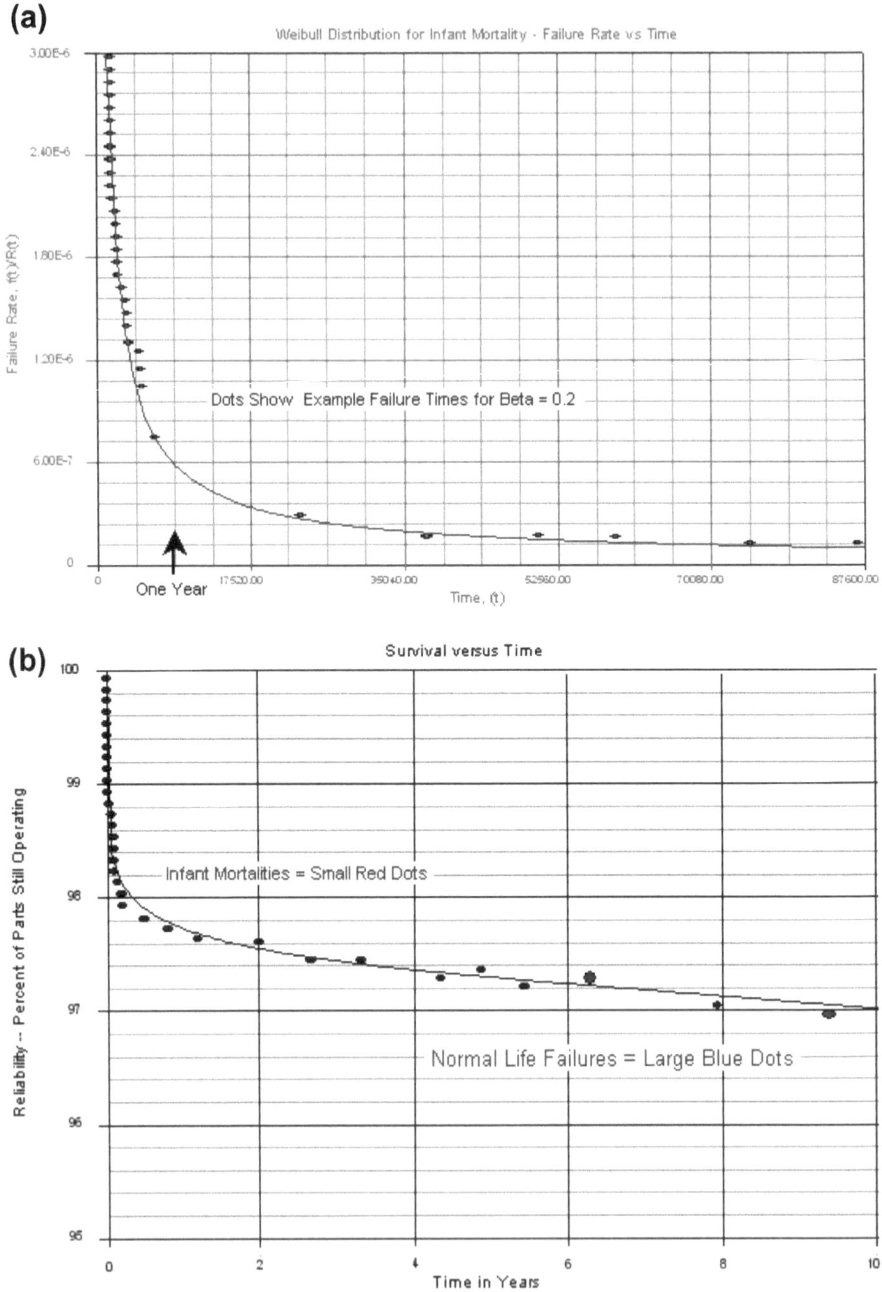

Figure 6.17 Weibull plots and the bathtub curve:[45] (a) infant-mortality period, (b) infant mortality and useful life period and (c) useful life and end-of-life period.

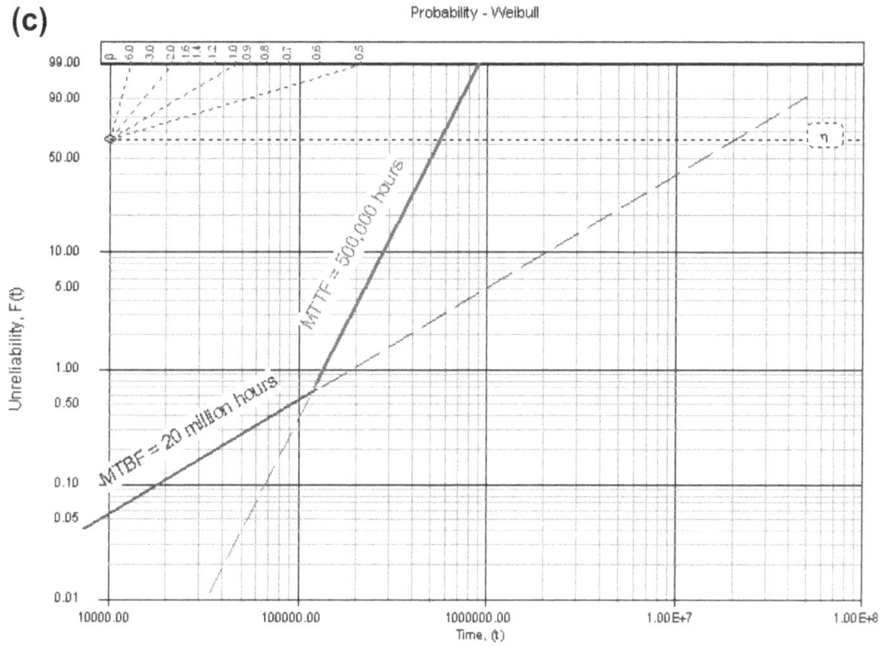

Figure 6.17 (*continued*).

6.3.2 Weibull analysis case study – anisotropic conductive film in flip-chip bonded devices

A Weibull analysis was performed by Kwon et al.,[47] using conductive adhesives. A test chip (14.7 mm × 8.5 mm) with gold bumps in a daisy-chain structure was used to evaluate anisotropic conductive film (ACF) adhesives in thermal cycling. Three adhesives were formulated with 0, 10, and 30% (w/w) of silica filler. Properties of the adhesives are given in Table 6.5. The ACF materials were placed on the substrate, the chip aligned to the substrate, and thermocompression bonded for 20 seconds with 5–6 kgf/cm^2 pressure during cure. Thermal cycling was performed between −40 °C and +125 °C, with a 15-minute dwell at each extreme. Failure criterion was a change in resistance of 50 mΩ and 100 mΩ. The Weibull plots (Fig. 6.18) show that ACF-C had the best performance (more cycles to failure) and ACF-A was the worst. These results are consistent with the fact that ACF-A had the highest CTE and ACF-C had the lowest. In comparison, values of modulus and T_g were comparable for all materials.

6.3.3 MIL-HDBK-217

Maintaining reliability and providing reliability engineering are essential needs with modern electronic systems. Reliability engineering for electronic equipment requires a means for a quantitative baseline, or a reliability prediction analysis. The most recognized and used reliability prediction handbook is Mil-HDBK-217. This standard

Table 6.5 Properties of anisotropic conductive film adhesives[47]

Adhesive name	Percent of silica filler	Coefficient of thermal expansion (ppm/°C)	Modulus (GPa)	Glass-transition temperature (T_g, °C)
A	0	67.1	1.8	106.6
B	10	55.9	2.5	107.9
C	30	40.4	3.0	105.4

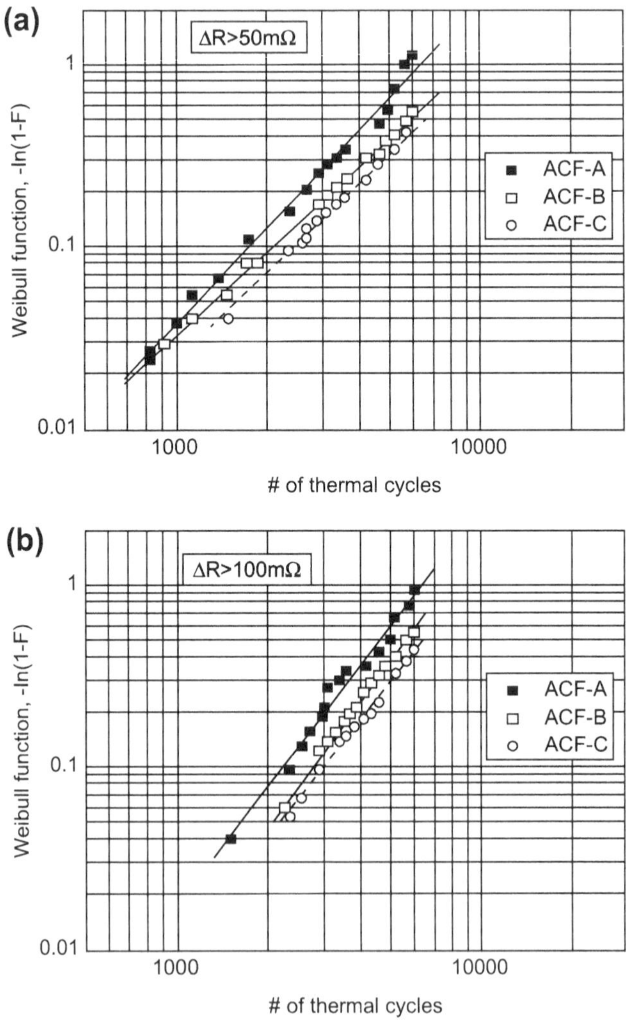

Figure 6.18 Weibull plot for ACF film adhesives in thermal cycling.[47] 50 mΩ failure limit (top) and 100 mΩ failure limit (bottom).

was developed for military and aerospace applications; however, it has also become widely used for industrial and commercial electronic equipment applications throughout the world. The most recent Revision F, Notice 2, was released in February of 1995. It contains failure-rate models for a wide range of electronic components such as integrated circuits, transistors, diodes, resistors, capacitors, relays, switches, and connectors. MIL-HDBK-217 requires a significant amount of data entered into the model. Typically, but not always, calculated results will show a higher failure rate than the corresponding commercial Telcordia document SR-332 that is a replacement for the Bellcore document TR-332 (Bellcore was purchased by SAIC in 1997 and the purchase agreement required Bellcore to change to Telcordia Technologies). This difference in the standards obviously results from the intended use of the MIL-HDBK-217 standard for harsh environments associated with aerospace applications.

The result of using the MIL-HDBK-217 standard for reliability prediction produces calculated Failure Rate and Mean Time Between Failures (MTBF) numbers for the individual components, equipment and the overall system. The final calculated prediction results are based on the roll-up, or summation, of all the individual component failure rates.

The handbook contains two methods of reliability prediction: *Part Stress Analysis* and *Parts Count Analysis*. The two methods vary in the degree of information required to be provided. The Part Stress Analysis Method requires a greater amount of detailed information and is usually more applicable to the later design phase.[48] The Parts Count Method requires less information such as part quantities, quality level, and application environment. It is most applicable during early design or proposal phases of a project. The Parts Count Method will usually result in a higher failure rate or lower system reliability, a more conservative result than the Parts Stress Method would produce.

part stress analysis. The Part Stress Analysis method is used the majority of time and is applicable when the design is near completion and a detailed parts list, or bill of materials, plus component stresses are available. By component stresses, the standard is referring to the actual operating conditions such as environment, temperature, voltage, current and power levels applied, for example. The MIL-HDBK-217 standard groups components or parts by major categories and then has subgroups within the categories. An example is the "low-frequency diode" subcategory of the "diode" group. Each component or part category and related subgroups have a unique formula or model applied to it for calculating the failure rate for that component or part.

The failure rate formulas include a base failure rate, λ_b, for the category and subgroup selected. The base failure rates apply to components and parts operating under normal environmental conditions, with power applied, performing the intended function(s), using base component quality levels, and operating at the design stress levels. The standard then applies many "π" factors multiplied against the base failure rates in order to factor in the actual operating conditions, environment, and stress levels. After the predicted failure rate at the actual operating conditions for each component in the project, the overall system level or equipment failure rate is found by adding up the calculated failure rates for each component.

An example of a MIL-HDBK-217 failure rate model for a simple semiconductor component is shown below:[49]

$$\lambda_p(\text{failures/million hours}) = \lambda_b \pi_T \pi_A \pi_R \pi_S \pi_C \pi_Q \pi_E \quad (6.11)$$

where base (λ_b) and π factors are the following: λ_b = Base failure rate, π_T = temperature factor, π_A = application factor (linear, switching, etc.), π_R = power rating factor, π_S = electrical (voltage) stress factor, π_C = contact construction factor, π_Q = quality factor, and π_E = operating environment factor

Each component or part group and its associated subgroup has a base failure rate plus numerous π factor tables, unique to that component or part, that list factors that are used in the model to adjust the base failure rate.

A low-frequency diode, for example, has a MIL-HDBK-217 model as follows[49]:

$$\lambda_p(\text{failures/million hours}) = \lambda_b \pi_T \pi_t \pi_R \pi_S \pi_C \pi_Q \pi_E \quad (6.12)$$

where λ_b = base failure rate, π_T = temperature factor (voltage regulator, voltage reference, and current regulator), π_t = temperature factor (analog, switching, fast recovery, power rectifier, transient suppressor), π_S = electrical (voltage) stress factor, π_C = contact construction factor, π_Q = quality factor, and π_E = environment factor.

Values for each of the "π" factors are given in tables of the handbook.

Parts count reliability prediction. The MIL-HDBK-217 Parts Count Reliability Prediction is normally used when accurate design data and component specifications are not determined. Typically, this occurs in the proposal and bid process or early in the design process. Minimal information is required for a Parts Count Reliability Prediction. The formula for a parts count analysis is the sum of the base failure rate of all components in the system. MIL-HDBK-217 provides tables for the same component groups in the Parts Stress Analysis, listing generic failure rates and quality factors for different environments. The predicted failure rate results will normally be harsher than those of the Part Stress Analysis approach.

It should be noted that in using MIL-HBK-217, the reliability predictions are based on empirical data for different component types and do not necessarily take into account the reliability of adhesive attached and interconnected components. The effect of the adhesive and its various possible failure modes and mechanisms on the reliability of devices under both operating conditions and long-term accelerated conditions should be considered a part of the equation.

6.4 Qualification, quality control, and specifications

Besides an understanding of the basic failure modes and mechanisms, important in the initial selection of adhesives, qualification and strict adherence to materials and processes specifications and work instructions are an integral part of the reliability process. The process begins with the design guidelines and requirements for a specific application. Qualification criteria and test specifications must then be generated and the adhesives selected must be qualified. In addition, materials specifications or procurement documents are required to assure that the adhesive meets a minimum of quality-assurance tests before acceptance from the supplier. A quality-control system should be in place that assures the material is properly stored and is still within its shelf life at the time of use. Once the adhesive is in use, process specifications, work instructions, and in-process controls are necessary to assure that the adhesive is properly dispensed and cured, and that the equipment used for processing and testing is calibrated. Finally, the completed assembly is inspected visually and some screen tests performed to assure that a known good product is delivered.

6.4.1 Qualification based on accelerated testing

Materials are qualified by passing a series of tests as severe or more severe than those of the intended application. Generally, accelerated environmental and mechanical tests are specified and performed on actual or simulated parts representative of the final product. Since the parts are subjected to accelerated tests, some of which are destructive, qualification parts are seldom deliverable items. Once an adhesive and process have been qualified, no changes in composition or processes are allowed without requalification. Testing at levels above the acceptance tests for an extended duration or number of cycles is necessary. The qualification tests for some applications, for example, for die and substrate attachment, have been specified in NASA and MIL specifications while those for other applications are being developed by industry and industry associations.

Qualification tests for adhesives are defined in Method 5011 of MIL-STD-883 and requirements for qualifying sealed hybrid microcircuits or multichip modules that contain adhesives are specified in MIL-PRF-38534. Space, automotive, and medical applications generally use the military specifications, but are modified or supplemented with additional tests and requirements for their particular application.

6.4.2 Qualification based on heritage data

Some adhesives have been used successfully for over 30 years in specific applications and environments and a reliability database (heritage data) already exists. These adhesives may thus be qualified for other applications by similarity of exposure, time, and operating conditions. In general, adhesives may be qualified and used if they had previously been demonstrated successful for an equal or shorter period of time under equal or less severe exposure and operating conditions. The reliability of silver-filled epoxy adhesives such as Henkel's Ablebond® 84-1 and Epoxy Technology's EpoTek® H-20 and of insulative adhesives such as Emerson & Cuming's Eccobond® 104 and Henkel's Ablebond 84-3 has been demonstrated through decades of use in consumer, military, space, and medical electronics.

6.4.3 Receiving inspection and acceptance testing

Once qualified, acceptance tests are necessary on a batch-by-batch basis to assure that the supplier has not changed his formulation, and the adhesive is as good as that initially qualified. Adhesives should comply with the quality-assurance (QA) provisions of materials specifications or procurement documents. Source inspectors may also be used. Some simple tests that users perform include lap-shear strength measurements on each batch or an infrared spectrum comparison with an original "gold" standard. A Certificate of Compliance (C of C) provided by the supplier guaranteeing that the material sold is traceable to a batch that has been tested and meets the requirements of the user is yet another QA approach. Most adhesives suppliers will conduct the tests specified in Method 5011 of MIL-STD-883 and will provide a C of C. Some companies may want to perform their own receiving-inspection tests before accepting the product.

6.4.4 In-process controls

In spite of all the precautions in selecting, qualifying, and procuring an adhesive, vigilance on the production line is still necessary to avoid failures. Process specifications and work-instruction documents provide step-by-step procedures to manufacture production quantities of parts in a manufacturing environment. A process specification often calls out sub-tier process documents such as cleaning, surface preparation, curing, and testing. The process should allow as much manufacturing freedom as possible commensurate with meeting the engineering requirements. For example, instead of specifying one cure schedule, a selection of several equivalent schedules should be permitted. Even better, a time–temperature curve of equivalent cures would give manufacturing greater freedom in meeting delivery schedules.

An essential part of the process specification that affects reliability, is the quality-assurance section. Quality-assurance provisions are usually in-process controls and tests for the more critical steps. For adhesives, these may consist of periodically checking the viscosity and pot life of the liquid adhesive, conducting bond-shear strength and hardness tests of cured samples, conducting tests to assure that surfaces are clean, and assuring that all processing and testing equipment are calibrated, especially those involving temperatures.

6.4.5 Screen tests and product acceptance

Some customers require that a series of screen (nondestructive) tests be applied to the final product before delivery. These tests may consist simply of a few temperature cycles or a regimen of sequential mechanical and electrical tests as for Classes H and K of MIL-PRF-38534 for hybrid microcircuits.[50] The sequence of tests for Class H microcircuits consists of the following:

- Internal visual
- Temperature cycling from −65 °C to 150 °C, 10 cycles or thermal shock, 0 °C to 100 °C, 15 cycles.
- Constant acceleration (3,000 g's in Y_1 direction) or mechanical shock (1,500 g's in Y_1 direction). The Y_1 direction is specified as the lid side of the package against the wall of the text chamber.

- Burn-in, usually 168 hours at 125 °C with power.
- Electrical test.
- Fine and gross leak testing.
- External visual.

For Class K (space-grade), the additional tests of nondestructive wire bond pull, PIND, and radiography are imposed. Screen tests assure the reliability of the entire electronic product of which the adhesive joints are a part.

6.4.6 Specifications and requirements

The first specifications for adhesives were generated by the staff of NASA and the DoD who were prompted by the high reliability that was required of microcircuits used in aerospace programs. These specifications covered primarily die and substrate attachments for hermetically sealed integrated circuits, hybrid microcircuits, and multichip modules. Subsequently, with the increased use of surface-mount adhesives in the assembly of commercial printed-wiring boards and underfills for flip-chip devices, industry associations took the lead in generating the requirements and test methods.

6.4.6.1 Specifications for die- and substrate-attach adhesives

Although the original NASA specification for the qualification and acceptance of die- and substrate-attach adhesives has been inactivated and Method 5011 of MIL-STD-883 is now the dominant specification, some of the NASA tests that were not carried over are still useful for improved reliability assurance. For example, even though Method 5011 does not specify an infrared (IR) spectrum for incoming material, the IR spectrum is a quantitative, easy test to run to assure that the chemical composition of the adhesive has not changed. Further, the corrosivity test in which the adhesive is applied to an aluminized Mylar® film and examined for etching of the aluminum is also a rather simple and valuable test. A comparison of the requirements and test methods of Method 5011 with the original NASA specification is given in Tables 6.6 and 6.7.

6.4.6.2 Specifications for surface-mount adhesives

With the advent and wide use of surface-mount adhesives, several industry and industry-association documents have been generated. Examples include the following:

- IPC CA-821, General Requirements for Thermally Conductive Adhesives
- IPC 3406, Guidelines for Electrically Conductive Surface-Mount Adhesives
- IPC SM-817, General Requirements for Dielectric Surface-Mount Adhesives
- Siemens Standard SN 59651

Table 6.8 compares the requirements and tests for IPC CA-821[51] with those of Siemens SN 59651,[52] both of which are applicable to surface-mount adhesives.

Table 6.6 NASA MSFC SPEC-592[a], requirements and test methods for die- and substrate-attach adhesives

Test	Requirement	Paragraph	Method	Paragraph
Visual	Uniform consistency, free of lumps or foreign matter	3.2.1	Visual examination at 30×	4.5.1
Pot life	>1 hour	3.2.2	Vendor-established method	4.5.2
Shelf life	>6 months	3.2.3	Properties meet requirements when tested at or before shelf life	4.5.3
IR spectrum	Disappearance/appearance of absorption bands (relative to baseline) criteria for requalification	3.2.4	IR spectrum of mixed or separate components	4.5.4
Cure schedule	<165 °C, <4 hours	3.3.1	N/A	N/A
Outgassing	<0.3% wt loss at 250 °C	3.3.2	Thermogravimetric analysis	4.5.5.1
	3,000 ppm_v moisture (maximum)	3.3.2	Packages with 10% of surface area of exposed adhesive analyzed after 24 hours and 1,000 hours at 150 °C by RGA	4.5.5.2
Bond-shear strength	>6.9 × 10^6 N/m^2	3.3.3	Shear test (die/substrate, @ 25 °C)	4.5.6
	>3.5 × 10^6 N/m^2		Shear test (die/substrate, @ 150 °C)	
	>10.3 × 10^6 N/m^2		Capacitors (@ 25 °C)	
	>5.2 × 10^6 N/m^2		Capacitors (@ 150 °C)	
	>0.7 × initial value, @ 25 °C >0.8 × initial value, @ 25 °C		After temperature cycling and solvent immersion After 1,000 hours at 150 °C aging	

Corrosivity	No change in light transmittance of aluminized Mylar®	3.3.4	Sample on aluminized Mylar® (48 hours)	4.5.7
Volume resistivity	$<5.0 \times 10^{-4}$ ohm-cm (silver), to $<15.0 \times 10^{-4}$ ohm-cm (gold)	3.3.5	4-wire probe measurement: 25 °C, 60 °C, 150 °C, and 25 °C after 1,000 hours aging at 150 °C	4.5.8.1
	$>1 \times 10^{13}$ ohm-cm (insulative, 25 °C) and $>1 \times 10^{8}$ ohm-cm (125 °C)		ASTM D257	4.5.8.2
Thermal conductivity	$\geq 3.5 \times 10^{-3}$ cal/cm-second °C (1.5 W/m K) (conductive) and $\geq 4.0 \times 10^{-4}$ cal/cm-second °C (0.17 W/m K) (insulative)	3.3.6	Measured @ 121 ± 5 °C	4.5.9
Coefficient of thermal expansion	≤65 ppm/°C (below T_g), ≤300 ppm/°C (above T_g, ≤150 °C)	3.3.7	Thermomechanical analysis (TMA)	4.5.10
Electrical stability	Resistance change <5% (conductive only)	3.3.8	Gold-plated Kovar® tabs (wire bonded)	4.5.11
Ionic impurities	<1,000 ppm NaCl; pH 4.0–9.0; Cl^- ≤300 ppm; Na^+ <50; K^+ <5 ppm	3.3.9	Federal Test Method Standard No. 406, Method 7071	4.5.14
Dielectric constant and dissipation factor	<6.0 (1 kHz and 1 MHz), <0.03 (1 kHz) and <0.05 (1 MHz)	3.3.10	ASTM D150	4.5.12, 4.5.13

NA – not applicable.
[a] This specification has been inactivated.

Table 6.7 Method 5011, MIL-STD-883, evaluation and acceptance procedures for polymeric materials

Test	Requirement	Paragraph	Method	Para graph
Visual	Uniform consistency, free of lumps or foreign matter	3.4.1	Visual examination at 30×	3.8.1
Viscosity	Range established by customer and included in the procurement document	3.4.2	User and supplier agree to a method on reporting viscosity	3.8.2
Pot life	>1 hour	3.4.3	User and supplier select the procedure to be used	3.8.3
Shelf life	≥12 months @ −40 °C for one-part systems; ≥12 months @ 32 °C for two-part systems. Shelf life is not extendable for Class K devices; may be extended once for Class H devices	3.4.4	Material, pot life, bond strength, and volume resistivity testing performed	3.8.4
Thermogravimetric analysis	<1% wt loss at 200 °C	3.5.2.1	10–20 °C/minute heating rate in nitrogen	3.8.5.1
	% Filler acceptance ±2% of certification lot	3.5.2.2	Heating from 25 °C to 600 °C at a 10–20 °C/minute heating rate in air	3.8.5.2
Outgassed materials	5,000 ppm_v moisture (maximum)	3.5.3	Packages with an equivalent exposed adhesive to the application shall be analyzed after 24 hours and 1,000 hours @ 150 °C	3.8.6
Ionic impurities	Total ionics ≤4.5 millisiemens/m, 4.0 < pH < 9.0, Cl^- ≤200 ppm, Na^+, K^+, and F^- ≤50 ppm	3.5.4	Extract analysis on 3 g of material (ground) added to 150 ml water and refluxed for 20 hours	3.8.7

Bond strength	1.0× requirement of Fig. 2019.4 Method 2019 of MIL-STD-883	3.5.5	Bond strength of 80-mil-square tabs to gold or nickel-plated Kovar® @ 25 °C and @ 25 °C after 1,000 hours @ 150 °C	3.8.8
Coefficient of thermal expansion	±10% of the value in the users material specification	3.5.6	Measured from −65 °C to +150 °C per ASTM D3386	3.8.9
Thermal conductivity	≥1.5 W/m K for electrically conductive ≥0.15 W/m K for insulative	3.5.7	ASTM C177 or ASTM C518	3.8.10
Volume resistivity	$\leq 5.0 \times 10^{-6}$ ohm-cm (silver-filled), $\leq 1.5 \times 10^{-5}$ ohm-cm (gold-filled)	3.5.8	Strip of epoxy cured onto a glass slide	3.8.11
	$>1.0 \times 10^{11}$ ohm-cm (insulative) at 25 °C and 1.0 × 106 ohm-m (125 °C)		ASTM D257	
Dielectric constant	≤6.0 at 1 kHz and 1 MHz	3.5.9	ASTM D150	3.8.12
Dissipation factor	≤0.03 at 1 kHz and ≤0.05 at 1 MHz	3.5.10	ASTM D150	3.8.13
Sequential-environment tests	No evidence of mechanical degradation after screens	3.5.11	Screen testing per MIL-STD-883 thermal shock (Method 1011), temperature cycling (Method 1011), mechanical shock (Method 2002), variable-frequency vibration (Method 2007), and constant acceleration (Method 2001)	3.8.14

6.4.6.3 Specifications for underfill adhesives

The use of underfill adhesives has resulted in the development of the draft version of J-STD-030, "Guideline for Selection and Application of Underfill Material for Flip Chip and Other Micropackages." The guideline covers critical material properties for underfill materials to assure compatibility in underfill applications for reliable electronic assemblies as well as selected process-related qualification tests such as thermal cycling. Table 6.9 summarizes selected materials' requirements for underfill adhesives from the proposed JEDEC J-STD-030.[53]

Specifications for underfill adhesives. Table 6.9 summarizes requirements for underfill adhesives from the proposed JEDEC J-STD-030.[53]

6.4.7 Standards

With the accelerated growth of electronics, thousands of specifications and standards have been generated by various government agencies in different countries and by private industry. This has resulted in much duplicity and great cost. In recent years, however, there has been a trend toward standardization even on a global level. Generally, documents are now preferred that have been generated, coordinated, approved, and monitored by some industry organizations on an international level such as ISO. Standards generated by ASTM, ANSI, and IPC have also been accepted by manufacturers and procurers of electronic materials and equipment and by military and space agencies. For example, the SIBR database (Standards Incorporated By Reference) of ASTM documents for procurement and regulation has been accepted by NASA.[54] There is also an increasing tendency for federal agencies to adopt private sector practices where possible to minimize the number of documents required for procurement.

6.4.8 Hazardous materials restrictions

The European Union has taken the lead in restricting the use of some hazardous materials in the manufacture of electrical and electronic hardware. The program, referred to as Restriction of Hazardous Substances (RoHS),[55] applies to all electrical and electronic products produced in Europe and products imported from other countries, thus has very broad application. The regulations put into force on July 1, 2006, apply to lead, cadmium, mercury, hexavalent chromium (Cr^{+6}), polybrominated biphenyl (PBB), and polybrominated diphenyl ether (PBDE). The RoHS guideline document specifies 0.1% or 1,000 ppm as the maximum amount permitted for lead, mercury, Cr^{+6}, PBB, and PBDE and 0.01% or 100 ppm for cadmium. The document also describes guidelines for sample preparation and analyses of the restricted materials.

These restrictions have presented some problems but also opportunities for adhesives manufacturers. A key problem has been finding or developing flame retardants to replace the brominated organic compounds that have been used successfully for many years. On the other hand, the restrictions on the use of lead in

Table 6.8 Requirements and test methods for surface-mount adhesives (IPC CA-821 and Siemens SN 59651) [51,52]

	IPC-CA-821 (Jan. 1995)		Siemens Standard SN 59651 (1997)	
Property	**Requirement (paragraph)**	**Method**	**Requirement (paragraph)**	**Method**
Appearance	–	–	Homogeneous, void free, ≤50 μm particle size (3.1)	Visual (8×)
Shelf life	Specified by supplier (3.2.4)	Meets specification requirements	≥5 months and storage conditions specified (4.1)	
Degree of fineness	Specified by supplier (3.2.5)	ASTM D1210	–	–
Viscosity	Range and speed agreed to by user and vendor (3.3.1)	ASTM D2556	Viscosity profile determined (3.2)	Casson viscosity (0–40 $second^{-1}$)
Work time	Per supplier recommendation (3.3.2)		>8 days and tack time >8 hours (4.2)	IEC 68-1
Spread or slump	–	Observe three dots dispensed (0.65 mm diameter and 0.25 mm thick) cured and uncured. Measure uncured dots after 70 minutes	<10% (5.6)	Average of 1.8 ± 0.2 mm dots on FR4 epoxy with solder resist. Check after 3 minutes uncured and cured.
Green strength	–	–	Device shift ≤0.15 mm (4.5)	Assembled and fixtured PWA slides down an incline 35 mm high and 400 mm long and strikes a stop.
Mass loss at temperature	–	–	<1%	Observed after 7 days at 85 °C
Thermal conductivity	>0.4 W/m K (3.3.3)	ASTM C177 or C518	–	–

(*Continued*)

Table 6.8 Requirements and test methods for surface-mount adhesives (IPC CA-821 and Siemens SN 59651)[51,52]—*cont'd*

	IPC-CA-821 (Jan. 1995)		Siemens Standard SN 59651 (1997)	
Property	**Requirement (paragraph)**	**Method**	**Requirement (paragraph)**	**Method**
Shear strength	<8.0 MPa (1,160 psi), removable grade; >8.0 MPa, permanent grade (3.3.4)	ASTM D1002 (A109 steel lap shear); conditioned per IPC-S-804 Methods 1, 2, 3, or 4	>5 N/mm^2 (825 psi) (5.7)	Die shear (25 °C) after cure
Solvent resistance	No degradation or change in surface roughness, swelling, tackiness, blistering, or color change (3.4.2)	–	–	–
Hydrolytic stability	Stencil-applied coating, shall show no evidence of bubbles or voids (3.4.3)	IPC-TM-650, Method 2.3.10	–	–
Dielectric strength	Specify minimum requirements (3.5.1)	Not specified ASTM D149 (1 kHz, 1 MHz)	–	–
Dielectric constant	Specify minimum requirements (3.5.2 and 3.5.3)	0.38 ± 0.05 mm sample, ASTM D149 (1 kHz, 1 MHz)	–	–
Volume resistivity	Specify minimum requirements (3.5.4)	ASTM D257	–	–

Surface resistivity	Specify minimum requirements (3.5.5)	ASTM D257	$>10^{10}$ ohms (5.3)	40 °C, 93% RH test
Moisture and insulation resistance	Specify minimum requirements (3.6.1)	IPC-TM-650, Method 2.6.3.1, 50 ± 5 °C and 90% RH	—	—
High temperature strength	No displacement (3.6.2)	IPC-TM-650, Method 2.4.42.1, solder float	>5 N/mm^2 (5.8)	Shear strength after solder float 255 ± 3 °C (1206 device)
Fungus resistance	Non-nutrient (3.6.3)	Exposure to mixed spore suspension, IPC-TM-650, Method 2.6.1	Do not exceed limiting quality A1.4 (5.2)	–
Electro-migration	No tendency to migration (3.6.4)	IPC-TM-650, Method 2.6.14	Electrolytic corrosion: No green or other discoloration or dendritic growth	IEC 426 and IEC 68-3
Outgassing	Limits agreed upon (3.6.5)	ASTM E595	–	–
Self shimming	Report results (3.6.6)	IPC-TM-650, Method 2.4.51	–	–
Coefficient of thermal expansion (CTE)	Report results (3.6.7)	ASTM D3386	–	–

Table 6.9 Selected materials requirements and test methods for underfill materials (JEDEC J-STD-030)[53]

Property Type	Parameter	Paragraph	Requirement description
Raw-material characteristics	Filler size	6.1.1	Report maximum filler size
	Filler type	6.1.2	Report filler type
	Filler loading	6.1.3	Report results (60–75%, typical)
	Underfill density	6.1.4	Report results (1.5–2.0 g/cm^3, typical)
	Viscosity	6.2	Report viscosity and thixotropic index (0.5/5 rpm); measure per ASTM D2556
	Viscosity change	7.4.1	Measure per ASTM D2556; a 20% increase (maximum) is acceptable
	Flow rate change (after 8 hours)	7.4.2	Usage-life limit classes: Class 1:50% reduced dispensing rate Class 2:40% reduced dispensing rate Class 3:30% reduced dispensing rate
	Settling test	7.4.3	Measure viscosity per ASTM D2556 at the top and the bottom of a syringe that has been stored vertically
	Gel time	6.2.2	Report results (5–20 minutes @ 121 °C, typical)
	Storage life	7.2	List optimal storage conditions (normally −20 °C to −50 °C) such the material remains usable in terms of viscosity, flow, and other performance criteria
	Underfill flow rate	8.4.2	Use glass slides to form a wedge shaped gap from zero to ~0.13 mm. The flow time is how long the front takes to reach the 1 cm mark
	Spread/slump test	8.6	Deposit three dots of adhesive 0.65 mm in diameter and 0.25 mm thick on two glass slides. Maintain sample 1 at 25 ± 5 °C and 50 + 25% RH. Cure sample 1 per vendor instruction. <10% increase in dot size is acceptable
	Pot life	8.9	Reduction in flow rate of 50% is considered the end of pot life (8 hours typical)

Mechanical (cured) characteristics	Flux compatibility	6.4	Comparable adhesion strength to fluxed and nonfluxed surfaces
Mechanical (cured) characteristics	Die shear	10.2.1	6.35-mm square gold-plated Kovar® tab, 0.05–0.13 mm standoff, and fillets with 25–50% of height of tab. Report shear strength
	Tensile shear (Stud Pull)	10.2.2	Silicon die mounted to substrate (0.05–0.13 mm standoff using copper wire as spacers)
	Lap shear/peel strength	10.2.3	Lap-shear testing of 1" × 4" aluminum or FR4 bars with and without solder mask (0.13 mm gap)
	Coefficient of thermal expansion	10.5	Report results
	Glass-transition temperature	10.6	Report results
	Young's modulus	10.4	Report results (5–10 GPA [flexural], typical)
Electrical (cured) characteristics	Surface insulation resistance	10.12	0.317 mm line and 0.159 mm space comb pattern tested per IPC TM-650, Test Method 2.6.3.3
	Electrochemical migration resistance	10.13	0.317 mm line and 0.159 mm space comb pattern, biased at 10 V dc + 1 V dc for 500 hours of 85 °C and 85% RH. Measure surface insulation resistance at 100 V dc + 10 V dc
	Volume resistivity	10.14	ASTM D 257 (25 °C, 500 volts); 10^{14} to 10^{15} ohm-cm typical
	Dielectric constant	10.15	ASTM D 150 (25 °C, kHz or mHz range); 3–4 typical
Environmental reliability	Non-nutrient (fungus)	10.11	Does not contribute to, support, or be degraded by biological growth when tested in accordance with IPC-TM-650, Test method 2.6.1
	Ionic content	12.1	121 °C extraction for 20 hours; 10 ppm maximum for each ion.

solders have been an opportunity to substitute electrically conductive, silver-filled epoxies.

References

1. Ardebili H, Pecht M. *Encapsulation Technology for Electronic Applications*. William Andrew Publishing Company; 2008.
2. Comstock R. Die attach epoxy's role in thin-package delamination. *Semicond Int*; Jul. 1993.
3. Britt L. The use of commercial components in a high reliability environment. *Surface Mount Technol*; Nov. 2000.
4. Loh K. Low-stress adhesive. *Adv Packaging*; Winter 1993.
5. Walwadkar S, Cho J. Evaluation of die stress in MEMS packaging: experimental and theoretical approaches. *IEEE Trans Comp Packaging Technol*. Dec. 2006; Vol. 29(Issue 4):735-742.
6. Chung KK, Drier G, Fitzgerald P, Boyle A, Lin M, Sager J. Tack-free flexible film adhesives. *Hybrid Circuit Technol*; May 1990.
7. Mooney CT, Bolger J. Die attach in Hi-Rel PDIPs: polyimide or low chloride epoxies. *IEEE Trans*; 1984.
8. Yoo HY, et al. Novel die attach adhesive for thin quad flat package, www.henkelna.com/industrial/white-papers-8082.htm. Accessed August 2010.
9. Swanson DW, Enlow LR. Stress effects of epoxy adhesives on ceramic substrates and magnetics. *Proc. IMAPS Conf*; 1999.
10. Swanson DW, Enlow LR. Stress effects of epoxy adhesives on ceramic substrates and magnetics. *Microelectron Reliability*. 2001;41:499-510.
11. Adams T. Using acoustic microimaging to inspect multichip modules. *Electron Engineer*; Apr. 1996.
12. Su B. Electrical, Thermomechanical, and Reliability Modeling of Electrically Conductive Adhesives. PhD Dissertation. Georgia Institute of Technology; May 2006.
13. Fracture Mechanics Design Approach, *The Adhesives Design Toolkit*, www.adhesivestoolkit.com. Accessed 2.09.10.
14. Stress Intensity, http://www.sv.vt.edu/classes/MSE2094_NoteBook/97ClassProj/anal/kim/intensity.html. Accessed 26.04.10.
15. http://www.efunda.com/formulae/solid_mechanics/fracture_mechanics/fm_lefm_Kc_Matl.cfm. Accessed 6.04.10.
16. Wu R, Jun K, Borghard TJ. An innovative die-attach for high thermal management. In: *Proc. Fourth Int. Conf. Electronic Materials, CIUMRS-ICEM-181*. Cheyu, Korea: Asian Tech. Information Program Report 98-098; Aug. 1998.
17. Abletherm 2600 Series Adhesives, Processing Tip Sheet, Ablestik Labs; Nov. 1999.
18. Self Shimming Thermally Conductive Adhesive Multi-Cure 9-20699, Preliminary Product Data Sheet, Dymax Corp., (Apr. 2004)
19. Product 3872, Henkel Technical Data Sheet; Aug. 2003.
20. Kim HK, Shi FG. Electrical reliability of electrically conductive adhesive joints: dependence on curing condition and current density. *Microelectron J*. 2001;32.
21. Ebel GH, Engelke HA. Polymeric materials and the new MIL-SPECS – a hybrid users view. In: *Proc. ISHM*; 1984:189-193.
22. Bolger J. Adhesive related failure mechanisms in military hybrid packages. *Int J Hybrid Microelectron*; Dec. 1984.

23. Tong QK, Markley DL, Fredrickson G, Kuder R. Conductive adhesives with stable contact resistance and superior impact performance. In: *Proc. 49th Electron. Comp. Technol. Conf. (ECTC)*; 1999:347-352. Jun. 1–4.
24. Wong CP, Lu D, Tong QK, Zhang E. Development of high performance surface-mount conductive adhesives. In: *Proc. 50th Electron. Comp. Technol. Conf*; 2000:892-898.
25. Jagt JC, Beric PJM, Lijten GF. Electrically conductive adhesives: a prospective alternative for SMD soldering. *IEEE Trans Comp, Packaging, Mfg Tech* 1995;18(2):292-298. Part B.
26. Gaynes MA, Lewis RH, Saraf RF, Roldan JM. Evaluation of contact resistance for isotropic electrically conductive adhesives. *IEEE Trans Comp, Packaging, Mfg Tech* 1995;18(2):299-304. Part B.
27. Lu D, Wong CP. Properties of conductive adhesives based on anhydride-cured epoxy systems. *J Electron Mfg* (Sept)1999;9(3):223-232.
28. Zwolinski M, Hickman J, Rubon H, Zaks Y. Electrically conductive adhesives for surface mount solder replacement. In: *Proc. Second Int. Conf. Adhesive Joining & Coating Tech. in Electronics Mfg*. Stockholm Sweden; 1996:333–340.
29. Kraus HS. Peculiar Behavior of Electrically Conductive Adhesives on Aluminum Surfaces, Ablestik Laboratories (undated).
30. Benson RC, Nall BH, DeHaas N, Romenesko BM, Charles HK, Colton RJ. Volatile and adsorbed species from microelectronic adhesives. In: *First Int. SAMPE Electron. Conf*; 1987:637-643. Jun.
31. Pecht MG, ed. *Quality Conformance and Qualification of Microelectronic Packages and Interconnects*. John Wiley & Sons; 1994.
32. Mooney CT. Electronic grade adhesives, reliability and performance. In: *Proc. 35th Electron. Comp. Conf*; 1985:326-330.
33. Unno Y. High density, low-mass hybrid and associated technologies, *Proc. Sixth Workshop on Electronics for LHC Experiments*, Krakow, Poland (Sep. 2000); Japan Institute of Electronic Packaging (JIEP); *Dictionary of Electronic Packaging Technology*, ISBN:4-7693-7089-X 63555.
34. Kohman GT, Hermance HW, Downes GH. Silver migration in electrical insulation. *Bell Systems Tech J.* (Nov) 1955;34(6).
35. Riemer DE. Material selection and design guidelines for migration-resistant thick-film circuits with silver-bearing conductors. In: *Proc. 31st Electron. Comp. Conf*; 1981.
36. Licari JJ, Enlow LR. *Hybrid Microcircuit Tech. Handbook*. 2nd ed. Noyes Publications; 1998: 137-142.
37. Perkins KL, Licari JJ. Investigation of organic adhesives for hybrid microcircuits. *NASA CR-150160*; Jul. 1975.
38. Interpretation of RGA Data Technical Brief, Oneida Research Services; 1994.
39. Particle Impact Noise Detection Test, MIL-STD-883, Method 2020.7.
40. King J. Qualification of Dow Corning Q3-6527 as a Particle Getter. *Singer Report YS57A834*; 1985.
41. Use of silicone gel particle getters for microcircuit packaging. In: *Proc. ISHM*; 1985.
42. Application Note 0503, Pike Technologies, www.piketech.com.
43. Ong MC, Zhao XL, Joman PP, Chin JM, Master RN. Lid adhesive failure study for flip-chip packaging. In: *15th Int. Symposium on Physical and Failure Analysis of Integrated Circuits*; 2008.
44. Jones JV. *Integrated Logistics Support Handbook*. 3rd ed. McGraw-Hill Professional; Jun. 8, 2006.
45. http://www.weibull.com/hotwire/issue21/hottopics21.htm. Accessed 30.03.10.

46. http://asqsandiego.org/articles/Reliability%20Eng-Part6.pdf. Accessed 30.03.10.
47. Kwon W, Yim M, Paik K, Ham S, Lee S. Thermal cycling reliability and delamination of anisotropic conductive adhesives: flip chip on organic substrates with emphasis on thermal deformation. *Trans ASME* (Jun) 2005;Vol. 127:86-90.
48. McLeish JG, Hillman C. Enhancing MIL-HDBK-reliability predictions with physics of failure models. *Adv Microelectron*; Mar./Apr. 2010;28-32.
49. MIL-HDBK-217F, Section 3, pp. 3–6.
50. Performance Specification, Hybrid Microcircuits, General Specification, MIL-PRF-38534; Jan. 1999.
51. IPC CA-821, General Requirements for Thermally Conductive Adhesives; Jan. 1995.
52. Adhesives for SMD Technology, *Siemens Standard SN 59651* (Ed. 1997-09).
53. Guidelines for Selection and Application of Underfill Materials for Flip-Chip and Other Micropackages, JEDEC J-STD-030; Dec. 2000.
54. http://standards.gov/sibr/index. Accessed 2.09.10.
55. Restriction of Hazardous Substances (RoHS), www.rohs.gov.uk. Accessed 2.09.10.

7 Test and Inspection Methods

Chapter Outline

Adhesives Technology for Electronic Applications. DOI: 10.1016/B978-1-4377-7889-2.10007-5

The reliability of electronic systems depends on extensive testing, inspection, and evaluation of all parts of the design and manufacturing process, including the adhesives used in the systems. Test methods for adhesives and adhesive-bonded assemblies may be generally classified as physical, electrical, environmental, thermal, mechanical or thermomechanical, and chemical (Table 7.1). This chapter summarizes most of the methods used to test and evaluate adhesives used in electronics. It is not intended to provide detailed instructions, but to lead the reader to the appropriate government or industry specifications, documents that are available from the various organizations listed in Table 7.2.

The test methods covered in this chapter are those most closely related to adhesives for die and substrate attachment, surface-mounting of components, underfill, and optoelectronic assembly. Some tests, such as bleedout and electrical stability, for which there are no standard test methods, are not covered but information for these procedures was already addressed in other portions of this book, for example, in Sections 2.1 and 6.1.

7.1 Physical tests

Physical-property tests are used to measure the properties of adhesives in the liquid or gelled states prior to curing and in the solid state after curing. Tests for the uncured state such as viscosity, visual examination, and surface energy or contact angle assure that fillers, if used, have not settled out, that the material has not exceeded its pot life or shelf life, and that the supplier has not changed the formulation. Visual examination and density after cure are performed to verify that voids are not present or, if present, meet specification requirements. Finally, light transmission and index of refraction measurements are important for adhesives used in optoelectronic applications.

7.1.1 Viscosity and thixotropy

Viscosity, a measure of the flow properties of a liquid, is important as an incoming receiving and acceptance test for adhesives. During automatic dispensing, measuring

Table 7.1 Classification of adhesive test methods

Physical	Electrical	Environmental	Thermal	Mechanical/ thermo-mechanical	Chemical analysis
Viscosity (ASTM D2196 and ASTM D2556)	*Volume resistivity* [MIL-STD-883, Method 5011 (3.8.11.1) (ASTM D257)]	*Solvent and chemical resistance* IPC-SM-817, paragraph 4.5.6	*Exotherm during cure*; *degree of cure*; glass-transition temp. (ASTM D2471, BSEN61006 Part 0)	*Hardness* (ASTM D2240)	*Weight loss during cure* (no standard method)
Spread and slump (IPC-SM-817, Method 4.5.5 and IPC-SM-870)	*Dielectric constant, dissipation factor* (ASTM D150-81)	*Corrosion or migration effects* (IPC-TM-650, Method 2.6.14)	*Gel time* (ASTM D2471)	*Modulus* [ASTM D790 (flexural), ASTM D638 (tensile), and ASTM D412 (Young's)]	*Degree of cure* Infrared spectrum (ASTM E1252 practice for general techniques for obtaining infrared spectra for qualitative infrared analysis) or Fourier Transform Infrared Reflection Spectroscopy (ASTM E1790-00 standard practice for near infrared qualitative analysis)

(*Continued*)

Table 7.1 Classification of adhesive test methods—*cont'd*

Physical	Electrical	Environmental	Thermal	Mechanical/ thermo-mechanical	Chemical analysis
Density [ASTM D1875 (uncured) ASTM D792-91 (cured)]	*Dielectric breakdown voltage* (ASTM D149)	*Radiation effects* (ASTM D1879-99 standard practice for exposure of adhesive specimens to high-energy radiation)	*Thermal conductivity* (ASTM C177 and ASTM C518)	*Elongation-at-break* [ASTM D790 (flexural) and ASTM D638 (tensile)]	*Gas chromatography, mass spectrometry* (ASTM D6420-99 standard test method for determination of gaseous organic compounds by direct interface gas chromatography-mass spectrometry)
Surface energy (wettability, contact angle) (ASTM D 5946-96)	*Arc resistance* (ASTM D495)	*Fungus resistance* (IPC-TM-650, TM 2.6.1)	*Weight loss (cured) TGA* (ASTM D3850)	*Coefficient of thermal expansion, glass-transition* (ASTM D3386)	*Ionic content* (total water extract conductance, pH, and specific ion contents) (Federal Test Method Standard 406, Method 7071)

Visual (MIL-STD-883, Methods 5011 and 2017)	*Electrical stability* (biased) (MSFC-SPEC-592[a], Method 4.5.11)	*Moisture absorption* (IPC-TM-650, TM 2.6.3.1, and ASTM D570)	*Degree of cure* (DSC)	*Bond strength tensile pull* (ASTM D1002), *die shear* (MIL-STD-883, Method 2019), *flip-chip strength* (MIL-STD-883, Method 2031), *substrate strength* (MIL-STD-883, Method 2017), *centrifuge* (MIL-STD-883, Method 2001)	*Residual gas analysis* (MIL-STD-883, Method 1018)
Light transmittance (ASTM D1003-00)	*Current-carrying ability, interconnect resistance* (N/A)	*Hydrolytic stability* (IPC-TM-650, TM 2.3.1)	*Glass-transition temp.* (DSC)	*Voids, delamination* (acoustic microscopy) (MIL-STD-883, Method 2030)	*Total mass loss, water vapor recovery, and volatile condensible materials* (ASTM E595)
Refractive index (ASTM D542)				*Radius of curvature* (no standard method)	

[a]NASA MSFC-SPEC-592 has been inactivated.

Table 7.2 Sources of industry and government documents

Organization	Address	Web site
ANSI	American National Standards Institute, 1430 Broadway, New York, N.Y. 10018	www.ansi.org
ASTM	American Society for Testing & Materials, 100 Barr Harbor Dr., West Conshohocken, PA 19428 (documents may be purchased from ANSI)	www.astm.org
BSI	British Standards Institution, 279 Lodge Causeway, Fishponds, Bristol, UK, BS163RA	StandardsUK.com; www.bsi-global.com
EIA	Electronic Industries Alliance, 2500 Wilson Blvd., Arlington, VA 22201	www.eia.org
EPA	Environmental Protection Agency, Ariel Rios Bldg., 1200 Pennsylvania Ave., NW Washington, D.C. 20460 *or* Superintendent of Documents P.O. Box 371954, Pittsburgh, PA 15250-7954	www.epa.gov
IEC	International Electrotechnical Commission, 3, rue de Varembe, P.O. Box 131, CH-1211, Geneva 20, Switzerland	www.iec.ch
IPC	IPC Association Connecting Electronics, 2215 Sanders Road, Northbrook, IL 60062-6135	www.ipc.org
ISO	Intl. Organization for Standardization, 1, rue de Varembe, Case postale 56, CH-1211 Geneva 20, Switzerland	www.iso.org
JEDEC	Solid State Technology Association, 2500 Wilson Blvd., Arlington, VA 22201	www.jedec.org
Military, Federal	DODSSP 700 Robbins Ave., Bldg 4D, Philadelphia, PA 19111-5098	http://dodssp.daps.mil/
NASA	National Aeronautics & Space Admin.	http://ndeaa.jpl.nasa.gov/nasa-mp/mp-hp.htm
NIST	National Institute of Standards and Technology, 100 Bureau Dr. Stop 1070, Gaithers-Burg, MD 20899-1070	www.nist.gov
UL	Underwriters Laboratories, 2600 N.W. Lake Rd., Camas, WA 98607-8542	www.ul.com

and monitoring, the viscosity of adhesives is also important to assure that the pot life has not expired and that reproducible high-yield results are achieved. Rechecking viscosity is also used as a criterion for extending the shelf life of an adhesive to verify that the material has not changed and will be within its working life when used.

Viscosity is the internal friction that results from intermolecular forces of attraction and interactions between fillers and resins in adhesives, a measure of resistance to flow. Most adhesives are non-Newtonian fluids that exhibit shear-thinning behavior, or decrease in viscosity with increasing shear rate. Exceptions to this general rule are the capillary-flow underfill adhesives that tend to be Newtonian in fluid behavior.

While various methods and equipment can be used to measure viscosity, the most common uses the Brookfield viscometer. ASTM D2196, *Rheological Properties of Non-Newtonian Materials by Rotational (Brookfield) Viscometer*,[1] covers this method for Model LVF and Model RVF viscometers. Brookfield "LV" viscometers are used for low-viscosity materials, "RV" for medium-viscosity formulations, and "HV" for high-viscosity formulations. Test Method ASTM D2556, *Apparent Viscosity of Adhesives Having Shear-Rate Dependent Flow Properties*,[2] also specifies the Brookfield RVF viscometer for measuring viscosity and thixotropic index. Selection guidelines for the type of spindle and speed (rpm) are given for viscosity ranges. High-viscosity (>500,000 cP) SMT adhesives, for example, typically use a Brookfield HAT with Helipath, and readings are taken at 1 and 10 rpm or at 2.5 and 20 rpm using Spindle "D".[3]

The method and type of equipment used also vary with the type of adhesive. As an example, the viscosity of electrically conductive adhesives is measured according to ASTM D1824, *Apparent Viscosity of Plastisols and Organosols at Low-Shear Rates by Brookfield Viscosity*.[4,5] The viscosity of typical electrically conductive die-attach adhesives is measured with a Brookfield HBT viscometer with Spindle TB and Speed 5.[6] For higher-viscosity conductive adhesives and for underfill adhesives, a Brookfield RVT or RVF viscometer is used[7] with Spindles 6 or 7[8] at speeds of 4–10.4 rpm.[9] Another Brookfield viscometer, the "Cone-and-plate" viscometer with a CP-51 spindle is used for low-to-intermediate viscosity adhesives.[10] Finally, the Brookfield HAT and HBT instruments are used for the high-viscosity (1–2 million cP at 1 rpm) adhesives typically used in SMT applications.[3]

Besides the Brookfield viscometers, the Haake rotoviscometer, PK10 M10/PK1 2 Cone, Casson Model is used to measure viscosity at shear rates of 0.4–30 second^{-1}. This instrument is used alone or in combination with a Brookfield viscometer for SMT adhesives. The equipment is used not only to measure viscosities, but also to provide flow properties in the form of graphs of shear stress versus shear rate.

Thixotropy is measured by several rheological tests. One involves obtaining a hysteresis loop by plotting the increase in shear rate from 0.01 to 10 second^{-1} versus viscosity and then back to 0.01 second^{-1}. The return curve tracing well with the original shows good thixotropic properties, well suited for screen printing or needle dispensing adhesives (see Chapter 4, Fig. 4.10). A second method involves measuring the viscosity by maintaining the shear-rate constant for up to 30 minutes.[11]

Thixotropy may also be expressed as the thixotropic index (TI) which is the ratio of viscosities measured at speeds that have a ratio of 1:10 rpm, for example at 2 rpm and at 20 rpm. For non-Newtonian adhesives used in automated dispensing, shear thinning is apparent from TI values of 2–6. In general, capillary-flow underfills have a thixotropic index of 1 or close to 1.

In addition to viscosity and TI, the *yield value* (also known as the yield point or yield stress) can be measured. The yield value can be measured by one or two methods. First, it may be obtained from the curve of shear stress to shear rate.[12] The curve is extrapolated to give the shear stress at zero shear rate and the yield value is read off the *y*-axis. In the second method, a Haake low-shear viscometer (CV-100) is used and the stress peak upon startup of rotation is a measure of the yield value.

7.1.2 Spread and slump

The degree to which an adhesive spreads after dispensing and the degree to which it slumps after curing are important parameters in achieving high yields in high-speed automated assembly. Adhesives must retain their dot or pattern shapes after dispensing. Excessive flow or spread in conductive adhesives may result in electrical shorting of adjacent conductors.

The spread and slump for surface-mount adhesives may be measured according to IPC-SM-817, Method 4.5.5.[13] According to this method, sets of adhesive dots (minimum three per set), 0.65 cm in diameter and 0.25-mm thick, are deposited (for example, by screen or stencil printing) on each of two frosted glass microscope slides. One sample is stored 50–70 minutes at 25 ± 5 °C and $50 \pm 5\%$ relative humidity and then measured for spreading of the uncured adhesive. The second sample is cured and measured for the increase in pattern from the original dimensions. The acceptance criterion is an increase of less than 10% of the original diameter of the dot or pattern.

7.1.3 Density and specific gravity

The *density* of a material is mass per unit volume reported as g/cm^3. The density of uncured adhesives is useful in verifying the correct portion of liquid resin and filler ingredients of a formulation. The density of a cured adhesive is an indicator of the specific formulation and of completeness of cure.

The density of uncured liquid formulations is measured as specified in ASTM D1875, *Density of Adhesives in Fluid Form*. A weight-per-gallon cup whose volume is exactly 83.2 ml at 25 °C is weighed to the nearest 0.1 g, filled with adhesive, and reweighed.[14] The net weight of adhesive divided by the volume of the cup gives the density.

The density of a cured, solid adhesive may be measured by taking a small sample of the adhesive (any shape), weighing it (in grams), then immersing it in a graduated cylinder containing a known volume of water (in cm^3). The volume of water displaced by the sample is equal to the volume of the sample, and density then equals mass/volume.

Test methods for the density and specific gravity of cured adhesives are specified in ASTM D792, *Density and Specific Gravity (Relative Density) of Plastics by*

Displacement.[15] Specific gravity is the ratio of the density of a material to the density of water. For all practical purposes, density and specific gravity are the same, however, for high precision, the density of water must be taken into account. Specific gravity may be measured by weighing a piece of cured adhesive (at least 1 cm^3) in air, submerging it in water at 23 °C, and reweighing it while submerged. Specific gravity is then calculated as:

$$sg = w_a/(w_a - w_s) \tag{7.1}$$

where sg is specific gravity, w_a is the weight of the sample in air, and w_s is its weight when submerged in water. Taking into consideration the weight of the sinker and wire, if used to immerse the part in water, the equation becomes:

$$sg = w_a/(w_a + w - b) \tag{7.2}$$

where w_a is the apparent weight of the sample, without wire or sinker, in air; w equals the apparent weight of a totally immersed sinker, if used, and of partially immersed wire; and b equals the apparent weight of the sample and sinker completely immersed in water and of the partially immersed wire.

7.1.4 Surface tension, surface energy, and contact angle

A knowledge of the surface tension of the liquid adhesive and the surface energy of the substrate to which it will be applied is important in predicting the wettability and subsequent adhesion of the cured adhesive. As a rule, the surface tension of the adhesive should be less than the surface energy of the substrate. Additives are often added to adhesive formulations to reduce their surface tensions, while cleaning processes are used to increase surface energies of substrates.

The contact angle that a liquid makes with a solid surface is a measure of the wettability of the liquid to that surface as well as a measure of the cleanliness or surface treatment of the surface. The height of a water droplet on a surface relative to the base area it covers is indicative of the surface's affinity to wetting. Contact angles decreasing from 90° to 0° indicate progressive improvement in wettability with 0° being completely wettable. Contact angles increasing from 90° to 180° indicate progressively poorer wettability with 180° being completely non-wetting. Contact-angle meters, known as *goniometers*, are useful quality-control tools to assure that a cleaning process has removed hydrophobic residues or that a liquid coating will wet and adhere to a surface. Instruments to measure surface tensions and contact angles are available from Paul N. Gardner Co., BYK Gardner, and Ramé-Hart.

7.1.5 Visual

Visual inspection of adhesives is performed on both uncured and cured samples. Uncured materials are inspected for uniformity of filler distribution, consistency, and adhesive coverage. Inspection methods are defined in paragraphs 3.4.1 and 3.8.3 of Method 5011.4 of MIL-STD-883F.[16]

Visual-inspection criteria for cured adhesives used for element or substrate attachment are specified in paragraph 3.1.2 of Method 2017, *Internal Visual (Hybrid)* of MIL-STD-883F.[17] Visual-inspection criteria for adhesives are classified according to element and substrate attachment and according to coverage and defects.

Elements bonded to substrates are visually examined at magnifications of 10× to 60× as stated in paragraph 3.1.4 of MIL-STD-883F, Method 2017. In order to meet Class H, adhesive attachment for non-end-terminated devices requires a minimum 50% filleting visible around the perimeter of the device unless the fillet is continuous on two full nonadjacent sides. To meet Class K, a minimum 75% of the fillet must be visible. This requirement also applies to end-terminated components such as capacitors. An exception to this rule is for an adhesive applied directly to more than 50% of the bond area on the substrate by using preform adhesives or screening or stencil printing paste adhesive directly to the bonding area. In such cases, mechanical strength measurements may be performed in lieu of direct visual inspection. Another exception to the rule is where the device or substrate is transparent and more than 50% coverage may be visually verified.

Failure criteria for element or die-attach defects include the following:

- Adhesive material that extends onto the top surface of the device.
- Cracks in the adhesive that are greater than 5 mils in length around the perimeter or 10% of the contact periphery, whichever is greater. Fissures and pulling back of the filler, however, are not considered cracks.
- Adhesive strings where the diameter at the attachment point is less than 50% of the maximum dimension of the string.
- Adhesive residue on a bonding pad.

Internal visual inspection for substrate attachment is also performed at magnifications of 10× to 60×. The 50% and 75% fillet rules also apply for element or die-attach adhesives.

Failure criteria for substrate-attach defects are as follows:

- Adhesive buildup that reduces separation to less than 1 mil between the mounting material and active metallization or between a mounting post not electrically common with the attachment.
- Any crack greater than 5 mils in length.
- Adhesive strings where the diameter at the attachment point is less than 50% of the maximum dimension of the string.
- Adhesive residue on a substrate bonding pad or package bonding post as evidenced by discoloration or adhesive residue on a substrate mounting post.

Inspection criteria for monolithic integrated circuits are defined in Method 2010.11 of MIL-STD-883F.[17] Failure criteria for die mounting are as follows:

- Adhesive that extends onto or vertically above the top surface of the die.
- Adhesive fillet not visible along 75% of each side of the die.
- Adhesive that bridges package posts or is on the post bond area.
- Separation, cracks, or fissures greater than 2 mils width in the adhesive, at the cavity wall or cavity floor.

- Transparent die with less than 50% of the area bonded.
- Flaking, lifting, peeling, or crazing of the die-attach material.
- Adhesive that is connected to the fillet or conductive cavity (e.g., metal package base or metallized floor of ceramic package) and extends up the cavity wall to within 1.0 mil of the package post.

7.1.6 Luminous transmittance

Luminous transmittance is a measure of the amount of light that passes through a transparent material. Adhesives used in electro-optical interconnects must have a high transmittance. The *total light transmittance* through a material is equal to the total incident light less the light that is absorbed and light that is scattered.

A related property is haze or the scattering of light as it passes through a transparent material. The standard test method for measuring both transmittance and haze is ASTM D1003-00, *Standard Test Method for Haze and Luminous Transmittance of Transparent Plastics.*[18] Instruments that can measure transmittance, haze, and clarity are available from BYK Gardner, Paul W. Gardner Co., and Hunter Associates Laboratories. A typical test sample consists of a disk 50 mm or 100 mm in diameter although almost any flat sample can be used.

Ultraviolet-visible spectrophotometers can also be used to measure light transmittance over a wavelength range of 200–2,000 nm when samples are cast as films.[19]

7.1.7 Index of refraction

When light traverses from air through a liquid, its *index of refraction* is the ratio of the sine of the angle of incidence to the sine of the angle of refraction. The index of refraction is a constant for each material and is greater than one when light enters a denser medium from a rarer medium as from air through a liquid. The index of refraction is important for adhesives used in the assembly of polymer optical fiber bundles and terminations where the index of the adhesive should match that of the silicon optical fibers. Furthermore, the index of refraction may be used as an indicator of the degree of purity of solids and liquids such as polymer resins, solvents, diluents, and mixed adhesive formulations. The refractive index is measured using a refractometer, of which the preferred method, ASTM D542, *Index of Refraction of Transparent Organic Plastics*, uses an Abbé refractometer.[20] Refractive indices ranging from 1.3 to 1.7 in increments of 0.001 can be measured with high accuracy.

7.2 Electrical tests

The main electrical properties related to organic materials are volume resistivity (for both conductive and nonconductive adhesives) and dielectric constant and dissipation factor (for insulative materials). Other electrical tests for conductive materials that are more application specific are electrical stability, current-carrying capacity, and interconnect resistance.

7.2.1 Volume resistivity

Volume resistivity is a measure of the electrical conductivity of adhesives or the insulation resistance of a nonconductive material. Volume resistivity is used to qualify adhesives for electronics and also to evaluate long-term reliability after aging. Volume resistivity for conductive adhesives is given by the equation:

$$\rho = (R \times w \times t)/l \tag{7.3}$$

where ρ = resistivity in Ω-cm, R = measured resistance (Ω), w = width in cm, and l = length between the inner pair of probes, in cm.

This test is specified in Method 5011.4 of MIL-STD-883F in 3.8.11.1.3.[21] Volume resistivity is usually measured at 25 °C, 60 °C, and 150 °C, and then measured at 25 °C after 1,000 hours of aging at 150 °C.

Insulation resistance is the ohmic relationship of applied voltage to the total current between two electrodes in contact with a specific material. Insulation resistance is directly proportional to the length and inversely proportional to the area as follows:

$$\mathrm{IR} = \rho l/A \tag{7.4}$$

where IR = insulation resistance (Ω), l = length (cm), ρ = volume resistivity, and A = area (cm^2). Specifically, volume resistivity, ρ is ohmic resistance of a cube of bulk dielectric material one cm/side expressed as Ω-cm. Requirements for insulative adhesives are given in paragraph 3.8.11.2 of Method 5011.4 of MIL-STD-883F.[21]

7.2.2 Dielectric constant and dissipation factor

Dielectric constant (ε) and dissipation factor (DF) are related to capacitance (C) and can be calculated from capacitance measurements. Capacitance, between a parallel-plate capacitor, is directly proportional to the dielectric constant of the insulator separating the conductors and the conductor area, A, and inversely proportional to the distance, d, between the conductors as shown in:

$$C = \varepsilon A/d \tag{7.5}$$

If the gap between the plates is filled with a dielectric material, the capacitance increases by a constant factor characteristic of that material. The *dielectric constant*, ε, is defined as:

$$C_{\mathrm{m}} = \varepsilon C_{\mathrm{v}} \quad \text{or} \quad \varepsilon = C_{\mathrm{m}}/C_{\mathrm{v}} \tag{7.6}$$

where C_{m} and C_{v} are capacitance of the material and a vacuum, respectively. Because dielectric material affects the force with which two oppositely charged plates attract each other, the dielectric constant may also be defined as the relative effect of the medium on this force of attraction, according to Coulomb's equation:

$$F = QQ'/\varepsilon d^2 \tag{7.7}$$

where F = force of attraction between the two plates, Q = charge on one plate, Q' = charge on the second plate, ε = dielectric constant, and d = distance between plates.

Thus, the higher the dielectric constant of the material between the plates, the more attenuated will be the force of attraction between the plates. The dielectric constant of vacuum is 1. Of air, it is slightly above 1, which for all practical purposes is also taken as 1. This simplifies the measurement somewhat. Details of sample preparation, test methods, and equipment for measuring dielectric constants are given in ASTM D150.[22] Measurements are made at a standard frequency of 1 MHz for comparative purposes. Measurements at higher frequencies extending into the gigahertz range are necessary for high-speed and high-frequency circuits.

The dissipation factor, DF, is the ratio of the resistive (loss) component of current I_r to the capacitive component of current, I_c, and equals the tangent of the dielectric loss angle, δ, as follows:

$$\mathrm{DF} = I_r / I_c = \tan\delta \tag{7.8}$$

As for dielectric constant, test methods for dissipation factor may also be found in ASTM D150.[22]

7.2.3 Electrical stability

Electrical-stability testing is essential for conductive adhesives used for electrical connections. Electrical conductivity can degrade at elevated temperatures, on aging with or without power, and on exposure to humidity and temperature. The specific test method used depends on the application. One test used for die-attach adhesives specified in NASA MSFC-SPEC-592 (now inactive) involves a series of gold-plated Kovar® tabs attached with conductive epoxy to metal pads on an interconnect substrate. In the test vehicle, a bias of 5 V and current density of 139 ± 3.9 A/cm^2 (900 A/in^2) are applied to a series of wire-connected tabs, and the resistance change is measured after exposure to 150 °C periodically up to 1,000 hours. The maximum allowable resistance change is 5%.

Electrically conductive adhesives that are used as solder replacements are evaluated for material characteristics and contact resistance on solder surfaces. In one study, the requirement for volume resistivity is given as 0.001 Ω-cm or less and for contact-resistance change as 20% or less after 500 hours of 85/85-exposure.[23]

7.2.4 Current-carrying capacity and interconnect resistance

Other characteristics of electrical interconnects are current-carrying capacity and interconnect resistance. Current-carrying capacity is important for the qualification of adhesives for flip-chip interconnects. Typical current densities for conductive epoxies used as flip-chip bumps are 10–20 A/mm^2 and typical bump resistances are 12 mΩ.

No standard test method exists for these tests. According to the general guidelines of IPC-3406,[24] junction resistance shall be measured before and after performing the

IPC aging tests on all surface materials that come in contact with adhesives. Bond-joint resistance is measured by depositing a strip of adhesive 1 in × 0.5 in (minimum) onto gold-plated stainless steel and then applying a second gold-plated stainless-steel strip to give a one-half inch overlap.[25]

7.2.5 *Metal migration*

Metal migration, especially silver migration, may be quickly tested by applying a drop of deionized water across two closely spaced silver or silver-filled conductors while a DC potential of 2–10 V is applied. Catastrophic bridging and shorting of the conductors lines are visible at low magnification. The onset of migration is also indicated by the decrease in insulation resistance between the lines.

Other standard tests include UL 796 (Underwriters Labs.) and ASTM F1996-01.[26] According to the latter test, samples are prepared that represent the minimum circuit spacing and are energized with a DC or AC potential equal to the anticipated voltage rating for the PWB or interconnect substrate. Prior to humidity exposure the samples are subjected to a 60-second dielectric withstanding voltage of 40 V/mil to a maximum of 1,000 V. The samples are then exposed at 23 °C, 95–100% RH, energized for 1,344 hours, and examined for metal migration at 10×.

7.3 Environmental tests

Adhesives may be exposed to solvents in subsequent assembly operations such as cleaning, and adhesive bonds are subject to elevated temperatures associated with wire bonding in packaged devices and surface-mounted parts. Adhesives in electronics are also bonded to different surfaces, and corrosion testing is required to verify the stability of cured adhesive interconnects. Adhesives alone or in combination with moisture can degrade the electrical stability over time. Adhesives on printed-wiring boards must also be resistant to fungus growth and be nonflammable. Finally, high-radiation and prolonged radiation exposures in space or satellite applications can result in reduced mechanical attachment and reduced high-temperature stability.

7.3.1 *Solvent and chemical resistance*

Cured adhesives must be resistant to the effects of solvents and chemicals used in subsequent cleaning and processing steps. Dissolving or swelling of adhesives degrades their mechanical, electrical, and physical properties. Adhesives' resistance to solvents is evaluated based on the function or application.

For example, solvent or chemical resistance of adhesives used in surface-mount assembly may be tested according to Method 4.5.6 of ANSI/ IPC-SM-817, *General Requirements for Dielectric Surface-Mounting Adhesives*.[13] Although this method specifies specific solvents including isopropanol; 1,1,1-trichlorotrifluoroethane; an azeotrope of 6% methyl alcohol and 94% trichlorotrifluoroethane; and a 10% alkaline detergent solution (about pH 13), other solvents and chemicals encountered during manufacturing, screening, and rework processes should also be evaluated. Deposited dots of SMT adhesives exposed to the solvents for various periods of time are examined visually for physical changes or degradation.

7.3.2 Corrosion

The potential for corrosion caused by adhesives requires evaluations in both cured and uncured states. In one test, the adhesive components or the cured adhesive are applied to aluminized-Mylar® film and examined after 48 hours for etching or transmittance of light through the Mylar. A second test involves embedding copper wire with the adhesive and measuring corrosion by resistance changes in copper wire. Because of the wide variety of tests and the fact that they are closely related to adhesive reliability issues discussed in Chapter 6, the reader is referred to that chapter.

7.3.3 Radiation effects

Adhesives are used in space applications that may be exposed to high-energy radiation. Evaluating the effect of total dose and dose rate on the mechanical, electrical, and chemical properties of adhesives is required for reliability. ASTM D1879, *Exposure of Adhesive Specimens to High-Energy Radiation*[27] defines conditions for exposure. Categories of dose rate exposure within this method are:

Gamma radiation, *x*-radiation:	10^5–10^7 rad/hour
Electrons (beta radiation)	
Radioisotopes:	10^5–10^7 rad/hour
Accelerators:	10^8–10^{10} rad/hour
Neutrons and gamma radiation:	10^5–10^7 rad/hour

The effects of a total dose of ionizing radiation from a cobalt-60 source may be measured by Method 1019.5 of MIL-STD-883F.[28] Although this method is intended for radiation effects on semiconductor devices, the radiation sources, doses, and procedures can be used to evaluate the effects on exposed adhesives, provided that changes in bond strength or other physical or electrical properties are measured.

7.3.4 Fungus resistance

Fungus-resistance tests are used to determine the reaction of adhesives to several species of fungi under high humidity and a warm environment in the presence of inorganic salts. Testing for fungus resistance is more appropriate for exposed adhesives as in printed-wiring assemblies than for hermetically sealed devices or modules. One test method specified as Method 4.5.14 in IPC-SM-817[13] references Method 2.6.1 of IPC-TM-650, *Test Methods Manual*.[29]

According to this method, a test sample and a known nutrient control sample are inoculated with fungus spores and stored at 30 °C and 90% RH for 20 hours followed by 4 hours at 100% RH. The samples are checked for fungus growth after 7 days. The control should show signs of fungal growth. If the test sample also shows fungus growth, the test can be stopped; if not, the test is extended to 28 days. The test adhesive is considered "non-nutrient" if no growth shows at the end of 28 days.

7.3.5 Moisture absorption

Moisture absorption in uncured and cured adhesives is important for reliable adhesive bonding. Absorbed moisture in adhesives, particularly those used in surface-mounted parts, may result in voids in the bond line due to moisture desorption or entrapment. Absorption of moisture after cure also affects reliability. Popcorning, the cracking of plastic parts during solder reflow, is due to the rapid vaporization and expulsion of moisture retained by the die-attach adhesive or the plastic encapsulant.

Water absorption of a cured adhesive is the percent of water absorbed in a given period of time at a given temperature. Adhesive suppliers generally measure and report water – absorption data based on exposure to water over 24 hours at room or elevated temperature based on ASTM D570, *Water Absorption of Plastics*.[30] Samples are usually preconditioned at 50 ± 3 °C prior to immersion. Moisture absorption may also be tested as specified in paragraph 3.3.10 of IPC-TM-650.[29]

7.3.6 Hydrolytic stability

Hydrolytic stability is the resistance of a cured polymer material to reverting to a semisolid or liquid form when exposed to high humidity and temperature. Testing for hydrolytic stability is important for adhesives and other polymers that may be directly exposed to high humidity for long periods of time, as in tropical environments. Hence, testing applies more to printed-wiring assemblies than to hermetically sealed circuits. The hydrolytic stability test used for solder masks (IPC-TM-650 Method 2.6.11 as specified in IPC-SM-817 Section 4.5.15) may also be used to evaluate adhesives. Test specimens are prepared by stenciling adhesive 0.25 ± 0.05-mm thick. The specimens should be free of air bubbles or voids when examined under backlighting. Test samples are then exposed to 95 ± 4% RH and 97 ± 2 °C for 28 days and examined visually for evidence of reversion indicated by softening, chalking, blistering, cracking, tackiness, loss of adhesion, or liquefaction.

7.4 Thermal tests

With increasing temperature, organic coatings, and plastic materials, in general, undergo a variety of changes from simple changes of state (for example, melting, softening, subliming, or glass-transition phase changes) to catastrophic decomposition due to fracture of the polymer structure, oxidation, and outgassing of decomposition products. The thermal limits of adhesives should be known or measured to assure that the material will remain stable under the actual operating conditions or accelerated conditions, if those are used as screening or qualifying tests. Thermal testing of formulated, but uncured-adhesive materials include the measurement of heat absorption (endotherm) or heat evolution (exotherm) during cure, gel time at various cure temperatures, and weight loss during cure. Thermal tests for cured adhesives include thermal conductivity, CTE, T_g, and weight loss.

Key test methods for thermal stability include differential scanning calorimetry (DSC) (also known as differential-thermal analysis [DTA]), thermal conductivity, thermogravimetric analysis (TGA), gas chromatographic and mass-spectrometric analyses, and changes in mechanical properties such as tensile strength and elongation. These analytical methods may also be used to study chemical and physical changes that occur from exposure to various types and intensities of radiation. The degradation products of radiation often parallel those from thermal exposure.

7.4.1 Differential scanning calorimetry (DSC)

During the process of curing adhesives (as well as other polymeric materials) physical and chemical changes occur due to absorption of heat (endotherm) or evolution of heat (exotherm). Heat changes generated by uncured adhesives as they polymerize may be measured and used by adhesives suppliers or users to establish optimum curing schedules and for material characterization and quality control (see also Chapter 4). The peak exothermic temperature is monitored using a temperature sensitive probe and can be used as a measure of the total energy to assure completeness of cure.

In DSC, material changes due to heat absorbed or evolved are monitored and measured continuously using commercially available calorimeters.[31,32] Small amounts of the material (approx. 10–20 mg) are subjected to a controlled-temperature increase. Heat changes correspond to phase changes that occur at specified temperatures such as softening, melting, sublimation, glass transition, oxidation, decomposition, or polymerization curing.[33,34] Heating rates are generally 5–20 °C/min but are affected by the filler; for example, silver-filled adhesives require slower scans due to the higher thermal conductivity of the silver. Thermal exposure may be conducted in air or an inert atmosphere.

Test methods are specified in several ISO, ASTM, IPC, and BSI documents as follows:

ISO 11357-1	Plastics – differential scanning calorimetry – general principles
ISO 11357-5	Determination of characteristic reaction cure temperatures and times, enthalpy of reaction and degree of conversion
IPC-TM IPC 650 2.4.25C	Glass-transition temperature and cure factor for laminates and printed circuit boards
ASTM D3417	Replaced by ASTM D3418
ASTM D3418	Standard test method transition temperatures and enthalpies of fusion and crystallization of polymers by DSC

(Continued)

ASTM E1356	Standard test method for assignment of the glass-transition temperatures by differential scanning calorimetry or differential-thermal analysis
BSEN61006, Part 0	Methods of test for determination of glass-transition temperature of electrical insulating materials

7.4.2 Thermal conductivity

Thermal conductance and conductivities are derived from the Fourier equation. The total amount of heat a material conducts is directly proportional to the surface area, the time of contact, and the temperature gradient and is inversely proportional to the thickness of the sample, according to:

$$Q \propto (T_2 - T_1)At/d \tag{7.9}$$

where Q = total heat flow in joules, J; A = surface area, m^2; d = thickness of sample, m; T_2 = temperature of the hot surface, °C or K; T_1 = temperature of the cold surface, °C or K; and t = time, second.

When a proportionality constant, k, is introduced, the thermal-conductivity equation becomes:

$$Q = \mathrm{k}(T_2 - T_1)At/d \tag{7.10}$$

And rearranging to solve for k

$$\mathrm{k} = Qd/(T_2 - T_1)At \tag{7.11}$$

and the units become W/m K.

The constant, k, is a material constant called the *coefficient of thermal conductivity*. It is the time rate of heat flow under steady-state conditions through a unit area and unit thickness per unit temperature gradient. Several methods for measuring thermal conductance are available, of which the *guarded-hot-plate* and *heat-flow-meter* methods are the most widely used. The most accurate method for plastics having thermal conductivities of 3.4×10^{-3} cal/(second)(cm)(°C) [1.4 W/m K] or less (most filled and unfilled polymers) is the guarded-hot-plate method as described in ASTM C177.[35,36] A comparable method is described in ISO 8302.[37]

7.4.2.1 Guarded-hot-plate method

Two types of guarded-hot-plate apparati can be used. They are similar in principle, but differ somewhat in construction. One example, shown in Fig. 7.1, consists of three sections: a central heating plate, auxiliary heating plates, and cooling plates. Two identical test samples are placed at opposite sides of the main heater. The guard portion insures unidirectional heat flow from the central heater through the specimens and eliminates any influences from the edge of the sample. Thermocouples are located at various positions in the guarded-hot plate to measure the difference between the

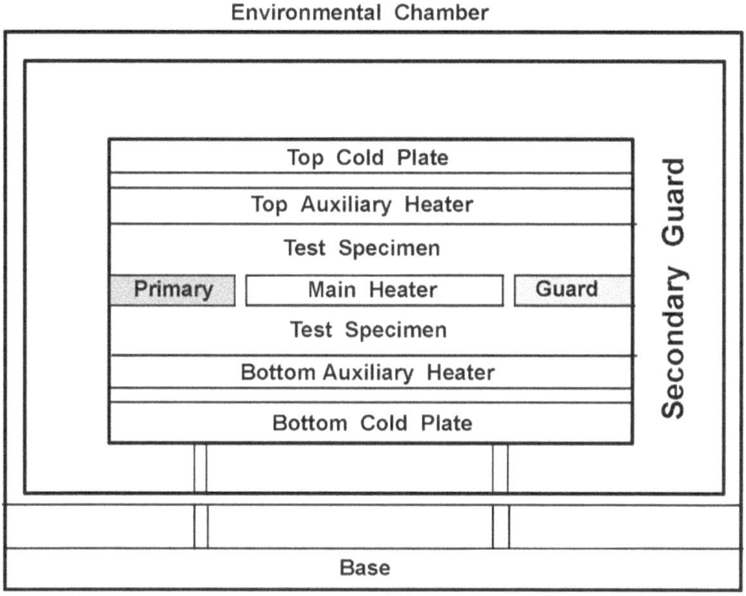

Figure 7.1 Schematic of guarded-hot-plate apparatus.

temperature of the guard ring and central heater and the temperature of the hot and cold surfaces. The central heating unit and cooling units are adjusted until the temperature drop through the two sides does not differ by more than 1% and the fluctuation over a 1-hour period is no greater than 0.5% of the temperature difference between the hot and cold plates. When equilibrium is reached, measurements are taken of the temperature differences across specimens; the hot-plate temperature, T_2; the cold-plate temperature, T_1; and the electrical power input, Q, to the central heater. Knowing the thickness and area of the sample, the thermal conductivity can be calculated using the thermal-conductivity equation given above. The sample thickness may vary with the maximum thickness based on calculations listed in ASTM C177.[35]

7.4.2.2 Heat-flow-meter method

The steady-state thermal transmission properties of an insulator can be measured using a heat-flow meter. This is a secondary or comparative method of measuring thermal conductivity since only the ratio of the thermal resistance of the specimen to that of a standard specimen is measured. According to this method, described in ASTM C518,[38] a sample is placed between a warm and cold plate. The meter measures heat flow through the specimen by generating a temperature difference across a slab of material of known thermal resistance. A calibration curve is generated using samples of known thermal conductivities. These samples should have thermal conductivities close to those expected of the samples tested. Sample thicknesses may range from 0.001 to 0.8 in but, for accuracy, the value of thermal conductivity divided by the sample thickness should be less than 400 W/m K. In the heat-flow-meter method, the spacing between the warm and cold plates is based on the theoretical maximum thickness of the specimens in the conventional guarded-hot-plate apparatus. Calculations of this spacing may be found in ASTM C518.

7.4.2.3 Laser-flash method

For high thermal-conductivity adhesives, such as the silver-glass compositions whose thermal conductivities are greater than 20 W/m K, the indirect laser-flash method is used. Unlike the steady-state methods, the flash method does not measure thermal conductivity directly, but measures thermal diffusivity, from which thermal conductivity is calculated as follows:

$$k = DC_pd \tag{7.12}$$

where k = thermal conductivity, D = thermal diffusivity, C_p = specific heat capacity, and d = density.

The flash method, described in ASTM E1461, *Standard Test Method for Thermal Diffusivity of Solids by the Flash Method*,[39] is based on measuring the temperature rise of the back surface of the sample when the front surface is irradiated with a high-energy pulse from a laser or flash lamp. Thermal diffusivity is usually carried out by rapidly heating one side of the sample and measuring the temperature rise on the other side. Essentially, the temperature rise follows the propagation of the pulse pattern. Small sample sizes of 12.5-mm diameter and 1–2-mm thick can be used. The flash method is applicable only for homogeneous solid materials. For materials that are substantially inhomogeneous such as composite or anisotropic materials, the thermal diffusivity values may be erroneous.

7.4.3 Thermogravimetric analysis

Weight loss with increasing temperature is performed as a measure of the thermal stability of adhesives. Thermogravimetric analysis (TGA) has been used for many decades as a quantitative measure of the total weight loss of outgassed products from polymers as they are heated at a controlled rate. A TGA apparatus consists of a precision electrobalance contained within a controlled-temperature oven such that the change in mass of a sample can be dynamically measured as a function of increasing temperature. The temperature is increased at a programmed rate or the sample may be kept at constant temperature and weight changes recorded as a function of time (isothermogravimetric analysis). Sample weights of several milligrams to several grams can be accommodated and weight losses or gains as little as 0.5 μg can be measured. The programmed rate of temperature increase is normally 10 °C/minute. Weight loss versus temperature curves (TGA curves, also known as *pyrograms*) is useful in determining the onset of outgassing, the weight loss at a specified temperature, and the temperature at which total decomposition occurs. TGAs may be conducted in nitrogen or, when thermal oxidative stability is being evaluated, in air.

Thermogravimetric analyzers may be connected to a variety of chemical analyzers to determine the exact composition of the outgassed materials as they are evolved.

Among chemical analysis methods are gas chromatography, infrared spectroscopy, and mass spectroscopy; for example, a TGA apparatus may be coupled with a Fourier Transform Infrared (FTIR) spectrophotometer to measure the thermal oxidative stabilities of several fluorinated polyimides.[40]

TGA is used to qualify adhesives for use at process temperatures such as those experienced during belt-furnace sealing. The standard test procedure for this method is ASTM D3850, *Rapid Thermal Degradation of Solid Electrical Insulating Materials by Thermogravimetric Method*.[41] According to this procedure, a manufacturer's system is calibrated to within ±0.01 mg and ±0.1 °C. The purge rate is 40–100 cm^3/minute and the heat rate is set to 5 °C/minute. Method 5011.4 of MIL-STD-883F, 3.5.2 also references this method. A maximum heat rate of 10 °C/minute is allowed in a nitrogen atmosphere with 20 cm^3/minute nitrogen flow.

7.4.4 Flammability testing

Adhesives may be tested for flammability by the same tests generally used for testing other polymeric materials. For most commercial and industrial applications, materials must pass standard tests specified by the Underwriters Laboratory UL 94[42] of which there are 12 classifications depending on the application. Three of these tests are generally used when stand-alone cured samples 127 mm × 12.7 mm and thicknesses representative of their intended use can be prepared. These are UL 94 V-0, V-1, and V-2, all of which are vertical burning tests. According to the tests, five cured polymer (adhesive) samples are ignited multiple times in a gas flame and the times for flame combustion to stop and glow to cease are recorded and averaged. In addition, cotton is placed beneath the sample and noted whether flaming drips ignite the cotton. Of the three versions, V-0 is the most severe requirement where the flame must cease within 10 seconds and glow within 30 seconds and no drips igniting the cotton. For V-1 and V-2 flaming combustion must cease within 30 seconds and glowing combustion within 60 seconds. Flaming drips are not allowed in V-1 but are allowed in V-2. Other tests are similar except that the sample is oriented horizontally. [43]

UL-746A[44] covers three other tests that are more representative of the actual use of materials in electronic/electrical applications where ignition can occur from electrical sources rather than a direct flame. Electrical sources include overloaded circuits and devices due to high currents, hot spots on circuit boards, or high voltage arcing. These tests are applicable to thin films or small samples generally those not capable of supporting themselves. These tests are: hot-wire ignition (HWI) test, high current/amp arc ignition (HAI), and high voltage arc tracking rate (HVTR) (Table 7.3).

Other tests similar to the UL tests are described in: ASTM D635, IEC 60695, IEC 60335-1, IEC 60112, IEC 60950, UL 1694, ISO 3795, ISO 9772, and ISO 9773.

7.5 Mechanical and thermomechanical tests

Mechanical and thermomechanical tests for adhesives are primarily used to evaluate the stress imparted by adhesives on bonded parts. Included in this category are

Table 7.3 Electrical ignition tests per UL 746A

Test	Type	Measurement	Summary
HWI	Hot-wire ignition	Measures resistance to ignition from high temperatures from a device or conductor failure carrying more than its rated current	Mean number of seconds to ignite a sample wrapped with a non-chrome resistance wire dissipating a specified level of energy
HAI	High amp arc ignition	Measures ability to withstand electrical arcing directly on or just above the sample surface	Number of arc exposures to ignite a specimen using standardized electrodes, geometry, and electrical supply circuit
HVTR	High voltage arc tracking rate	Measures arc tracking rate on sample surface	Rate in in/minute that a tracking path can be produced on the sample surface under standardized electrical stress

physical characteristics such as hardness and modulus that are a measure of rigidity of the cured adhesive. Other mechanical properties used to evaluate and classify materials are modulus and elongation-at-break, as well as the coefficient of thermal expansion above and below the glass-transition temperature. A third general area of tests is bond strength that may include tensile shear as well as die-shear and stud-pull tests for devices and centrifuge for both devices and substrates. Finally, radius of curvature is used to characterize residual stress of devices imposed by adhesives as related to cure conditions and post-processing, such as wire bonding and plastic molding or encapsulation.

7.5.1 Hardness

Hardness is used both as a qualification and acceptance test for adhesives. It is a simple, low-cost test that can be used as a rough indicator of the degree of cure. There are several instruments for measuring hardness, but the most widely used are based on a spring-loaded indentor that is pressed into a sample. Small portable instruments such as the Shore Durometers give an analog or digital reading on a 0–100 scale. The Shore A Durometer is used for soft materials, Shore B for moderately hard, Shore C for medium hard, and Shore D for very hard materials. This method is defined in ASTM D2240.[45]

7.5.2 Modulus and elongation (flexural)

Flexural modulus is a measure of the strength of adhesives. Modulus data are most often used in stress analysis (one-dimensional or as an input to 3-D modeling). In addition to flexural modulus, elongation-at-break is also recorded. The standard test method is ASTM D790, *Flexural Properties of Unreinforced and Reinforced Plastics and Electrical Insulating Materials*.[46] Two methods are specified. The first uses a three-point bending test; two supports hold the part and a load is applied with a single point to induce flexure. In the second method (four-point bending), the force is applied using two heads or points. The flexural modulus is calculated from the following equation:

$$E = L^3F/4bh^3y \tag{7.13}$$

where L = length of span; y = distance covered by a load F during flexure, measured from the initial position; F = load (in Newtons); b = width of test bar (in mm); and h = thickness of test bar (in mm).

7.5.3 Modulus and elongation (tensile)

Young's modulus, also referred to as elastic modulus, tensile modulus, or modulus of elasticity in tension is the ratio of stress-to-strain and is equal to the slope of a stress–strain diagram for the material. In the standard test method, ASTM D412, a force is applied to a "dog-bone-shaped" sample of the cured adhesive. The force at elongation (strain) is measured. Most often, the elongation is 25–30% although for elastomeric materials it may be 100% or greater.

Tensile and elongation characteristics of materials may also be evaluated according to ASTM D638-96, *Tensile Properties of Plastics* (*Metric*).[47] According to this method, bi-axial strain gages are used to monitor axial and transverse strain in dog-bone test samples. Ultimate strength, modulus, Poisson's ratio, and percent strain at failure are determined.

7.5.4 Coefficient of thermal expansion and glass-transition temperature

Thermomechanical-analysis (TMA) testing is used to measure a material's expansion coefficient above and below the glass-transition temperature or T_g. Thermomechanical analysis continuously monitors the expansion of a probe on a sample as a function of temperature. The standard test method for TMA is ASTM D3386, *Coefficient of Linear Thermal Expansion of Electrical Insulating Materials*.[48] In addition to the glass-transition temperature or T_g, the expansion coefficients above and below T_g are reported.

7.5.5 Bond strength

Bond strength is measured to assure that bonding materials retain adhesion at temperature and after aging or exposure to thermomechanical stresses. Bond strength may be measured in tensile, peel, or shear. Bond strength is measured as tensile strength for SMT adhesives and as shear strength for component and substrate adhesives. Specialized adhesive test methods are also used for flip-chip and substrate strength, and centrifuge testing is used to evaluate both die and substrate adhesion.

7.5.5.1 Tensile strength

Tensile strength is evaluated by pull testing on metal "lap-shear" coupons, most often aluminum. A standard test method is ASTM D1002, *Apparent Shear Strength of Single-Lap-Joint Adhesively Bonded Metal Specimens by Tension Loading (Metal-to-Metal)*.[49] Aluminum coupons are overlapped by 1/2 in (typical) and a wire is used as a spacer to control bond line thickness. Method 5011, paragraph 3.8.8.2 of MIL-STD-883F references this specific method. Other tensile lap-shear strength measurement tests are described in BSEN 1465, Adhesives Determination of Tensile Lap-Shear Strength of Rigid-to-Rigid Bonded Assemblies (1995) and ISO 4587, Adhesives Determination of Tensile Lap-Shear Strength of Rigid-to Rigid Bonded Assemblies (2003).

7.5.5.2 Peel strength

Peel strength is the average load per unit width of bond line required to separate adhesive-bonded materials when pulled at a specific angle. The width is generally 1 inch and the angle 180° but may also be 45 or 90° depending on the application and the test data desired. Peel strengths, reported as pounds per inch width (ppi), are especially useful in assessing the adhesion of laminates, tapes, and films used for interconnect substrates, flex cable, and packaging.

According to the peel test, the sample is attached to a movable crosshead via grips or clamps. The test speed is kept constant, typically at 10 or 12 in/minute and the sample pulled until separation or break occurs. Most peel tests require a load of 10 pounds or less. The peak load at separation, average load, and energy at break are measured and stored in the equipment's controller memory and subsequently transferred to a computer where loads are plotted versus time. Test methods and standards include ASTM D903-98 and ASTM D3807.

Peel testers consist basically of a test stand, calibrated force gauges, digital readouts, various sizes and configurations of grips/clamps, and software programs. Peel testers are available from many instrument suppliers.

7.5.5.3 Die-shear strength

Die-shear testing is performed both for qualification and quality assurance testing of adhesive-attached semiconductor die and other chip devices. MIL-STD-883F, Method 2019.7[50] defines the test method most commonly used. A uniform force is applied to the edge of the die using a die-contact tool. The tool loads against the edge of the die perpendicular to the mounting plane (substrate). A lateral force is applied sufficient to shear the die from the substrate. Failure criteria are given as the minimum force in kilograms required to detach the die based on the area of the die. Graphs of failure criteria are provided in Method 2019 for minimum strength, 1.25 times minimum, and two times minimum strength as a function of die area. Most wire-bond pull testers are now also equipped for die-shear testing.

7.5.5.4 *Die and substrate attachment strength*

Another method for measuring either die- or substrate-attach adhesion strength involves attaching a contact tool to the top of the die or substrate with an adhesive that is stronger than that of the die-attachment adhesive. The tool applies a vertical force until the die detaches from the substrate or until a minimum specified strength is achieved. The test method and failure criteria are specified in Method 2027 of MIL-STD-883F.[51]

7.5.5.5 *Centrifuge testing (constant acceleration)*

Centrifuge testing, also known as constant acceleration, is applicable to both die and substrate adhesion and is defined in Method 2001 of MIL-STD-883F.[52] According to this method, parts assembled with adhesives are centrifuged in the Y_1 direction with the die-side against the wall of a spinning test chamber (centrifuge). The centrifugal force pulls against the attached devices. Parts may be centrifuged to a known "G" load or until failure occurs. Centrifuge testing is widely used as a 100% screen test for hermetically sealed hybrid microcircuits and multichip modules for both Class H (military) and Class K (space). In both cases, according to MIL-PRF-38534E[53] modules must pass 3,000 G without any detachments. In addition, for qualification, modules must pass 5,000 G in the Y_1 direction.

Mechanical shock Method 2002.4 of MIL-STD-883[54] may be used as an option instead of constant acceleration as the screen test, but both mechanical shock and constant acceleration must be used for qualification.

7.5.6 Acoustic microscopy

Scanning acoustic microscopy (SAM) is an ideal nondestructive method for revealing internal flaws within materials or between material interfaces. SAM is extensively used in detecting voids, delamination, and other separations that can occur in adhesive-attached parts, especially after thermal cycling. SAM is particularly useful in the analysis or evaluation of many types of electronic parts, including ceramic and plastic-encapsulated ICs, plastic-encapsulated microcircuits (PEMs), hybrid microcircuits, CSPs, PBGAs, and printed-wiring boards.

The acoustic microscope produces an image by mechanically scanning with an ultrasonic transducer in an *X*, *Y* pattern over the test specimens. This "scan" is performed in a vessel filled with deionized water or other liquid that couples the ultrasonic energy from the transducer to the part. Water is the most effective medium but, where the electronic part is sensitive to moisture, other liquids such as alcohols can be used. The transducer emits a short ultrasonic pulse and then detects the echo. If the part is free of defects, there will be two signals, from the near and far surface. If there is an internal defect, such as a void or delamination, the transducer will detect an additional return signal. This signal has a direct relationship to the depth of the defect. Moreover, acoustic images from several depths can be acquired simultaneously, revealing defects at each interface in an electronic device.

Images are typically displayed as either C-scans or B-scans. A C-scan is an overall X, Y view of the test specimen at a particular depth or interface. A color scale or black and white image displays signal amplitude. B-scans display an image of a cross-sectioned view at a chosen location with the scan indicating signal amplitude. Acoustic microscopes have the flexibility to adapt different transducers for different applications from low-frequency transducers (1–15 MHz), used for detecting large flaws in thick materials (through which sound travels slowly) to 50 MHz transducers that can examine thin Kovar® of 30 mils or aluminum up to 200 mils thick. Higher frequency transducers are needed for thin substrate materials such as alumina or gallium arsenide. The high-frequency transducers allow finer resolution and greater detail, but do so at the expense of penetrating power.

The acoustic microscope has the software to analyze the image once the scan is completed. Analysis tools include: histogram (for percentage of voids), vertical and horizontal profile, zoom, manipulation of color scale, measurements, and multiple scans. The system also has the capability to print either in color or black and white, or to transmit images electronically over the network.

Although there is no standard test method for acoustic microscopy, Method 2030, *Ultrasonic Inspection of Die Attach* of MIL-STD-883F[55] is used to nondestructively detect unbonded regions and voids in IC die-attachment materials. Parts are rejected if a single void larger than 15% of the contact area is present (>10% at a corner) or more than 70% of any given quadrant is detached.

7.5.7 Radius of curvature

The ability of an adhesive to absorb stresses associated with bonding large devices to substrates with mismatched expansion coefficients is directly measured by the radius of curvature. Radius of curvature of a device bonded to a leadframe is determined in order to predict relative stress absorption capabilities of materials. There is no industry standard for this test, and measurement techniques vary by vendor. The radius of curvature is used in determining the bending stress as defined below:

$$\text{Bending stress} = Eh/2Yr \tag{7.14}$$

where E = Young's modulus, h = thickness of the specimen, Y = Poisson's ratio, and r = radius of curvature.[56]

Typically, an automatic dispenser is used to apply enough adhesive to completely cover the underside of a die. Bond line thicknesses of 1.0–1.5 mils and fillets of 50–75% of the die thickness are recommended. The star pattern of dispensing is suggested to achieve optimum coverage. A surface-roughness tester, such as a Brown & Sharpe Surface Roughness Measuring Instrument Model Surfcom 120A or equivalent, is used to measure the profile of the cured sample.[57]

Radius of curvature, r, is determined as (see Fig. 7.2):

$$r^2 = (r - x)^2 + (L/2)^2 \tag{7.15}$$

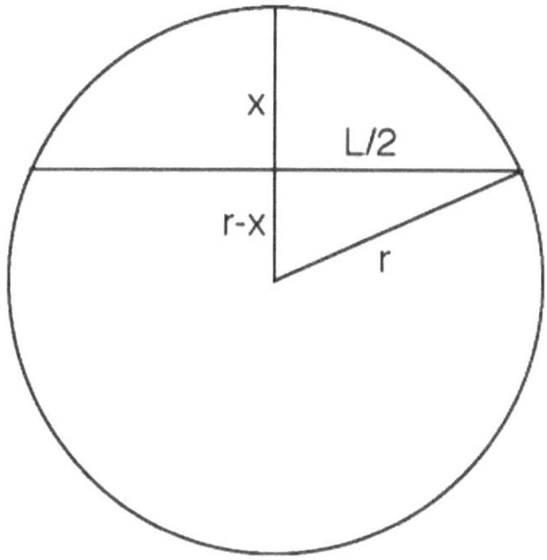

Figure 7.2 Radius of curvature diagram.

And rearranging:

$$r = [x^2 + (L/2)^2]/2x \tag{7.16}$$

where r is the radius of curvature in millimeters, x is the maximum amplitude (μm) from the printout, and L is the traversed length of the specimen.

Laser interferometry can also be used to measure the radius of curvature. Changes in surface deflection are detected from the angle of reflection, and the radius of curvature is determined.

7.6 Chemical analysis

Many chemical and analytical or instrumental methods are available for analyzing adhesives for chemical structure and composition, outgassing, decomposition products, and ionic impurities.

7.6.1 Weight loss during cure

Weight loss during cure is a measure of the amount of volatiles given off during cure. High-weight loss is a concern in contaminating curing ovens and subsequently redepositing onto critical surfaces of circuits, such as wire-bond pads. There is, as yet, no industry standard for this test. Some adhesives' suppliers use "simulated" silicon die attached to a glass microscope slide and accurately weigh the glass slide, the silicon chips, and the adhesive-bonded samples; then, by difference, calculate the initial weight of uncured adhesive used. After curing, the test coupons are reweighed and, by difference from the original sample weight, the weight loss and percent loss are calculated.[58]

7.6.2 Infrared spectral analysis

Infrared spectroscopy (IR) is an excellent method for determining chemical changes that occur in organic materials through decomposition, oxidation, polymerization, reactions with other compounds, or changes in the formulation. Infrared spectroscopy is also a valuable tool for detecting and identifying organic compounds, polymer types, impurities in compounds, and for quantitative analysis of mixtures of compounds. Infrared spectra result from different modes of vibration and rotation within a molecule. Pure rotational spectra occur at the long wavelengths (greater than 667 cm^{-1}). At shorter wavelengths, the radiation has sufficient energy to cause changes in vibrational levels of the molecule. Thus, the frequencies of vibration are related to the masses and binding forces of the atoms that become the basis for infrared in determining the structure of a compound. The region of most interest for the analysis of organic materials lies between 5,000 cm^{-1} (2 μm) and 667 cm^{-1} (15 μm) where specific groups absorb at definite frequencies or wavelengths.

The percent absorption or transmission of infrared radiation by a material provides a "fingerprint" of the material that is useful in determining the exact nature of the compound. Absorptions at specific wavelengths in the near infrared are indicative of various groups within a molecule such as amino, carboxyl, keto, or hydroxyl groups, while absorption in the far infrared is indicative of the nature of the molecule as a whole. The spectrum obtained may also be compared with spectra of known compounds as recorded in the literature.[59] A standard test method for IR analysis is ASTM E1252.[60]

7.6.3 Gas chromatography

In gas chromatography, the components of a mixture are separated by volatilizing the sample and passing the gases through a column of solid porous material of 20–200 mesh or onto thin layers of finely divided solids. In one method, gas–liquid chromatography, the surface of the packing is coated with a nonvolatile liquid (the stationary phase). If the liquid is not present, the process is called gas–solid chromatography. The principle of chromatography is based on the different rates at which the chemical components pass through the packing material, and their detection as separate species as they exit the column. The components may be detected by thermal conductivity changes, density differences, or by ionization detectors.

7.6.4 Mass spectrometry and residual gas analysis (RGA)

The main source of moisture and other outgassed compounds in hermetically sealed ceramic or metal packages is from adhesives and other polymer materials used in the assembly of the electronic devices. Even though the assemblies are extensively vacuum baked prior to sealing and even though they are sealed in dry nitrogen, outgassing of moisture, greater than the 5,000 ppm_v allowed by MIL-PRF-38534, can occur after burn-in or after temperature cycling.

Mass spectrometry is an excellent analytical method to quantitatively measure moisture and trace amounts of low-molecular-weight organic species. In RGA, the outgassing from adhesives and other organic materials such as moisture, carbon dioxide, carbon monoxide, and low-molecular-weight hydrocarbons can be measured in low parts per million.

By mass spectrometry, molecules are first activated and dissociated into ionic fragments that are then separated according to their charge-to-mass ratios. Mass spectrometers essentially consist of four parts: equipment to ionize the sample, an ion accelerator, an ion separation system, and a detection system. The sample may be ionized by bombardment with electrons from a heated filament or by several other means. The ions produced are then accelerated and focused electrostatically and separated according to their atomic masses while traveling through a magnetic field.

A variation of magnetic-separation equipment, widely used today, is the quadrupole mass spectrometer. This equipment consists of four cylindrical rods arranged in an orthogonal array. Opposite rods are electrically connected and opposite voltages applied. The positively charged pair acts as a low-mass pass filter while the negatively charged pair acts as a high-mass pass filter. By adjusting the voltages on the two pairs of rods, the quadrupole is made into a narrow, bandpass-mass filter. The separated ions are then detected electrically. The ion beams are scanned across a collector where they pick up electrons and generate an electric current that is amplified and analyzed by computer. As with infrared spectra, an extensive database of mass spectra of chemical compounds is available and can be used to compare with the mass spectrograph of an unknown to verify its nature.[61]

A standard test method for RGA is Method 1018.4 of MIL-STD-883F[62] according to which mass spectrometry is used to detect moisture and other volatilized compounds present in sealed microcircuit packages. The sealed modules are first preconditioned to remove external contaminants and adsorbed moisture by prebaking for 16–24 hours at 100 ± 5 °C then placed in a vacuum holding chamber that is attached to the spectrometer. The device and chamber are then pumped down and stabilized at 100 ± 5 °C, after which the lid of the device package is punctured and internal gas from the package allowed to enter the mass spectrometer.

In qualifying adhesives for outgassing, an amount of adhesive and surface area representative of that to be used in actual devices or modules is cured in gold-plated or nickel-plated lead-less packages. After vacuum baking, the samples are sealed in dry nitrogen, leak tested, heated for up to 1,000 hours at 150 °C and analyzed. Moisture must be no greater than 5,000 ppm_v for both Class H and Class K parts. If a moisture getter is used, the moisture requirement is less than 3,000 ppm_v after the 1,000-hour exposure.

Because of the expensive equipment involved and the expertise required for accurate analyses, moisture and gas analysis is generally outsourced to laboratories that are certified.

7.6.5 Total-mass loss and water regained

Another variation on outgassing is total-mass loss of cured-adhesive materials. Outgassing and subsequent condensation of volatiles is a concern, for example, in optical and space applications. A standard test method is ASTM E595, *Total-Mass Loss and Collected-Volatile Condensible Materials from Outgassing in a Vacuum Environment*.[63] Three measurements are made in this test: total-mass loss (TML), collected-volatile condensible materials (CVCM), and water vapor regained (WVR). Historically, maximum values of 1% TML and 0.1% CVCM have been used to qualify materials for spacecraft applications. TML and CVCM are determined after 24 hours exposure at 5×10^{-5} Torr and 125 °C. Water vapor regained is determined by reweighing after storage in a 23 °C and 50% RH environment. Samples are normally mixed in 10-g batches and cured specimens are cut into cubes 1.5–3.0 mm/side. Adhesive films may be applied to surfaces to simulate use conditions.

7.6.6 Ionic content

The total ionic content and the amounts of specific metal ions such as sodium and potassium greatly affect the electrical properties of an insulative adhesive and of the electronic function of a circuit. A simple test to determine the total ionic content of a polymeric material is to digest a weighed, powdered sample in deionized water and to measure the resistivity of the water extracts after refluxing for 20 hours. This test gives a rough idea of the extent of ionic contaminants in the sample. For example, a drop in resistivity from 10 MΩ-cm to 10,000 Ω-cm indicates a considerable amount of ionic species. The total ionic content is reported as ppm of equivalent NaCl:

$$\text{Total ionic content} = 23.47(L_2 - L_1)/W \qquad (7.17)$$

where L_2 = specific conductance in micromho/cm of a sample, L_1 = specific conductance in micromho/cm of a blank, W = sample weight in grams, 23.47 = the constant A/L_s where A = grams of NaCl in 50 cm^3 of a 0.0005N NaCl solution, and L_s = specific conductance of the 0.0005N NaCl solution.

Exact elemental analysis can then be performed by evaporating the water and analyzing the residue by atomic absorption spectroscopy or flame photometry. These are emission spectrographic methods whereby the atoms or molecules of the sample, on being excited by energy input from an arc, spark, or flame, emit wavelengths of radiation as they return to the normal state. These wavelengths indicate qualitatively and quantitatively the atomic species present in the sample. The spectra of the elements consist of lines of distinctive color, for example, the yellow D line for sodium. The spectra of molecules consist of groups of lines called bands. The number of lines present in an emission spectrum depends on the number and position of the outermost electrons and the degree of excitation of the atoms. Element analysis may also be performed by ion chromatography.

Standard test methods and requirements for total ion content and specific ions such as chloride, sodium, fluoride, and potassium are given in Method 5011.4, paragraph 3.8.7.3.[21]

A simple qualitative test for chloride ions consists of immersing or swabbing the part in deionized water and adding a few drops of 0.1N silver nitrate solution to the water extracts. A white turbidity or precipitate indicates that an insoluble silver chloride has formed and that chloride ions may have been present. The test, however, is not entirely specific for chlorides since a few other anions, such as sulfate, also form white insoluble precipitates that must then be distinguished by other qualitative or quantitative methods.

References

1. Standard Test Methods for Rheological Properties of Non-Newtonian Materials by Rotational (Brookfield) Viscometer, ASTM D2196, *Am Soc.* Testing and Materials. 2003.
2. Standard Test Method for Apparent Viscosity of Adhesives Having Shear-Rate Dependent Flow Properties, ASTM D2556, *Am Soc.* Testing and Materials. 1997.
3. Loctite 3607 (Oct. 1995); Loctite 3610 (Nov. 1998); Loctite 3612 (Nov. 1996); Loctite 3614 (Nov. 1996); Technical Data Sheets, Loctite.
4. Standard Test Method for Apparent Viscosity of Plastisols and Organosols at Low Shear Rates by Brookfield Viscosity, ASTM D1824, *Am Soc.* Testing and Materials. 2002.
5. Epibond 7002 and Epibond 7200, Technical Data Sheets, Vantico, Inc. (formerly Furane). Feb. 2001.
6. Hysol KO110 Die-Attach Adhesive, Technical Data Sheets, Henkel; Oct. 2001.
7. Thermoset DS MD-110, Technical Data Sheet, Thermoset Lord Chemical Products. Apr. 1999.
8. Hysol FP 4652, Technical Data Sheet, Henkel; Jul. 2010.
9. Hysol FP 4460, Technical Data Sheet, Henkel; Jul. 2010.
10. Hysol KO120, Technical Data Sheet, Henkel; Oct. 2001.
11. Durairaj R, Ekere NN, Salam B. Thixotropy flow behavior of solder and conductive adhesive pastes. *J Mater Sci: Mater Electron.* 2004;15:677-83.
12. Bell GC, Rosell CM, Josline ST. Rheology of silver filled glass die attach adhesive for high speed automatic processing, Du Pont semiconductor materials. In: *Proc. 37th Electronic Components Conference*; May 11–13, 1987.
13. General Requirements for Dielectric Surface Mounting Adhesives, ANSI/ IPC-SM-817, Method 4.5.5; Nov. 1989.
14. Standard Test Method for Density of Adhesives in Fluid Form, ASTM D1875, *Am Soc.* Testing and Materials. 2003.
15. Standard Test Method for Density and Specific Gravity (Relative Density) of Plastics by Displacement, ASTM D792, *Am Soc.* Testing and Materials. Dec. 12, 2000.
16. Evaluation and Acceptance Procedures for Polymeric Materials, MIL-STD-883F, Method 5011.4, paragraphs 3.4.1 and 3.8.3; Oct. 1995.
17. Internal Visual (Hybrid), MIL-STD-883F, Method 2017 (Jun. 2004), Internal Visual (Monolithic), Method 2010.11. Jun. 2004.
18. Standard Test Method for Haze and Luminous Transmittance of Transparent Plastics, ASTM D1003-00, *Am Soc.* Testing and Materials. Oct. 6, 2000.
19. Dymax OP-30 and OP-32, Technical Data Sheets, Dymax Inc.
20. Standard Test Method for Index of Refraction of Transparent Organic Plastics, ASTM D542, *Am Soc.* Testing and Materials. Oct. 11, 2000.
21. Evaluation and Acceptance Procedures for Polymeric Materials, MIL-STD-883F, Method 5011.4; Oct. 1995.
22. Standard Test Method for AC Loss Characteristics and Dielectric Constant (Permittivity) of Solid Electrical Insulating Materials, ASTM D150, *Am Soc.* Testing and Materials. 2004.

23. Tong QK, Markley DL, Fredrickson G, Kuder R. Conductive adhesives with stable contact resistance and superior impact performance. In: *Proc. 49th Electronic Components and Technical Conference*; Jun. 1–4, 1999.
24. Guidelines for Electrically Conductive Surface Mount Adhesives, Institute for Interconnecting and Packaging Electronic Circuits, IPC 3406. Jul. 1996.
25. Bond Joint Resistance Test, Method TM-PT-15, Ablestik Electronic Materials (now Henkel). Sep. 15, 1993.
26. Standard Test Method for Silver Migration for Membrane Switch Circuitry, ASTM F1996-01, *Am Soc.* Testing and Materials. 2003.
27. Standard Test Method for Exposure of Adhesive Specimens to High-Energy Radiation, ASTM D1879, *Am Soc.* Testing and Materials. Oct. 10, 1988.
28. Ionizing Radiation (Total Dose) Test Procedure, MIL-STD-883F, Method 1019.6; Mar. 2003.
29. Test Methods Manual, IPC-TM-650, Institute for Interconnecting and Packaging Electronic Circuits.
30. Standard Test Method for Water Absorption of Plastics, ASTM D570, *Am Soc.* Testing and Materials.
31. Cassel R.B. Enhanced DSC instrumentation design and characterization of adhesives, Adhesives and Sealants Industry, Jan. 31, 2002.
32. Thermal Analysis Applications in the Semiconductor Packaging Industry, www.perkinelmer.com.
33. Murphy CB. Differential thermal analysis. *Anal Chem.* May. 1960;32:168R-71R.
34. Murphy CB. Differential thermal analysis. *Anal Chem.* Apr. 1962;34:298R-301R.
35. Standard Test Method for Steady-State Heat Flux Measurements and Thermal Transmission Properties by Means of the Guarded Hot-Plate Apparatus, ASTM C177, *Am Soc.* For Testing and Materials. Jan. 11, 2004.
36. Anderson DR. Thermal conductivity of polymers. *Chem Rev.* 1966;Vol. 66.
37. ISO 8302, Thermal insulation – determination of steady-state thermal resistance and related properties – guarded hot plate apparatus, International Organization for Standardization. 1991.
38. Standard Test Method for Steady-State Thermal Transmission Properties by Means of the Heat Flow Meter Apparatus, ASTM C518, *Am Soc.* Testing and Materials.
39. Standard Test Method for Thermal Diffusivity of Solids by the Flash Method, ASTM E1461, *Am Soc.* Testing and Materials. Oct., 2, 2001.
40. Turk MJ, Ansari AS, Alston WB, Gahn GS, Frimer AA, Schieman DA. Evaluation of the thermal oxidative stability of polyimides via TGA techniques. *J Polymer Sci. Part A: Polym Chem.* 1999;Vol. 37.
41. Standard Test Method for Rapid Thermal Degradation of Solid Electrical Insulating Materials by Thermogravimetric Method, ASTM D3850, *Am Soc.* Testing and Materials. 2000.
42. UL 94, Standard for Safety of Flammability of Plastic Materials for Parts in Devices and Appliances Testing, Underwriters Labs, www.ul.com.
43. ASTM D 635-06, Standard Test Method for Rate of Burning and/or Extent and Time of Burning of Plastics in a Horizontal Position.
44. UL 746A, Standard for Safety of Polymeric Materials – Short Term Evaluations.
45. Standard Test Method for Rubber Property – Durometer Hardness, ASTM D2240, *Am Soc.* Testing and Materials. Jan. 7, 2004.
46. Standard Test Method for Flexural Properties of Unreinforced and Reinforced Plastics and Electrical Insulating Materials, ASTM D790, *Am Soc.* Testing and Materials. Jan. 3, 2003.

47. Standard Test Method for Tensile Properties of Plastics (Metric), ASTM D638, *Am Soc.* Testing and Materials. Jan. 12, 2003.
48. Standard Test Method for Coefficient of Linear Thermal Expansion of Electrical Insulating Materials, ASTM D3386, *Am Soc.* Testing and Materials. Oct. 12, 2000.
49. Standard Test Method for Apparent Shear Strength of Single-Lap-Joint Adhesively Bonded Metal Specimens by Tension Loading (Metal-to-Metal), ASTM D1002, *Am Soc.* Testing and Materials. Oct. 10, 2001.
50. Die Shear Strength, MIL-STD-883F, Method 2019.7; Mar. 2003.
51. Substrate Attach Strength, MIL-STD-883F, Method 2027; Mar. 1989.
52. Constant Acceleration, MIL-STD-883F, Method 2001; Aug. 1977.
53. Performance Specification, Hybrid Microcircuits, General Specification, MIL-PRF-38534E; Jan. 2003.
54. Mechanical Shock, MIL-STD-883F, Method 2002.4; Aug. 1998.
55. Ultrasonic Inspection of Die Attach, MIL-STD-883F, Method 2030; May 1987.
56. Loh KI. Low stress adhesive. *Adv Packaging*; Winter 1993.
57. Radius of Curvature, Test Method TM-MT-15, Ablestik Electronic Materials (now Henkel). Nov. 1994.
58. Weight Loss During Cure, Test Method PT-80, Ablestik Electronic Materials (now Henkel). May 31, 1996.
59. Bellamy LJ. *Infrared Spectra of Complex Molecules*. 3rd ed. London: Chapman and Hall; 1975.
60. Standard Practice for General Techniques for Obtaining Infrared Spectra for Qualitative Analysis, ASTM E1252, *Am Soc.* Testing and Materials. Oct. 3, 1998.
61. NIST Standard Reference Database, No. 69, NIST Chemistry WebBook, http://webbook.nist.gov; Jul. 2001.
62. Internal Water-Vapor Content, MIL-STD-883F, Method 1018.4; Jun. 2004.
63. Standard Test Method for Total Mass Loss and Collected Volatile Condensible Materials from Outgassing in a Vacuum Environment, ASTM E595, *Am Soc.* Testing and Materials. 1999.

Appendix

Conversion factors

Table A.1 Thermal conductivity units and equivalents

Cal/(second · cm · °C)	BTU/(hour · ft · °F)	W/(m·K)	W/(cm · °C)	W/(in · °C)
1	241.9	418.6	4.186	10.63
4.13×10^{-3}	1	1.73	0.0173	0.044
2.39×10^{-3}	0.578	1	0.01	0.0254
0.239	57.8	100	1	2.54
0.094	22.74	39.4	0.394	1

Table A.2 Dimension units and approximate equivalents

Mil	Inch	Centimeter	Millimeter	Angstrom	Micrometer	Microinch	Nanometer
1	0.001	0.00254	0.0254	254000	25.4	1,000	2.54×10^4
1000	1	2.54	25.4	2.54×10^8	25,400	1×10^6	2.54×10^7
394	0.394	1	10	1×10^8	$1 x 10^4$	3.94×10^5	1×10^7
39.4	0.0394	0.1	1	1×10^7	1×10^3	3.94×10^4	1×10^6
3.94×10^{-6}	3.94×10^{-9}	1×10^{-8}	1×10^{-7}	1	1×10^{-4}	3.94×10^{-3}	0.1
0.0394	3.94×10^{-5}	1×10^{-4}	0.001	10,000	1	39.4	1,000
0.001	1×10^{-6}	2.54×10^{-6}	2.54×10^{-5}	254	0.025	1	250
4×10^{-6}	4×10^{-9}	1×10^{-7}	1×10^{-6}	10	1×10^{-3}	0.0394	1

Table A.3 Pressure and force conversion factors

To Convert From:	To:	Multiply By:
Pascals (Pa)	Pounds per in^2 (psi)	1.45×10^{-4}
Pounds	Newtons	4.44
Dynes/cm^2	Pounds per in^2 (psi)	1.45×10^{-5}
Newtons/mm^2	GigaPascals (GPa)	1×10^{-3}
MPa	Newtons/mm^2	1
Pascals (Pa)	Dynes/cm^2	10

Table A.4 Viscosity conversion factors

To Convert From:	To:	Multiply by:
Poise	Centipoises	100
Centipoises	Poise	0.01
Poise	lb/(second · ft)	0.0672
Poise	lb/(hour · ft)	242
Poise	Pa second	0.1
Pa second	Centipoises	1,000
Poise = gm/(sec)(cm)		

Abbreviations, Acronyms, and Symbols

Table A.5 Abbreviations, acronyms, and symbols

ACA	Anisotropic Conductive Adhesive
ACS	American Chemical Society
AEW	Amine Equivalent Weight
ALGAP	Aluminum Gallium Phosphide
amp	ampere
ANSI	American National Standards Institute
ANTEC	Annual Technical Conference (of SPE)
ASIC	Application Specific Integrated Circuit
ASTM	American Society for Testing and Materials
BCC	Bumped Chip Carrier
BGA	Ball-Grid Array
BSI	British Standards Institution
BT	Bismaleimide Triazine
CAF	Conductive Anodic Filament
CALCE	Center for Advanced Life Cycle Engineering
C of C	Certificate of Compliance
CCD	Charged-coupled Device
CERDIP	Ceramic Dual-In-Line Package
CFC	Chlorofluorocarbon
CHMT	Component, Hybrid, and Manufacturing Technology
CMOS	Complementary Metal Oxide Semiconductor
COB	Chip-on-Board
CCD	Charged Coupled Device
COG	Chip-on-Glass
COL	Chip-on-Lead
COS	Chip-on-Substrate
COTS	Commercial Off The Shelf
CPGA	Ceramic Pin-Grid Array
CPMT	Components, Packages, and Manufacturing Technology
cP	Centipoise

Table A.5 Abbreviations, acronyms, and symbols—*cont'd*

CSP	Chip Scale Package
CTE	Coefficient of Thermal Expansion
CVCM	Collected Volatile Condensible Materials
DADS	Diaminodiphenyl Sulfone
DCB	Double Cantilevered Beam
DGEBA	Diglycidylether of Bisphenol A
DGEBF	Diglycidylether of Bisphenol F
Dicy	Dicyandiamide
DI	Deionized
DIP	Dual-In-Line Package
DoD	Department of Defense
dph	dots per hour
DRAM	Dynamic Random Access Memory
DSC	Differential Scanning Calorimetry
DSCC	Defense Supply Center, Columbus Ohio
DTA	Differential Thermal Analysis
EDX	Energy Dispersive X-Ray
EEPROM	Electrically Erasable Programmable Read Only Memory
EEW	Epoxy Equivalent Weight
EIA	Electronic Industries Alliance
EMI	Electromagnetic Interference
EMTA	Emerging Markets Trade Association
EPA	Environmental Protection Agency
ESD	Electrostatic Discharge
FCIP	Flip-Chip in Package
FCOB	Flip-Chip on Board
FCOF	Flip-Chip on Flex
FDA	Food and Drug Administration
FEP	Fluorinated Ethylene Propylene
FPD	Flat-Panel Display
FTIR	Fourier Transform Infrared
G	Gravity
g or gm	grams
GaN	Gallium Nitride
GaP	Gallium Phosphide
GPS	Global Positioning Satellite
HAST	Highly Accelerated Stress Test
HCFC	Hydrochlorofluorocarbon
HDI	High Density Interconnect, *also* Hexamethylene Diisocyanate
HFC	Hydrofluorocarbon
HFE	Hydrofluoroether
HHPA	Hexahydrophthalic Anhydride
HTC	High Temperature Cofired Ceramic
IC	Integrated Circuit
ICA	Isotropic Conductive Adhesive

(*Continued*)

Table A.5 Abbreviations, acronyms, and symbols—*cont'd*

ID	Inside Diameter
IEC	International Electrotechnical Commission
IEEE	Institute for Electrical and Electronics Engineers
IEMT	International Electronics Manufacturing Technology
InGaN	Indium Gallium Nitride
IMAPS	International Microelectronics and Packaging Society
I/O	Input/Output
IPA	Isopropyl Alcohol
IPC	Institute for Interconnecting and Packaging Electronic Circuits (formerly Institute of Printed Circuitry)
IR	Infrared (also Insulation Resistance)
ISBN	International Standard Book Number
ISHM	International Society for Hybrid Microelectronics (now IMAPS)
ISO	International Standards Organization
ITO	Indium Tin Oxide
JEDEC	Joint Electronic Device Engineering Council (now Solid State Technology Association)
JIS	Japanese Industrial Standards
K	Kelvin, also potassium
KB	Kauri-Butanol
LC	Liquid Crystal
LCD	Liquid-Crystal Display
LED	Light Emitting Diode
LMI	Low Mobile Ions
LOC	Lead on Chip
LSI	Large-Scale Integration
LTCC	Low Temperature Cofired Ceramic
MCM	Multichip Module
MCOT	Modified Cyclo-olefin Thermoset
MELF	Metal Electrode Leadframe
MEMS	Micro Electromechanical System
MEPTEC	Microelectronics Packaging and Test Engineering Council
MOE	Modulus of Elasticity
MPa	Megapascal
MPC	Microphase Cleaning
MSDS	Material Safety Data Sheet
MSFC	Marshall Space Flight Center
MSL	Moisture Sensitivity Level
MVTR	Moisture Vapor Transmission Rate
N	Newton
NC	No Clean
NCMS	National Center of Manufacturing Science
NFFUF	No-Flow Fluxing Underfill
NFU	No-Flow Underfill
NIST	National Institute of Standards and Technology
NMP	N-Methyl Pyrrolidone

Table A.5 Abbreviations, acronyms, and symbols—*cont'd*

OA	Organic Acid
OD	Outside Diameter
ODA	4,4′ Oxydianiline
ODS	Ozone-depleting Solvent
OE	Optoelectronics
OSHA	Occupational Safety and Health Administration
OLED	Organic Light Emitting Diodes
OSP	Organic Solderability Preservative
Pa	Pascal
P	Poise
PA	Phthalic Anhydride
PBB	Polybrominated Biphenyl
PBDE	Polybrominated Diphenyl Ether
pbw	parts by weight
PCB	Printed-Circuit Board
PCMCIA	Personal Computer Memory Card International Association
PDA	Personal Data Assistant
PDIP	Plastic Dual-In-Line Package
PEEK	Polyaryletherether ketone
PEM	Plastic-Encapsulated Microcircuit
PFI	Polymer Film Interconnect
PGA	Pin-Grid Array
phr	parts per hundred of resin
PIND	Particle Impact Noise Detection
PLCC	Plastic Leadless Chip Carrier
PMDA	Pyromellitic Dianhydride
PoF	Physics of Failure
ppi (width)	Pounds per inch (width)
ppm	parts per million
ppm_v	parts per million by volume
PRF	Performance
PTFE	Polytetrafluoro ethylene
PTMEG	Polytetramethylene ether glycol
PWA	Printed-Wiring Assembly
PWB	Printed-Wiring Board
QFP	Quad Flat Pack
RGA	Residual Gas Analysis
RH	Relative Humidity
RIE	Reactive Ion Etch
R	Rosin
RA	Rosin Activated
RFID	Radiofrequency Identification
RMA	Rosin Mildly Activated
ROHS	Restriction of Hazardous Substances
ROSE	Resistivity of Solvent Extracts

(*Continued*)

Table A.5 Abbreviations, acronyms, and symbols—*cont'd*

rpm	revolutions per minute
RT	Room Temperature
RTV	Room Temperature Vulcanizing
s or sec	second
SAIC	Science Applications International Corporation
SAM	Scanning Acoustic Microscopy
SAMPE	Society of Aerospace Materials and Processes Engineers
SCF	Super Critical Fluid
SEER	Strain Energy Release Rate
SEM	Scanning Electron Microscopy
SIP	System-in-Package
SIR	Surface Insulation Resistance
SMA	Surface-Mount Adhesive
SMT	Surface-Mount Technology
SOIC	Small Outline Integrated Circuit
SOT	Small Outline Transistor
SOP	Small Outline Package
SPE	Society of Plastics Engineers
SRAM	Static Random Access Memory
TAB	Tape Automated Bonding
TCR	Temperature Coefficient of Resistance
TDCB	Tapered Double Cantilevered Beam
TDI	Toluene Diisocyanate
TGA	Thermogravimetric Analysis
TI	Thixotropic Index
TO	Transistor Outline
TMA	Thermomechanical Analysis
TML	Total Mass Loss
TSOP	Thin Small Outline Package
TSV	Through-Silicon Vias
TTT	Time-temperature-transformation
UL	Underwriters Laboratories
UV	Ultraviolet
VCM	Volatile Condensable Material
VFM	Variable Frequency Microwave
VLSI	Very Large Scale Integration
VOC	Volatile Organic Compound
WI	Wetting Index
WLU	Wafer-Level Underfill
WS	Water Soluble
WVR	Water Vapor Regained

Index

FSC
www.fsc.org
MIX
Papier | Fördert gute Waldnutzung
FSC® C083411

Kolibri 360 GmbH
Ferdinand-Jühlke-Straße 7
99095 Erfurt, Deutschland
produktsicherheit@kolibri360.de